The Indian Ocean and the Portuguese-Speaking World

RECONFIGURING IDENTITIES IN THE PORTUGUESE-SPEAKING WORLD

Edited by

Paulo de Medeiros and Cláudia Pazos-Alonso

VOL. 20

PETER LANG

Lausanne - Berlin - Bruxelles - Chennai - Oxford - New York

Ana Mafalda Leite, Elena Brugioni,
Jessica Falconi and Marta Banasiak (eds)

The Indian Ocean and the Portuguese-Speaking World

Literary and Cultural Intersections

PETER LANG
Lausanne - Berlin - Bruxelles - Chennai - Oxford - New York

Bibliographic information published by the Deutsche Nationalbibliothek.
The German National Library lists this publication in the German National Bibliography; detailed bibliographic data is available on the Internet at http://dnb.d-nb.de.

A catalogue record for this book is available from the British Library.

Library of Congress Cataloging-in-Publication Data

Names: Leite, Ana Mafalda, editor. | Brugioni, Elena, editor. | Falconi, Jessica, editor. | Banasiak, Marta, 1985– editor.
Title: The Indian Ocean and the Portuguese-speaking world : literary and cultural intersections / Ana Mafalda Leite, Elena Brugioni, Jessica Falconi, Marta Banasiak.
Description: Oxford ; New York : Peter Lang, 2025. | Series: Reconfiguring identities in the Portuguese-speaking world, 2235-0144 ; volume no. 20 | Includes bibliographical references and index.
Identifiers: LCCN 2024031381 (print) | LCCN 2024031382 (ebook) | ISBN 9781800790964 (paperback) | ISBN 9781800790971 (ebook) | ISBN 9781800790988 (epub)
Subjects: LCSH: Indian Ocean Region—In literature. | Portuguese literature—History and criticism. | Mozambican literature (Portuguese)—History and criticism. | Timor-Leste—Literatures—History and criticism. | Indic literature (Portuguese)—History and criticism. | LCGFT: Literary criticism. | Essays.
Classification: LCC PQ9022.I53 I53 2025 (print) | LCC PQ9022.I53 (ebook) | DDC 869.09/32171269—dc23/eng/20240729
LC record available at https://lccn.loc.gov/2024031381
LC ebook record available at https://lccn.loc.gov/2024031382

Texts 2–5, 8, 12–14, 16 are translated by Hilary Owen and Patricia Ober

This volume results from the research developed within the project NILUS – Narrative of the Indian Ocean in the Lusophone Space, financed by FCT (Portugal).

Cover image: 'Fishermen and the Ocean' by Subodh Kerkar. Reproduced with permission from the artist.
Cover design by Peter Lang Group AG

ISSN 2235-0144
ISBN 978-1-80079-096-4 (print)
ISBN 978-1-80079-097-1 (ePDf)
ISBN 978-1-80079-098-8 (ePub)
DOI 10.3726/b17729

Published by Peter Lang Ltd, Oxford, United Kingdom
info@peterlang.com – www.peterlang.com

This publication has been peer reviewed.

Contents

ANA MAFALDA LEITE, ELENA BRUGIONI, JESSICA FALCONI AND MARTA BANASIAK

1 Lusotopic navigations across the Indian Ocean

This volume derives from research carried out under the aegis of the project *NILUS – Narratives of the Indian Ocean in the Lusophone Space*,[1] working from the premise that the Indian Ocean shapes a new 'transnational imaginative geography' (Ghosh & Muecke, 2007). The aim was to analyse how visual and written narratives from Lusophone – or rather 'Lusotopic' – spaces (Portugal, Mozambique, East Timor and Goa) establish a critical dialogue with the existing theories of Indian Ocean Studies and how this engagement may allow a set of original theoretical developments within the area. We therefore draw on the concept of *Lusotopy*, which appeared in the 1990s as the title of an academic journal as well as a tool enabling the Social Sciences to approach cultural, historical and political phenomena related to 'the area drawn by Portuguese history and colonization' (Cahen & Dos Santos, 2018). This concept is closely connected to the analysis of diaspora, emigration and circulation. Addressed from a literary and visual studies perspective, Lusotopy comprises a more useful tool for conceptualizing the Indian Ocean that emerged from Portuguese colonialism than the notion of Lusophony, supported mainly by the idea of a common language and culture. Furthermore, the concept of Lusotopy allows us to include narratives in English from contemporary India, and better address the multilingual and multicultural contexts in which the written and visual

1 The *NILUS* project (PTDC/CPC-ELT/4868/2014) was funded by the Portuguese Foundation for Science and Technology and coordinated by Ana Mafalda Leite at the CEsA – Centre for African and Development Studies of the Lisbon School of Economics and Management, University of Lisbon.

narratives under scrutiny in this work are inscribed, proposing a set of literary and cultural counterpoints which, in our view, correspond to the very specificity of the Indian Ocean world.

As for Lusophone studies as a field within literary and cultural studies, the conceptual and epistemological revisions developed in the *NILUS* project and set out, in part, in this volume have been fundamental for the emergence of new theoretical and analytical frameworks that are not based on the binary opposition of coloniality and the postcolonial condition, nor grounded upon the concept of linguistic or national identity. Rather, they trace new critical outlines that we consider crucial for reading and situating contemporary visual and written narratives from *Lusotopic* space(s). *The Indian Ocean and the Portuguese-speaking world : literary and cultural intersections* therefore brings together the findings of the *NILUS* research team, as well as dialogues and collaborations with scholars, researchers, writers and artists whose work has been extremely significant for the achievement of the project outcomes.

Echoing liquid spaces of the contemporary age – from the Black Atlantic (Gilroy, 1993) to the South Atlantic (Santos, 2001; Almeida, 2000; Shoat & Stam, 2012) – *NILUS* was based on a theoretical and disciplinary articulation between Indian Ocean Studies and Lusophone literary, visual and cultural studies. The aim was to fill a significant disciplinary gap occasioned by the almost complete absence of a critical dialogue between these two areas of study, especially in Lusotopic contexts. In respect of scholarly production within Indian Ocean Studies, the works that stand out are those of a historical nature combining political and anthropological studies, especially regarding the period before the arrival of Europeans in the Indian Ocean and the pre-modern era, while the modern and contemporary periods continue to be underexplored (Pearson, 2011). In light of these considerations, we believe that the disciplinary dialogue proposed by the project and presented in this volume indicates analytical, conceptual and epistemological potential of great merit and timeliness, significantly extending the scope of the fields of study under consideration.

It is important to underline that while scholarship in history, anthropology and the social sciences in general has been extremely productive in establishing comparative approaches between Lusotopic spaces and the

Indian Ocean world, literary and cultural studies have been more modest in setting up transnational and translinguistic paths through which to address and analyse literatures and visual narratives from different spaces and places of the Indian Ocean World (Pearson, 2011). According to the acclaimed historian Michael Pearson, critical and epistemological approaches based on heteroglossic, transnational and transdisciplinary paradigms, such as the Indian Ocean, represent a crucial step in promoting alternative and situated readings, simultaneously contributing to the consolidation of a more complex set of epistemologies for the study of the Indian Ocean World (Pearson, 2011: 82). The transdisciplinary approach emerging from these disciplinary encounters will therefore help shape Indian Ocean Studies as a specific and organic field of studies, providing alternative critical paradigms and epistemologies that enrich humanistic critique and re-position literary practice within a vital field of knowledge, to better embrace and understand the challenges of contemporaneity. The main goal of this volume, then, is to pursue these comparative and transdisciplinary dimensions, adding new layers to the study of written and visual narrative from Portuguese-speaking contexts as well as opening up new critical possibilities within the area of Indian Ocean Studies.

At the crossroads of comparative literatures, area studies and postcolonial critique, in keeping with the objectives of the project, this volume sets out to draw a new cartography of Lusotopic cultural and literary production. One of the most original aspects of this book is the articulation between the type of analysis that is dominant within Lusophone Studies, mainly based on the paradigm of the nation and new readings guided by the transnational perspective of the Indian Ocean. Engaging with the complex relationship between time and space, narratives and other representations from the Lusotopic Indian Ocean point to a contextual dimension that blurs the boundaries of the national space, unveiling relationships and dynamics that must be observed from a critical perspective that goes beyond linguistic cartographies and national geopolitical contexts. Taking it as given that the Indian Ocean shapes a 'transnational imaginative geography' (Ghosh & Muecke, 2007), the purpose of this book is to reflect on the role of the Lusotopic space and its narratives in the construction of the 'transnational imaginary' of the Indian Ocean. As both the Indian Ocean and the Lusotopic space are multiple and complex cultural constructions, a comparative approach between national and regional contexts that share a common colonial past is fundamental for a multifaceted analysis of analogies and differences in the perceptions and representations of the Indian Ocean imaginary. In other words, critical and epistemological approaches based on heteroglossic, transnational and transdisciplinary paradigms, such as the Indian Ocean, represent a crucial step in enabling alternative and situated readings within Lusotopic cultural, visual and literary studies, enhancing the study of the Indian Ocean World (Pearson, 2011).

As established by the last decade of Mozambican literary scholarship, the Indian Ocean offers a new critical perspective from which to analyse and (re)address literary and visual narratives from the Mozambican context (Falconi, 2008, Leite, 2004, Noa, 2012, Brugioni, 2019, among others). Therefore, the study of Mozambican literature from the Indian Ocean perspective has been a main object of our research, introducing new comparative approaches that rely on the concept of the Indian Ocean as 'a network of structured relations' (Chaudhury, 1992), as an inter-regional arena (Bose, 2006) and a 'world-system'[2] tackling visual and written narrative as the *realm* where the Indian Ocean world takes shape and acquires meaning. In this respect, the notion of *sea/ocean* – namely the (Indian) Ocean – serves to underline the *relation* (Glissant, 1990) between literary texts produced, for instance, in Mozambique and in Mauritius, and foreground the importance of the *transnational oceanic/sea imaginary* within contemporary Swahili poetry published and produced in Mozambique. Therefore, addressing the Indian Ocean from a theoretical perspective as well as a theme and a narrative device is a fundamental step in grasping both the importance of the Indian Ocean within visual and written narratives from the Portuguese-speaking world and the general relevance of Lusophone studies for the area of Indian Ocean Studies. As has been demonstrated by Isabel Hofmeyr, among others, literatures from the Indian Ocean comprise and explore various cultural, historical and social specificities that demonstrate the complexity of this 'interregional arena' (Bose, 2006), and offer the possibility of presenting 'new transnational paradigms for the Global South'

2 Moorthy and Jamal (2010).

(Hofmeyr, 2007). As far as Lusophone studies are concerned, the paradigm of the Indian Ocean would provide as yet underdeveloped 'relational perspectives'. From these it would, for example, be possible to approach Mozambican literature in a transnational dimension, determining a redefinition of topics and imaginaries specific to Mozambican literary writing. Therefore, this book represents an attempt to (re)define the semantic and cultural repertoires of Mozambican literature, while contributing to a reconfiguration of the relationship between literary medium and contextual specificities across the Indian Ocean world.

The wide-ranging conceptual discussion encompassed in the different parts of this volume – itself one of the stated goals of the *NILUS* project – demonstrates how concepts are explored not in light of their literary origins and meaning (cf. *coolitude* from the *coolie*) but according to the way they describe, signify and scrutinize proximity regarding material conditions and social challenges. This approach, in turn, opens up new relational possibilities (Glissant, 1990), or rather, new comparisons and theoretical counterpoints. It is therefore important to clarify that discussion in this volume demonstrates, paradigmatically, how concepts represent different yet juxtaposed responses to questions and concerns shared by different subjects and places within the Indian Ocean world. Consequently, the chapters and sections of this book focus on the way that 'theories travel far beyond the context of their origins' (Said, 2001), re-orienting their political power and meanings.

As was previously mentioned, historiographical studies on the Indian Ocean comprised the starting point and the main disciplinary basis for the constitution of the transnational and transdisciplinary field of Indian Ocean Studies. Following the research on the Mediterranean proposed by Fernand Braudel and the French *Annales*, as well as the perspectives of world-system analysis developed by Immanuel Wallerstein, new historiographical approaches to the Indian Ocean began to appear in the 1980s, contributing to the emergence of a 'new thalassology' (Vink, 2007). Among the issues addressed within the historiographical scholarship, the inclusion of East Africa in general, and Mozambique in particular, as a region within the sphere of influence of the Indian Ocean, was a controversial element throughout the consolidation and institutionalization of Indian Ocean

Studies, as demonstrated by Kirti N. Chaudhuri' (1992) and Sugata Bose (2006), among others. At the same time, the incorporation of perspectives centred on the Indian Ocean as *a unit of analysis* in the historiography of Mozambique and other disciplinary fields – such as anthropology and, more recently, literary and cultural studies – has been a relevant epistemological turn in recent decades (Falconi, 2013; Gupta, 2018; Meneses, 2009, 2012).

Mindful of these premises, the *NILUS* project set up a fruitful dialogue between literary studies and the history of the Indian Ocean, particularly through the discussion of a vast bibliographical repertoire produced in different languages by scholars whose work addresses different regions and locations of the Indian Ocean. On the one hand, this dialogue was seen by the team as an essential framework for addressing written and visual narratives from contexts formerly colonized by Portugal in the Indian Ocean and other spaces connected to them by the liquid geography of the sea. On the other hand, the corpus produced within the scope of historiography has come to constitute an enormous and useful source of conceptual tools and critical paradigms for the analysis of cultural imaginaries that literary and artistic narratives encompass. Here it is worth noting that seminal concepts emerged from historical scholarship, and its application in the field of literary and cultural studies has been fundamental to broadening the disciplinary focus on the Indian Ocean and deepening its relevance as a unit of analysis within the domain of cultural and artistic production. We refer to concepts such as 'inter-regional arena', or 'littoral society', formulated by Sugata Bose and Michael Pearson, respectively, to define structuring elements of a possible unity-in-multiplicity of the Indian Ocean world. Such concepts produced significant critical developments for the analysis of the aesthetics and poetics of the Indian Ocean, such as the categories of 'transnational imaginative geography' (Ghosh & Muecke, 2007) and 'coastal form' (Samuelson, 2017) representing theorizations of great value for our project and for the outline of this volume. Several essays in this book point up historiographical issues that have proved to be fundamental in contextualizing the relationship between the Indian Ocean, Portuguese-speaking spaces and other shores of this ocean. One of these issues is, precisely, the integration of Mozambique into the network of exchanges and circulations in the Indian Ocean, addressed by Joana

Pereira Leite and Ana Mafalda Leite. These scholars adopt a long-term perspective within the broader framework of this African space's 'openness to the world' through transoceanic trade routes which also brought intense circulation and exchanges of ideas and cultures. These exchanges resulted in an identity matrix that is still relevant for equating the social and cultural dynamics of contemporary Mozambique. As Pereira Leite and Leite emphasize, until recently, this integration was barely explored, if not deliberately erased, both by historiographical studies and discourses and the political narratives of construction and affirmation of Mozambican national identity. Therefore, we strongly believe that equating this matrix becomes increasingly relevant and urgent also in light of the recent conflicts that have emerged in the North of Mozambique and in the Cabo Delgado region, rooted in the complex articulation of religious, social, cultural and economic factors, on a local and global scale.[3] This is crucial for an understanding of the movement of people and ideas along the East African coast, as well as their relations with the interior of the African continent (Forquilha & Pereira, 2022).

The dialogue that the project and this book have established with historiographical reflection seeks to account for the possible intersections between history, memories and narratives, thus expanding the *corpus* of analysis to incorporate diverse productions and practices, such as travel literature and 'remediations' of cultural memory. Manuel Lobato's chapter on travel narratives focuses on African and Asian territories, while Falconi and Brugioni's chapter on the cultural memory of the Indian Ocean, enshrined in the Commemorations of the Portuguese Discoveries, dialogues with history in order to chart cultural representations from various shores of the Indian Ocean. Within these perspectives, the ocean is addressed not only as a historical, aesthetic and – possibly – utopian space but also as an 'uneven place' with its centres and peripheries, its fractures and conflicts.

3 On armed attacks and violence in Cabo Delgado, among a vast bibliography, see Eric Morier-Genoud, 'A insurgência jihadi em Moçambique: origens, natureza e início', *Cadernos IESE*, 21 (2021), 1–56. The article is also available in English in *Journal of Eastern African Studies*, 14.3 (2020), 396–412.

The essays in this book aim to account for the multiplicity of representations that have accumulated over this area of the world; its borders remain fluid, challenging hegemonic notions of space and geography. Carmen Tindó Secco's study of the presence of the Islamic religion in Mozambican literature, as well as Clarissa Vierke's work on Swahili poetry, bring together the historical dimension and the Indian Ocean's 'openness to the world' down the ages, resulting in a 'diverse space of multiple connections' on to which different cartographies have been projected, drawn by different transregional and translinguistic contacts and affiliations.

With regard to literary and artistic dialogues, it is important to underscore both the concept of 'relational poetics' introduced by Ute Fendler in her chapter and the definition of Indian Ocean as 'narratable space' formulated by the Mozambican writer João Paulo Borges Coelho in his essay: '[N]arratable space is the structural element that provides the link with reality in that undertaking which we call fiction. It is not so much a plot or a storyline but rather, as I see it, a *place* where plots and storylines may be constructed.' Borges Coelho defines the Indian Ocean as one of these narratable spaces; a 'narratable space' that accommodates a range of different forms and shapes and establishes tensions between them, frequently tending towards conflict and rupture. This relational and narratable definition of the Indian Ocean has been also theorized by the Mauritian poet Khal Torabully (1999, 2020) through the concepts of 'coolitude and thalassography'; his creative research has given rise to a kind of 'oceanography' that places transoceanic voyages at the heart of the poetics of the indentured worker and the coolie, breaking with land-centred narratives, investing in maritime transnational space and connecting memories, stories, cultures and imaginaries of the world.[4]

4 Later, Khal Torabully (2002, 2014) developed the idea of 'coral identity', underlining the semiotic/symbiotic crossing of the indentured worker and the coolie and putting forward the idea of a open, plural, structuring poetic in which the Indian ocean is the fundamental binding element. Breaking with the geographical, anthropological and memorial marginalization of representations and circulations in the Indian Ocean, he redefines the History of the Indian Ocean as a relational history between people and the sea; according to Khal Torabully the ocean should become a space of deconstruction, conceived as bounding and diluting element, a

In her chapter, Ute Fendler, drawing on the theories elaborated by Torabully in dialogue with the concept of 'tout-monde' developed by Édouard Glissant (1997), proposes a critical approach to the *poetics of a relational imaginary*, with a special focus on the iconological elements in photos, sculptures and installations. She considers that the process of relating images could create a surplus of meaning that enriches the imaginary of the Indian Ocean. The relational construction of a visual imaginary brings together the various layers of production and distribution with the colonial and postcolonial implications it carries as a historical burden. Fendler states that this 'creative and symbolic dimension' can be linked to the perspective proposed by Moorthy and Jamal (2010) and particularly to the ideas of Stephen Muecke (2010): 'Fabulations are the kind of storytelling that takes off on a flight of becoming rather than repeating the available discourses and genres.' Fendler analyses the connections between the work of Goan artist Subodh Kerkar, who describes himself as an 'ocean artist', and the work of two Mozambican artists, the photographer Sergio Santimano and the sculptor Pekiwa. Examining the representation of the Indian Ocean by the three artists through images of boats and fishermen, we are able to discern relational connections and convergent 'narratable' intentions. Santimano's photograph of fishermen pushing the boat into the sea trying to surmount the strong waves breaking on the shore is a masterly depiction of the daily confrontation between human beings and the sea; for his part, the artist Pekiwa in his boat sculptures merges experiences and stories, transforming the boat into a conduit between the human being and the ocean. Likewise, and with a similar purpose, Subodh Kerkar, among many other creative plastic strands, works with performative installations. His series 'Fishermen' is a performative work in which fishermen enact daily practices, and in fact, we observe the fishermen gathering on the beach in such a way that their bodies form the silhouette of a boat, performing the act of rowing, or fishing. According to Fendler, the performance of the

central paradigm of cultural imaginaries and interactions, from a polycentric – no longer Eurocentric – point of view; therefore, within this perspective, writing (on) the sea is 'the equivalent of holding up a mirror to the writing of history, of hi/stories' (2020).

fishermen may be considered a 'reenactment of daily practices and of oral history'. In this way, the human body turns into the medium of a communicative memory, of a shared and lived history.

Working in the context of photographic iconology, in their chapter Meg Samuelson and Pamila Gupta present a study on the Capital Art Studio in Zanzibar, and discuss how its founder, Ranchhod Oza, framed Zanzibar's circulating subjects before and after the Revolution. They reflect on the photographic practice designed by his son, Rohit Oza, which involves re-taking scenes that his father had framed the decades before. Capital Art Studio thus presents a significant methodological archive; its moving frames and circulating subjects suggest profound ways of thinking through Zanzibar's pivotal position between the African continent and the Indian Ocean, and theorizing the potential uses of the photographic medium itself. Gupta and Samuelson observe how Art Studio's distinctive practices of 'moving frames' can convey the idea of 'circulating subjects', a specific sense of fluid motion that sheds new light on circularity in the Indian Ocean, as explained by Sugata Bose in his definition of Indian Ocean as an 'interregional arena of human interaction' built on 'circular migration' (Bose, 2006).

Another important contribution in the field of visual arts, as an artistic dialogue and as an example of the Indian Ocean as 'transnational imaginative geography', is Jessica Falconi's and Giulia Spinuzza's chapter on 'Fishing for stories: Audiovisual narratives of the Indian Ocean: Goa, Mozambique. East Timor'. The chapter addresses different audiovisual narratives centred on territories of the Indian Ocean, namely Goa, Mozambique and East Timor, with the aim of exploring some of the elements that unite and divide this area. In this itinerary, guided by Michael Pearson's reflection on *littoral societies* (1985, 2007, 2017) Falconi and Spinuzza undertake a reading in the counterpoint of these narratives that have in common the topic of fishing, an activity with important characteristics and social and cultural meanings in each of the territories under scrutiny. This topic allows the authors to explore historical, social and environmental dynamics and themes related to preservation and safeguarding, while audiovisual representation provides the opportunity to address the coasts, islands and oceans – in this case, the Indian Ocean – as spaces for imaginaries and aesthetic creation (Samuelson, 2017; Verne & Verne, 2017) whose future is uncertain, impacted by the routes and maps of the global economy. Taking up the concepts formulated by Pearson and Samuelson, the two authors observe that the littoral examined in the audiovisual narratives may be understood as an *ecotone*, a space of transition, in this case between sea and land, and also a space of meeting (Pearson, 2019) that bears witness to the persistence of 'amphibian lives', where the communities represented life in symbiosis with the sea. Falconi and Spinuzza state that these documentaries, through aesthetic and narrative choices, point to an amphibian aesthetic, as well as establishing an affective link between the inhabitants of the coast and the sea.

In this context, visual, audiovisual and artistic dialogues can be seen as an extensive network of representations within the *relational imaginary* of the Indian Ocean, putting forward multiple iconic narratives and memories on the interrelation between the land and the sea, the human and the environment. This approach allows us to identify imaginaries, perceptive sensibilities and cultural practices that, paraphrasing Mozambican poet Virgílio de Lemos, are 'the very making of the sea'.[5]

As already mentioned, when it comes to the literatures from the Lusotopic contexts located in the basin of the Indian Ocean, the theoretical approach proposed by Indian Ocean Studies opens up a multiplicity of new insights. On one hand, we are dealing with consolidated fields of literary studies, on the other hand, most of the critical work that has been done within these literatures promotes rather telluric and national readings; new critical frameworks seem to emerge particularly by the sea. Whether the Indian Ocean space is defined as an 'interregional arena' (Bose, 2006), or as 'one, among many liminal spaces of hybrid evolution, whose area boundaries are both moveable and porous' (Moorthy & Jamal, 2010), the insertion of Lusophone literatures into this transnational and translinguistic paradigm allows 'new literary and cultural cartography based on "Afro-Asian" counterpoints, that is, between different territories formerly colonized by Portugal in the Indian Ocean' (Falconi, 2021: 30).

5 Cf. Virgílio de Lemos, *Para Fazer um Mar*. Lisbon: Instituto Camões, 2001. On this see Falconi (2013).

It further also provides an opportunity to establish multiple points of convergence with the Anglophone and Francophone literatures of the region. While translinguistic comparative perspectives can unveil common cultural and intellectual linkages within the literary Indian Ocean, the transdisciplinarity of Indian Ocean Studies, and its theoretical openness allows us to reorient a set of critical interests. Changing the customary analytical approach by revisiting the national literary canons from a broader theoretical and critical standpoint, enables multiple (re)interpretations of canonical works, as well as new critical readings of recent literary works. In both cases, the unearthing, or rather the 'unwatering' of new declensions of these literatures moves away from established social and cultural telluric bonds and cartographies with a Lusophone linguistic and (post)colonial imprint, as well as from national(istic) critical frameworks, and moves towards a new (re)defined and more complex matrix to address the Indian Ocean as an 'index of a bigger history' [o indício de uma história maior] (Brugioni, 2015: 103). A comparison of writing and visual narrative is, in our view, one of the most valuable ways to broach the most important and peculiar feature of the Indian Ocean world: its wide diversity of languages, subjects, histories and cultural practices whose reading requires – inevitably – a wide range of linguistic skills and disciplinary tools.

Given that maritime crossings of humans and nonhumans are decisive for the comprehension of history in and of the oceans (Pearson, 2003: 194), the essay by Marta Banasiak, Joana Passos and Kamila Krakowska Rodrigues analyses multiple trajectories and travel routes, understood as political, commercial and tourist trajectories, which have brought about contact and interaction between the different peoples of the region, across different historical contexts. This starting point questions the primacy afforded to nation-based models of thought, enclosed within their frontiers, that underpinned critical work on the Mozambican literature, as well as the international debate that has tended to privilege the colonizer/colonized binary while disregarding regional dynamics. By choosing writings that transgress circumscribed national limits, while at the same time promoting a platform for regional comparison, the authors draw a conceptual map of this collective space which recreates its specific history and dynamics, investigating the reciprocal exchange of regional influences in the present day, and the perceptions of self and other that have been disseminated along the different shores of the Indian Ocean. Spurred on by the perceived need to expose the unofficial alliances and/or animosities and mistrust between societies along the margins of the Indian Ocean, in the lusotopic sphere, this essay draws together the examples of Goan and Mozambican literary texts. This comparative approach allows the reader to discover the affective ties created by the sea routes that linked the East Coast of Africa with the Malabar coast on the western seaboard of the Indian subcontinent, and the emergence of (proto)national and (proto)nationalist awareness, on both sides of the Indian Ocean, as an intrinsically transnational phenomenon. Finally, a focus on more contemporary crossings – tourist movement in the basin – shows how modern tourist contacts mirror the relations which structure the world system, and the potential of Indian Ocean Studies to deconstruct these monolithic relations.

Returning to Pearson's words quoted above, and having explored the human crossings within the Indian Ocean, our interest should turn to the nonhuman elements, both material and non-material that, along with human interaction, compose the *genius loci* of the Indian Ocean (Borges Coelho, 2019). Given the fact that the Indian Ocean and its margins are strongly marked by the 'wrath and fury expressed in the climatic manifestations (…) with naturally frequent phenomena such as cyclones, typhoons and earthquakes being exponentially aggravated by anthropogenic actions and climate change', as Gabriela Beduschi Zanfelice puts it in her essay, it seems only natural to combine Indian Ocean Studies with ecocriticism. Highlighting once more the transdisciplinarity of the Indian Ocean Studies, and aware of being 'indebted to the Warwick Research Collective's proposition of world literature as the literature of the modern capitalist world-system, or the literature that registers, both in thematic and formal levels, the rhythm of its lived experience, whether their authors are conscious of it or not', Zanfelice applies the Mozambican *genius loci*, the concept of 'postcolonial environment' developed by Pablo Mukherjee (2010). Emphasizing the importance of the Indian Ocean and different maritime cultures in the process of constructing the Mozambican literary imaginary, she explores in the texts of João Paulo Borges Coelho the ecosystems where human and nonhuman subjects are deeply integrated with narratives that register the social and material conditions of Mozambican society 'without losing sight of its connections to the world-system and the world-ecology at large'.

When discussing the new critical approaches to the Lusotopic narratives of the Indian Ocean, it should be remembered that we are dealing with visibly marginalized literary production, or, as Paulo de Medeiros puts it, with literatures that suffer from 'redoubled forms of invisibility' (2020: 23) within the global literary market. Therefore, the last section of this volume includes literary works from writers of the Indian Ocean basin, namely Khal Torabully (Mauritius), João Paulo Borges Coelho, Luís Carlos Patraquim (Mozambique), and Jessica Faleiro (Goa, India). We cannot overstate the importance of Torabully's and Borges Coelho's theoretical and historical reflections on the Indian Ocean, as expounded through their artistic production. Hence the two texts selected to open this section are hybrid and liminal meditations on the Indian Ocean world by these writers. Not only do their words resonate in the majority of the chapters gathered in this volume, but they also demonstrate how the Indian Ocean emerges, first and foremost, through its creative, literary voices. These texts are followed by the poetry of Torabully and Luís Carlos Patraquim, and a lyrical prose work by Jessica Faleiro.

The concepts and theorizations discussed by the scholars who have contributed to this volume, the creative works and authors analysed in the different sections of the book and finally, the wide bibliographical repertoire discussed in each one of the chapters illustrate, paradigmatically, the vast multiplicity of the Indian Ocean world and, concomitantly, the demanding task faced by scholars who dedicates their efforts to the study of this oceanscape. As Michael Pearson noted in his review essay on Indian Ocean Studies published in 2011, the very diversity of the Indian Ocean World constitutes a critical challenge that can only be met through collaborative strategies and shared methodologies:

> The Indian Ocean World, if indeed there is such a thing, is notoriously diverse. Over forty states today, many languages, cultures, religions, topographies and a vast body of histories, often difficult to access. No person can hope to master all this. Obviously, the need is for collaboration, especially between people with different language skills. Then, in a decade or so maybe some later day Braudel will synthesize what I hope will be a vast body of new and provocative work. (Pearson, 2011: 82)

A key aim of the *NILUS* project, and by extension this book, is to prompt new collective and collaborative studies of the Indian Ocean. It is certainly our hope that the navigations offered by this volume will help re-position Lusotopic contexts among the relevant places of the Indian Ocean, promoting a more comprehensive – and less uneven – definition of the Indian Ocean world, not as a world apart but as a location whose specificity offers valuable answers to the questions faced today inside and outside Indian Ocean places.

Bibliography

Almeida, M. Vale de. (2000). *Um Mar da Cor da Terra: Raça, Cultura e Política da Identidade.* Oeiras: Celta.

Borges Coelho, J. P. (2019). 'O Índico como lugar'. In A. M. Leite, E. Brugioni and J. Falconi (eds), *Estudos sobre o oceano Índico: Antologia de textos teóricos*, pp. 13–28. Lisbon: Edições Colibri.

Bose, S. (2006). *A Hundred Horizons: The Indian Ocean in an Age of Global Empire.* Cambridge, MA: Harvard University Press.

Braudel, F. (1985). *La Méditerranée, vol. 1: L'espace et l'histoire, vol. 2: Les hommes et l'héritage.* Paris: Flammarion.

Brugioni, E. (2015). 'Por detrás de tantos nomes, o mar. Reflexões em torno da relação entre Moçambique e o Oceano Índico: narrativas, imaginários e representações'. *Via Atlântica*, 16(1), 93–110.

——. (2019). *Literaturas africanas comparadas. paradigmas críticos e representações em contraponto.* Campinas: Editora da Unicamp.

Cahen, M. and Dos Santos, I. (2018). 'Lusotopie, Lusotopy: What Legacy, What Future?'. *Lusotopie, Concepts et Pratiques*, 17(2), 187–203. Available at: lusotopie-2838-1.pdf.

Chaudhuri, K. N. (1990). *Asia before Europe: Economy and Civilization of the Indian Ocean from the Rise of Islam to 1750.* Cambridge: Cambridge UP.

Falconi, J. (2008). *Utopia e conflittualità. Ilha de Moçambique nella poesia mozambicana contemporanea.* Roma. Aracne.

——. (2013). 'Para fazer um mar. Literatura moçambicana e Oceano Índico'. *Diacritica*, 27(3), 77–92.

——. (2021). 'Declinações da Nação nas literaturas africanas de língua portuguesa'. *Abriu* 10, 9–38.

Forquilha, S. and Pereira, J. (2022). 'Dinâmicas da migração e o desenvolvimento da insurgência jihadista no Norte de Moçambique'. In C. N. Castel-Branco, R. Ali, S. Chichava, S. Forquilha and C. Muianga (eds), *Desafios para Moçambique 2022*, pp. 38–56. Maputo: IESE.

Ghosh, D. and Muecke, S. (eds). (2007). *Cultures of Trade: Indian Ocean Exchanges*. Newcastle, UK: Cambridge Scholars Publishing.

Gilroy, P. (1993). *The Black Atlantic, Modernity and Double Consciousness*. London: Verso.

Glissant, E. (1990). *Poétique de la relation*. Paris: Gallimard.

——. (1997). *Traité du Tout-Monde. Poétique IV*. Paris: Gallimard.

Gupta, P. (2018). *The Portuguese Decolonization in the Indian Ocean World*. London: Bloomsbury Academic.

Hofmeyr, I. (2007). 'The Black Atlantic Meets the Indian Ocean: Forging New Paradigms of Transnationalism for the Global South – Literary and Cultural Perspectives'. *Social Dynamics*, 33(2), 3–32.

Leite, A. M. (2004). *Literaturas Africanas e Formulações Pós-Coloniais*. Lisbon: Colibri.

Medeiros, Paulo de. (2020). 'Mia Couto and the Antinomies of World-Literature'. In K. van Haesendonck (ed.), *The Worlds of Mia Couto*, pp. 11–31. Oxford: Peter Lang.

Meneses, M. P. (2009). 'Food, Recipes, and Commodities of Empires: Mozambique in the Indian Ocean Network'. *Oficinas do CES*, 335, 1–37.

Moorthy, S. and Jamal, A. (eds). (2010). *Indian Ocean Studies: Cultural, Social, and Political Perspectives*. New York: Routledge.

Morier-Genoud, E. (2021). 'A insurgência jihadi em Moçambique: origens, natureza e início'. *Cadernos IESE*, 21, 1–56. The article is also available in English in *Journal of Eastern African Studies*, 14(3), 396–412.

Muecke, S. (2010). 'Fabulation: Flying Carpets and Artful Politics in the Indian Ocean'. In S. Moorthy and A. Jamal (eds), *Indian Ocean Studies: Cultural, Social, and Political Perspectives*, pp. 32–44. London: Routledge.

Mukherjee, U. P. (2010). *Postcolonial Environments: Nature, Culture and the Contemporary Indian Novel in English*. London: Palgrave Macmillan.

Noa, F. (2012). *Perto do Fragmento, a Totalidade. Olhares sobre a literatura e o mundo*. Maputo: Ndjira.

Pearson, M. N. (1985). 'Littoral Society: The Case for the Coast'. *The Great Circle*, 7(1), 1–8.

——. (2003). *The Indian Ocean*. London: Routledge.

——. (2006). 'Littoral Societies: The Concept and the Problem'. *Journal of World History*, 17(4), 353–373.

——. (2007). 'Studying the Indian Ocean World: Problems and Opportunities'. In H. P. Ray and E. Alpers (eds), *Cross Currents and Community Networks: The History of the Indian Ocean World*, pp. 15–33. Oxford: Oxford University Press.

——. (2011). 'History of the Indian Ocean: A Review Essay'. *WASAFIRI*, 26(2): Indian Oceans, 78–99.

——. (2017). 'Places in the Indian Ocean World'. *The Journal of Indian Ocean World Studies*, 1(1), 4–23.

Samuelson, M. (2017). 'Costal Form: Amphibian Positions, Wider Worlds, and Planetary Horizons on the African Indian Ocean Littoral'. *Comparative Literature*, 69(1), 16–24.

Santos, B. de S. (2001). 'Entre Próspero e Caliban: colonialismo, pós-colonialismo e inter-identidade'. In M. I. Ramalho and A. Sousa Ribeiro (eds), *Entre ser e estar. Raízes, percursos e discursos da identidade*, pp. 23–85. Oporto: Afrontamento.

Shohat, E. and Stam, R. (2012). *Race in Translation: Culture Wars Around the Postcolonial Atlantic*. New York: New York University Press.

Torabully, K. (1999). *Chair Corail: Fragments Coolies*. Petit-Bourg, Guadeloupe: Ibis Rouge.

——. (2014). 'Between Fixity and Fluidity: Agglutination and the Poetics of the Coral'. In O. Ettea and U. Wirth (eds), *Nach der Hybridität: Zukünfte der Kulturtheorie*, pp. 93–12. Berlin: Edition Tranvía.

Verne, J. and Verne, M. (2017). 'Introduction: The Indian Ocean as Aesthetic Space'. *Comparative Studies of South Asia, Africa and the Middle East*, 37(2), 314–320.

Vink, M. P. M. (2007). 'Indian Ocean Studies and the "new thalassology"'. *Journal of Global History*, 2(01), 41.

Wallerstein, I. (2004). *World-System Analysis: An Introduction*. Durham: Duke University Press.

WReC. (2015). *Combined and Uneven Development: Towards a New Theory of World-Literature*. Liverpool. Liverpool University Press.

PART I

Time and Crossings: Histories, Memories, Travels, Theories

JOANA PEREIRA LEITE AND ANA MAFALDA LEITE

2 The narratives of history and the uncovering of the Indian Ocean

The long history of the opening to the world of Mozambique, key space in the Western Lusophone Indian Ocean, is intertwined with the History of the Indian Ocean, although this narrative of History has been profoundly erased, especially from the twentieth century, with the shifting of economic interests to the south of the then colony and through the creation of the cultural elites that centralized, from that south, the identity themes of the anticolonial struggle, and literary and artistic postcolonial practices.

The *NILUS* project (Narrativas Literárias e Visuais do Índico no Espaço Lusófono/Literary and Visual Narratives of the Indian Ocean in the Lusophone Space) is comprised within the framework of a vast amount of theoretical research in the sphere of the Social Sciences, namely Indian Ocean Studies, which have developed since the end of the twentieth century, seeking to identify the historical and cultural specificities of the countries of the Indian Ocean. This project takes as its point of departure the study and interrogation of the sharing, intermeshing and presence of the cultures and identities of the Indian Ocean, within the framework of the literary and artistic creation of the countries that belong to the Portuguese colonial presence, namely by examining the representations of Indian Ocean multiculturality based on the literary and visual narratives of the respective cultural heritages. Consequently, researching the Indian Ocean as a theoretical point of departure, as well as a narrative theme and device, will make it possible to emphasize the importance of the Indian Ocean in the visual and written narratives of the Lusophone space, as well as the importance that the progressive revealing of this narrative space may have in the framework of the inter-regional arena of the Indian Ocean. The paradigm of the Indian Ocean will allow us to create as yet undeveloped relational perspectives in order to address the transnational dimension of the Lusophone countries' narratives, redefining themes, semantic and cultural repertoires, and imaginaries, as well as uncovering possible comparisons.

The brief historical narrative we present about the Western Indian Ocean is constructed over a long duration, both on this side of and beyond the shores of the Western Indian Ocean. It is also considered in the long and medium term, underscoring the new relations that result from the Portuguese presence, and the bonds woven with the set of territories that constituted its colonial Empire in the twentieth century, which sees the enfeoffment (embedding) of Mozambique in the economic pole of Southern Africa. It also signals some of the permanencies and ruptures operated in the time of the Independence up to the present day. For this reason, the historical narrative presented here is fundamental for the necessary 'uncocealment' of the literary and visual narratives of poets, directors and novelists, and for the theoretical framing of Indian Ocean Studies in the Lusophone space. Whenever historical and political circumstances are muted, it is up to creativity, to theory, to Social Sciences narratives and to reveal significant links, dormant or interdicted, by past or present history. They are equally the ones who authorize us to think the unthought, the unthinkable or even the reverse of history.

Mozambique: A history shared between the Indian Ocean and Southern Africa

Evoking Mozambique when questioning the construction of the identities of the Portuguese-speaking Indian Ocean necessarily leads us to seek answers to its long history of openness to the world.

On the other hand, it involves remembering that the country belongs to an ancient history, the inexorable legacy of its Geography, which goes back to the ties established by the Swahili Corridor, as acknowledged by recent historical archaeology (Horton & Middleton, 2000; Horton 1987,

2001, 2010), both with the world system of the Orient, up to the ninth century AD, and later with the Mediterranean, through the Red Sea, from the tenth century. Today it is known that from the East coast of Northern Africa (present-day Somalia) to the southern shores of the Indian Ocean, to the littoral of Southern Africa (Mozambique in our times), in their origins the African communities established there from the beginning of the first millennium and sustained in a material life linked to the sea, opened up to the world through the path of transoceanic commerce (wood, slaves and ambergris, gold, ivory, quartz) and added to their original identities elements of the Arab-Islamic material, political and religious culture (Trimingham, 1980; Constantin, 1987; Horton, 2001).

The history of the confrontation of East Africa with the European expansion and, namely, with the deeds of the vulnerable Asian Portuguese empire (1500–1700), confirmed by abundant historiographical material (from Boxer, 1969 to Subrahmanyam, 1993), not free from controversy, today explains to us how these coastal communities established a bridge between the outside world, extremely turbulent and greedy for resources, and a continent shut in on itself (Alpers, 2009). From the sixteenth century, slaves captured in Mozambique were transferred by the Portuguese to the Indian subcontinent as far as Ceylon (Pinto, 1992). Along the coast, between Mombasa and the River Rovuma, throughout all of the seventeenth century and up the mid-eighteenth century, the tension between the Omani, Portuguese and Dutch expansionism was lasting and intense. In the interior, from the neighbourhood of the Great Lakes to the territories of Southern Africa, networks of long-distance caravanners (Nyamwezi, Yao, Tonga), just as in antiquity, assured the flow of strategic products (ivory and slaves). The Portuguese, in competition with Afro-Islamic (Arab-Swahili) traders, Arabs and also Indians, already established there, progressed as far as they could along the course of the Zambeze. They unsuccessfully challenged the Shona power in the Upper Zambeze and were obliged to interrupt their progress at the end of the seventeenth century (Changamira revolt 1693 and restructuring of the Shona kingdom under the Rozwi dynasty; Randels, 1975; Austen, 1987; Newitt, 1995; Beach, 1980). Between Quelimane and Tete, the Portuguese established on both sides of the river a settlement system based on enfeoffment, establishing unprecedented forms of female power in Zambezia (the Women Owners of Zambezia: Isaacman, 1972; Capela, 1995; Newitt, 1995; Rosário, 2021), as well as the hegemony of subjects of Indo-Portuguese origin, who from Goa, the Asiatic Empire had legitimated (Boxer, 1980; Subrahmanyam, 1993). From the first decades of the eighteenth century, the economic order of mercantile capitalism mobilized the slave trade from the ports of Northern Mozambique (Island of Mozambique and Quirimbas archipelago) to work on the plantations of Madagascar and the French islands of the Indian Ocean (Mascarenhas), at the time Île-de-France (Mauritius) and Borbon (Réunion) (Austen, 1987; Bento, 1993; Capela, 2002; Newitt, 1995). From that century, and during the nineteenth, slaves were taken there by the Portuguese, who settled in Gujarat, Maharashtra, Daman, Diu and Karnataka (the Siddis, Indians of African origin, just like the Habsh, settled from the beginning of the European expansion; Catlin-J & Alpers, 2004; Alpers, 2000, 2015). In the mid-eighteenth century (1752), when Portugal was already focusing most of its attention on the South Atlantic Empire, the then designated Portuguese East Africa acquired administrative autonomy from the State of India. It was equally at that moment in time when – after more than a century of intense confrontation with the Omani power in the Indian Ocean (loss of Muscat in 1650) and on its East African shores (definitive loss of Mombasa in 1729) the Portuguese were led to limit their presence to south of the Rovuma. The eighteenth century saw the consolidation of the mercantile power of the Asiatic diasporas, whose presence made itself felt in the territory from the end of the seventeenth century (the Company of the Mazanes of Diu obtained the Trade Monopoly with the Island of Mozambique in 1686). Effectively, mobilized by the Marquis of Pombal's legislative policy of free trade (Hoppe, 1970), the Banean mercantile expansion extended to the south of the territory, crossing the Zambeze until it reached Inhambane at the end of the eighteenth century (Antunes, 1992, 1998). In this era, the Indian communities, Hindu and Muslim, were present in significant numbers throughout the southern littoral of Mozambique. Established communities trading to the rhythm of the monsoon, their descendants, the children of African mothers were educated there and would live there forever (Lobato, 1970; Rita-Ferreira, 1980; Newitt, 1995). In the mid-eighteenth century, the Portuguese administration imposed heavy taxes on ivory, long since transported to India, which the Yao traders took to the Island of Mozambique. However, the aim of favouring the finances of the colony had adverse consequences once there was a redirecting of the white gold route to Kilwa for the benefit of the economic dynamism of the Arab-Swahili sultanate. It was the beginning of the lasting economic decline of the capital city of Portuguese East Africa, situated on the Island of Mozambique (Birmingham & Martin, 1992; Newitt, 1995; Lobato, 1996b). In effect, the nineteenth century brought profound changes to the World and to East and Southern Africa, a history shared with Mozambique. Nineteenth-century globalization associated with the emergence of Industrial Capitalism and the European Abolitionist order that accompanied it, had as a consequence of the banning of the slave trade south of the equator (1830), the displacement of the shameful trade, now illicit, from the Atlantic to East Africa. This irony of East African history (Curtin et al., 1987) made itself felt along the whole Swahili coast in no uncertain terms, through the escalation of the business and the multiplicity of agents involved, on the northern littoral of Mozambique, from the 1840s up to the last quarter of the century. In effect, between the Sultanate of Quitangonha (Bay of Nacala) and the Sultanate of Angoche (littoral north of the Zambeze), trading at its peak would attract Africans (Yao and Macua), Afro-Islamics (Swahili), Indians and even functionaries in the Portuguese administration. Slave ships docked there, both French (involved from early on in the Indian Ocean trafficking) and those who were then seeking in the Western Indian Ocean to satisfy the growing demand for labour in the plantations of the New World (Brazil, the Antilles, North America), which abolition had made difficult. Associated with the capture and trade in slaves was the ivory trade, stimulated by new habits of consumption in industrialized Europe (Rocha, 1989; Capela, 2002; Medeiros, 1988). The activism of the long-distance trading networks, Yao and also Arab-Swahili, caused agitation and extreme violence in the Macau territory and Upper Zambezia, with the intervention of the *Senhores dos Prazos* who then made and upheld the law in the region (see Baltasar, 2016). This was a place of predatory men and nature that was felt far beyond the Zambezian upland (Pélissier, 1984; Isaacman, 1972; Lobato, 1996a). This movement of extending economic, social, religious (dissemination of Islam) and political, happened throughout East Africa, from Mozambique to the area of influence of the Trade Empire of Zanzibar. There, in the context of nineteenth-century globalization, Omani proto-colonialism led, under the hegemony of the Bu-Said dynasty, the Arab-Swahili expansion to the lands of Central Africa (Sheriff, 1987). The historiography established from the last decades of the twentieth century allows us to flag up and question the consequences of this major fact of East and Central African history: that is to say, the opening of African societies to the exterior, the strengthening of communication between them, the introduction of new technologies and cultural values and a new way of dealing with economic realities. It took around a millennium for the interior of East Africa to open up definitively to the Indian Ocean (Kagabo, 1988; Horton, 2010).

The nineteenth century was also the moment for reconstructing the Indian diasporas in East and Southern Africa (Clarence-Smith, 1989). Free passengers and coolies, Muslims (Sunni and Ismali) and Hindus left British India at the end of this century and crossed the Indian Ocean in search of a better life, in small businesses, as workers contracted to the plantations (*coolies/indentured labourers*) and also in the colonial administration. They settled in British possessions (Uganda, Kenya, Natal, Transvaal), in Zanzibar under Omani power (but it became a British protectorate at the end of the century, 1890), in German East Africa (in Tanganika and even in the region of the Great Lakes, Ruanda-Urundi, a vast territory that in the aftermath of the First World War was shared out between Great Britain and Belgium) and in Mozambique (Leite, 1996, 2001; Khouri, 2008). Mozambique was equally a destination shared by many Goans who, from the last decades of the nineteenth century, left the Portuguese colony of India for the East African coast (Frenz, 2014). During the colonial twentieth century, and up to our own days, these communities confirmed their presence as irreplaceable agents in the commercial dynamization of Mozambique (Khouri & Leite, 2008, 2014; Leite & Khouri, 2008, 2012), as well as in all of East, Central-East and Southern Africa.

Finally, in the context of the Portuguese colonial occupation, which the Conference of Berlin established in 1885 (Alexandre, 1979; Clarence-Smith, 1985; Newitt, 1981), economic relations then established with South African Union (South Africa today) following the expansion of the powerful Mining Economy in the Transvaal – definitively set Mozambique's destiny throughout the twentieth century. The change of capital from Island of Mozambique, in long economic decline since the mid-eighteenth century, to Lourenço Marques (in 1898) was a mark of the strong symbolic value of that moment (Lobato, 1970, 1996b). The economic integration operated between the two territories, executed through the transit between Transvaal and the Port of Lourenço Marques, as well as the emigration of workers for the Rand Mine (First, 1983), financially assured the external accounts of the colony of Mozambique up to the end of the 1950s (Leite, 1990). This was how they compensated for the structural deficit of the trade balance which resulted from the institutionalization, in the 1930s, of the Colonial Pact (exchange of colonial cotton for wine and metropolitan textiles) instituted by the Administration of the Estado Novo (Leite, 1993a). It is certain that from the 1960s, the new economic policy of late colonialism (*Política de Integração do Espaço Económico Português/Policy of Integration of the Portuguese Economic Space*) came to frame relations between Mozambique, Portugal and the rest of the world (Leite, 1993b: 199), in the context of the challenge to colonialism that led to the outbreak of the liberation struggle in 1964 (Monteiro, 1993). However, ties with the neighbouring economy are maintained essentially unchanged and make the region to the south of the Save emerge as a fundamental and more visible pole of the economic and social development of the colony. This embedding of Mozambique in Southern Africa, a decisive event in its contemporary history, from the dawn of the twentieth century, operated, in our opinion, from the systemic point of view, a socio-economic, political and cultural structuring of consequence, perhaps more important, profound and lasting than the mark left by Portuguese colonization. It produced a geographical, economic, political and cultural hierarchization of and in Mozambican society. In the last instance, it had the power to marginalize and conceal the diversity of worlds that constitute a national identity under construction, inexorably inspired by the historical legacies proper to the populations of the North and Centre. This is what happened, in the view of many people, to the World of the Indian Ocean and to its age-old history, of which Mozambique is part, as we have had the opportunity to argue in this essay. But even so, that broad littoral territory, socially and culturally diverse, real and equally imaginary, that social analysis recently designated, not without propriety, the 'old Mozambique' (Chichava, 2007), constitutes today more than ever before, a profound mark of Mozambiqueness. Ironically, everything indicates that this oceanic component of Mozambican identity, an integral part of one of the first 'World Economies' (Braudel, 1949, 1958), was ignored until the beginning of the twenty-first century by the post-colonial political elites, attracted by the cultural and economic influence of the South African model and actors of the domination of North by South. The adoption, through unprecedented modes, of some of the ways of opening Mozambique to the Indian Ocean and the World is owed today to the recent globalization, moved by neoliberal and financial capitalism, and the resurgence of China (Broadman, 2007). The current discovery of the wealth of gas in the north of Mozambique has contributed to this. Formerly in the shadows, though historically close, today these new routes mobilize intensely and not always peacefully, men, opposing interests, material and cultural resources and capitals and competences.

As in the short story 'The Enchanted Cloth' by João Paulo Borges Coelho, these new directions and routes retake diversely and also 'unconceal' ancient geographies and sharing, religiosities and exchanges that the Indian Ocean, in its long and erased history, peopled by phantasmic narratives the literatures of the Indian Ocean. The *NILUS* project is intended to 'unconceal' the representations of that Indian Ocean narrative network, whose cultural, historical and social specificities give us the measure of the complexity of an 'inter-regional arena' (Bose, 2006)[1] offering the possibility of presenting a new transnational paradigm for the Global South (Hofmeyr, 2007).

Bibliography

Alexandre, V. (1979). *Origens do colonialismo português moderno*. Lisbon: Sá da Costa.

1 'arena inter-regional'.

Alpers, E. (2000). 'Recollecting Africa: Diasporic Memory in the Indian Ocean World'. *African Studies Review*, 43 (1), 83–99.
——. (2009). *East Africa and the Indian Ocean*. Princeton: Markus Wiener Publishers.
——. (2015). 'Maritime Mozambique'. Tsingy: Revue de l'OIES (Océan Indien, Espace et Sociétés), *Université de La Réunion. Dossier: Le Mozambique*, (18), 15–34.
Antunes, L. F. D. (1992). *A actividade da companhia de comércio dos Baneanes de Diu em Moçambique (1686–1777)*. Masters thesis, Universidade Nova de Lisboa, Lisbon.
——. (1998). 'Os mercadores baneanes guzerates no comércio e a navegação da Costa Oriental Africana (século XVIII)'. *Actas do Seminário 'Moçambique: navegações, comércio e técnicas'*, Faculdade de Letras da Universidade Eduardo Mondlane de Maputo, 1996, pp. 67–93. Lisbon: Comissão Nacional para as Comemorações dos Descobrimentos Portugueses.
Austen, R. (1987). *African Economic History: Internal Development and External Dependency*. London: James Currey.
Baltasar, J. A. d. S. (2016). *Rumo ao hinterland: a evolução social dos prazos do vale do Zambeze (séculos XVII e XVIII)*. Masters thesis, Universidade Nova de Lisboa, Lisbon.
Beach, D. N. (1980). *The Shona and Zimbabwe, 900–1850: An Outline of Shona History*. Gwelo: Mambo; London: Heinemann; New York: Africana.
Bento, C. L. (1993). *As ilhas de Querimba ou de Cabo Delgado. Situação colonial, resistências e mudanças (1742–1822)*, 2 vols. PhD Thesis, Instituto Superior de Ciências Sociais e Políticas da Universidade Técnica de Lisboa, Lisbon.
Birmingham, D. and Martin, P. (eds). (1992 [1983]). *History of Central Africa*, 2 vols, 4th edn. London: Longman.
Bose, S. (2006). *A Hundred Horizons: The Indian Ocean in the Age of Global Empire*. Cambridge: Harvard University Press.
Boxer, C. R. (1969). *The Portuguese Seaborne Empire, 1415–1825*. London: Hutchinson.
——. (1980). *Portuguese India in the Mid-seventeenth* Century. London: Oxford University Press.
Braudel, F. (1949). *La Méditerranée et le Monde Méditerranéen à l'époque de Philippe II*. Paris: Armand Colin.
——. (1958). 'Histoire et sciences sociales: la Longue Durée'. *Annales: Histoire, Sciences Sociales*, 13 (4), 725–753.
Broadman, H. G. (2007). *Africa's Silk Road: China and India's New Economic Frontier*. Washington, DC: The World Bank.
Capela, J. (1995). *Donas, Senhores e Escravos*. Porto: Ed. Afrontamento.

——. (2002). *O Tráfico de Escravos nos portos de Moçambique*. Porto: Ed. Afrontamento.

Catlin-Jairazbhoy, A. and Alpers, E. (eds). (2004). *Sidis and Scholars: Essay on African Indians*. Noida, UP: Rainbow Publication.

Chichava, S. I. (2007). *Le «vieux Mozambique». Étude sur l'identité politique de la Zambézie*. PhD Thesis, l'Institut d'Etudes Politiques de Bordeaux, Université Montesquieu-Bordeaux IV.

Clarence-Smith, W. G. (1985). *The Third Portuguese Empire 1825–75: A Study in Economic Imperialism*. Manchester: Manchester University Press.

——. (1989). 'Indian Business Communities in the Western Indian Ocean in the Nineteenth Century'. *The Indian Ocean Review*, 2 (4), 18–21.

Constantin, F. (ed.). (1987). *Les voies de l'islam en Afrique Orientale*. Paris: Karthala.

Curtin, P. D., Feierman, S., Thompson, L. and Vansina, J. (1987). *African History*, 10th edn. London: Longman.

First, R. (1983). *Black Gold: The Mozambican Miner, Proletarian and Peasant*. Brighton: Harvester Press.

Frenz, M. (2014). *Community, Memory and Migration in a Globalizing World: The Goan Experience, c.1890–1980*. New Delhi: Oxford University Press.

Hofmeyr, I. (2007). 'The Black Atlantic Meets the Indian Ocean: Forging New Paradigms for Transnationalism for the Global South: Literary and Cultural Perspectives'. *Social Dynamics: A Journal of African Studies*, 33 (2), 3–32.

Hoppe, F. (1970). *A África Oriental portuguesa no tempo do Marquês de Pombal, 1750–1777*. Lisbon: Agência Geral do Ultramar.

Horton, M. (1987). 'The Swahili Corridor'. *Scientific American*, 257 (3), 86–93.

——. (2001). 'The Islamic Conversion of the Swahili Coast 750–1500; Some Archaeological and Historical Evidence'. In Amoretti, B. S. (dir.), *Islam in East Africa: New Sources* (Archives. Manuscripts and Written Historical Sources. Oral History, Archaeology). International Colloquium (Rome, 2–4 December 1999), pp. 449–469. Rome: Università degli studi di Roma 'La Sapienza', Herder.

——. (2012). 'Did Asia Underdevelop Africa? An Archaeological Perspective on African Poverty'. In *Understanding African Poverty in the Longue durée*, Accra, International Institute for Advanced Study of cultures, Institutions, and Economics Enterprise (ILAS).

Horton, M. and Middleton, J. (2000). *The Swahili; the Social Landscape of a Mercantile Society*. Oxford: Blackwell.

Isaacman, A. (1972). *Mozambique: The Africanisation of a European Institution, the Zambezi Prazos, 1750–1902*. Madison: University of Wisconsin Press.

Kagabo, J. (1988). 'Les réseaux marchands arabes et swahili en Afrique orientale'. In D. Lombard and J. Aubin (eds), *Marchands et hommes d'affaires asiatiques dans l'océan Indien et la mer de Chine, 3e-20e siècles*, pp. 237–252. Paris: Éditions de l'EHESS.

Khouri, N. (ed.). (2008). 'L'Afrique des Indiens', *Outre–Mers Revue d'Histoire*, 360–361.

Khouri, N. and Leite, J. P. (2008). 'La presse coloniale portugaise du Mozambique et les Indiens 1930–1975'. *Lusotopie*, XV (1), 27–35.

——. (2012). 'The Ismailis of Mozambique: History of a Twofold Migration (Late 19th Century–1975)'. In E. Morier-Genoud and M. Cahen (eds), *Imperial Migration: Colonial Communities and Diaspora in the Portuguese World*. New York: Palgrave Mcmillan.

——. (eds). (2014). *Ismailis Khojas du Mozambique colonial à la Globalisation*. Paris: Harmattan.

Leite, J. P. (1990). 'La reproduction du réseau impérial portugais:quelques précisions sur la formation du circuit d'or Mozambique/Portugal'. *Estudos de Economia*, X (3), 365–401.

——. (1993a). 'Mozambique 1937–1970: Bilan de l'évolution de l'économie d'exportation. Quelques réflexions sur la nature du "Pacte colonial" '. *Estudos de Economia*, XIII (4), 387–410.

——. (1993b). 'Colonialismo e industrialização em Moçambique: Pacto colonial, dinamização das exportações e "import substitution" (1930–74)'. *Ler História*, 24, 53–70.

——. (1995). 'A economia do caju em Moçambique e as relações com a Índia: dos anos 20 ao fim da época colonial'. *Ensaios de homenagem a Francisco Pereira de Moura*, pp. 631–653. Lisbon: ISEG/UTL.

——. (1996). 'Diáspora Indiana em Moçambique. Em torno da presença indiana em Moçambique – Séc. XIX e primeiras décadas da época colonial'. *Economia Global e Gestão*, 2 (96), 67–108.

——. (2001). 'Indo-britanniques et indo-portugais: présence marchande au Sud de Mozambique au moment de l'implantation du système colonial, de la fin du XIX ème siècle jusqu'aux années 1930'. *Outre-Mers Revue d'Histoire*, 13–37.

——. (2007). 'A Administração belga face à presença asiática na região dos Grandes Lagos – do período da ocupação aos anos 30'. In A. Paiva (ed.), *Sob o signo de Horus – Ensaios de Homenagem a Eduardo Sousa Ferreira*. Lisbon: Edições Colibri.

Leite, J. P. and Khouri, N. (eds). (2008). 'Indiens des cinq colonisations de l'Afrique Orientale: mobilités et identités en diaspora de 1870 à nos jours'. *Lusotopie*, XV (1), 27–35.

Leite, J. P. and Khouri, N. (2012). *Os Ismailis de Moçambique-Vida económica no tempo colonial*. Lisbon: Edições Colibri.

Lobato, A. (1970). *Lourenço Marques, Xilunguíne. Biografia da cidade.* Lisbon: Agência Geral do Ultramar.

Lobato, M. (1996a). 'Redes mercantis e expansão territorial. A penetração portuguesa no vale do Zambeze e na África Austral durante o século XIX (1798–1890)'. *Studia*, 54–55, 165–210.

——. (1996b). 'A ilha de Moçambique antes de 1800. Algumas notas sobre a história e o património de uma capital colonial'. *Oceanos*, 25, 10–26.

Medeiros, E. (1988). *Etapas da escravatura no norte de Moçambique.* Maputo: Arquivo Histórico de Moçambique. Núcleo Editorial da Universidade Eduardo Mondlane.

Monteiro, F. A. (1993). *O Islão, o Poder e a Guerra (Moçambique 1964–74).* Porto: Universidade Portucalense.

Newitt, M. (1981). *Portugal in Africa: The Last Hundred Years.* London: C. Hurst & Company.

——. (1995). *History of Mozambique.* London: C. Hurst & Company.

Pélissier, R. (ed.). (1984). N*aissance du Mozambique. Résistance et révoltes anticoloniales (1854–1918)*, 2 vols. Orgeval: Éditions Pelissier.

Pinto, J. (1992). *Slavery in Portuguese India (1510–1842).* Goa: Himalaya Publishing House.

Randles, W. G. L. (1975). *L'empire du Monomotapa du XV au XIX* siècle. Paris: Mouton.

Rita-Ferreira, A. (1980). 'Moçambique e os naturais da Índia portuguesa'. *Seminário Internacional de História Portuguesa*, pp. 615–648. Lisbon: IICT./CEHCA.

——. (1982). *Presença Luso-Asiática e Mutações culturais no sul de Moçambique.* Lisboa: IICT/JICU.

Rocha, A. (1989). 'A resistência em Moçambique: o caso dos Suaíli, 1850–1913'. *Actas da I ª Reunião Internacional de História de África*, pp. 581–612. Lisbon: ICT/CEHCA.

Rosário, C. (2021). *Donas da Terra (Female Owners of the Land). Decolonizing Historical Representations through an Ethnography of Memories of Women of Power and Authority in Zambezia, Mozambique.* PhD Thesis, The University of Bergen.

Sheriff, A. (1987). *Slaves, Spices & Ivory in Zanzibar: Integration of an East African Commercial Empire into the World Economy, 1770–1873.* London: James Currey.

Subrahmanyam, S. (1993). *The Portuguese Empire in Asia 1500–1700: A Political and Economic History.* London: Longman.

Trimingham, J. S. (1980). *Islam in East Africa.* New York: Arno Press (Oxford: Oxford University Press, 1964).

MANUEL LOBATO

3 Travel literature and the colonial matrix on the Indian Ocean periphery: From Mozambique to Timor

Etiópia Oriental e Cousas de Vária História do Oriente is considered an unusual work that is difficult to place in any one literary genre. It is essentially a collection of geographical, historical and missionary narratives that have been published in one place, combining the register of memoire with the more mundane content typical of travel literature in that era. The edition that Luís de Albuquerque produced is a complex, heterogeneous work. The text of *Etiópia Oriental* itself, published in what constitutes the first part of this lengthy work, generally follows a coherent, clearly delineated plan with explicit thematic links. The second part, in contrast, and as the title *Cousas de Vária História* suggests, does not provide any consistent narrative links at all, highlighting rather an absence of thematic unity and a lack of organization in the way the materials are arranged. This variety encompasses a very wide range of themes and literary genres, from accounts of shipwrecks and battles to narratives collected from Dominican martyrology.

This heterogeneity in respect of literary genre, was probably designed to reach as broad a readership as possible. However, in a way that is perhaps less immediately obvious, other unacknowledged interests are also discernible in this publishing project. These relate primarily to the general layout of the work and to the selection of the materials included in it, as well as certain materials that one might expect to find there, which are absent. We believe that a series of very particular circumstances led to the publication of *Etiópia Oriental.* Not least among these is the fact that it contains no references at all to the activity in Mozambique of the Jesuits, who were not only rivals of the Dominicans but were actually pioneers in missionizing the interior of the East African coast, where they had arrived before the Order of São Domingos. In effect, the Jesuits had founded two short-lived missions, the first in the hinterland that adjoined Inhambane, in the centre-south of Mozambique,[1] and the second at the court of the 'emperor' Mwene Mutapa. When these missions failed,[2] the Jesuits decided to abandon the captaincy of Mozambique and Rios, but they were to return there in 1608, having previously announced in Goa that this was their intention.[3]

Read in this context, the publication of *Etiópia Oriental* appears to have responded very much to the moment, with the Dominicans seeking to re-assert their control over the missionary space which their order had become accustomed to considering their sole monopoly. When the Jesuits re-established themselves in Mozambique, they represented a threat to the exclusive nature of the missionary endeavour over which the Dominicans had presided for forty years. Thus, the publication of the text will certainly have been influenced by this context of uncertainty regarding the direction that missionization in Mozambique and Rios de Cuama was about to take. It is an episode that fits readily into the general climate of covert, and sometimes even overt, competition between the different missionary orders that

1 In 1560, the Jesuits established themselves in Tanga, in the kingdom of Gamba, on the left bank of the Inharrime River, 150 miles to the southwest of Inhambane. See J. F. Feliciano, 'Os encontros e desencontros das Missões. Uma leitura sociológica sobre o Leste Africano no século XVI', *Actas do Congresso Internacional de História. Missionação Portuguesa e Encontro de Culturas*, II, Braga: 1993, pp. 67–76.

2 The failure of the mission, after the death of its principal protagonist, led the Jesuits to support the conquest of the Zambezi valley, where there were supposedly a large number of unexploited gold mines, in an undertaking inspired by the Hispano-American model of territorial conquest. See Bertha Leite, *D. Gonçalo da Silveira*. Lisboa: AGC, 1946; 'Relação da Viagem que fizeram os Padres da Companhia de Jesus com Francisco Barreto na conquista de Monomotapa no anno de 1569 feita pelo padre Monclaro da mesma companhia', *Boletim da Sociedade de Geographia de Lisboa*, 4ª Série, Lisboa, 1885, pp. 10–11: 492–508, 542–562.

3 J. A. Guerreiro, 'A acção missionária e a sua organização canónica, em Moçambique, no período filipino (1581–1640)', *Actas do Congresso Internacional de História dos Descobrimentos*, V, Parte II, Lisboa: Comissão Executiva das Comemorações do V Centenário da Morte do Infante D. Henrique, 1961, p. 184.

constituted Portuguese patronage in the East. This also explains why certain areas of missionary activity in Africa, in what was known as Ethiopia, as João dos Santos calls it, receive scant attention in *Etiópia Oriental*. This is the case, for example, with 'Etiopia, a Alta', or Abyssinia, where Jesuits were in the majority. So this work, for all its countless merits, wound up in some respects becoming a massive propaganda exercise for the missionary activities of the Dominican Order in Mozambique, a region the Jesuits had previously abandoned but to which they were now returning, precisely at the point when this text was coming off the press in Évora.

Whatever the intentions of the Dominican publishers in Évora may have been, João dos Santos himself did not simply narrate the missionary successes of his own order or the military feats of the Portuguese. What he ultimately achieved was an unparalleled contribution to reformulating the image of Africa and the 'Ethiops' which had been handed down from the Early Middle Ages, and had subsequently remained in circulation, alongside newer information provided by the empirical *in situ* observations of Portuguese travellers.[4]

Among the authors from the later period in what is known as the 'geographical discoveries', João dos Santos is justly recognized for his finely honed powers of observation. Indeed, his forays into the natural world are some of the most detailed we possess. In this respect, his work was a precursor for later naturalists and botanists, some of them also members of religious orders, who worked on Brazil in the seventeenth and eighteenth centuries. As the editor of his personal memoirs, as well as a huge body of ethnographic material which he gathered directly on the ground, Santos emphasized the superiority of eye-witness accounts over written manuscripts and printed sources and he justified this by referring to the inaccuracies and mistakes he had found in printed works in circulation at the time, particularly the *Crónica do felicíssimo rei D. Manuel*, by Damião de Góis.

However, the geographical area which João dos Santos conceptualized as 'Ethiopia' is not what one would anticipate. Instead of limiting

4 On the image of Africans in Portuguese travel literature, see J. da Silva Horta, *A 'Guiné do Cabo Verde': produção textual e representações (1578–1684)*. Lisboa: FCG-FCT, 2011.

Ethiopia *stricto sensu* to Abyssinia, which he calls 'Etiópia, a Alta', as you would expect to see in a written work at the beginning of the seventeenth century, he used the term in a kind of generic sense, extending the designation 'Ethiopia' to cover that huge part of the African continent which we would now label 'black' or 'Sub-Saharan Africa'. Thus, what he calls 'East Ethiopia' corresponds to present-day 'East Africa' and this includes 'Abyssinia' or 'Ethiopia' proper. This represents something of a step backwards in respect of the term 'Africa' adopted half a century earlier by João de Barros, who had abandoned the Ptolemaic geographical system which still featured in Portuguese texts about the discovery of the Gulf of Guinea and its coast,[5] following instead the toponymy of Persian-inspired[6] Arab geographers. Thus, João dos Santos refused to include East Africa in the vague, generic Ptolemaic concept of the Indies, which was sometimes more precisely divided into 'Middle India' and 'Ethiopic India',[7] terms which had come down from Medieval geography, both Arab and Christian. He opted instead for the late Medieval definition which gave Ethiopia huge dimensions stretching from the Atlantic to the Indian Ocean, rather than following the classical authors, some of whom did also use the term 'Ethiopia' in the sense referred to above, to mean 'black' or 'Sub-Saharan' Africa.

In respect of Asia, João dos Santos's idea of geography is eccentric to say the least, and at times disconcerting. He sets out a long list of the regions which make up Asia, according to the geographical conceptualizations of the Greeks but to this he adds other regions which had become known through Portugal's maritime voyages. However, the criteria he adopts for updating the classics seem decidedly narrow. From the countries and regions visited by the Spanish and the Portuguese, he singles

5 M. S. Lopes, 'Portugal: uma fonte de novos dados. A recepção dos conhecimentos portugueses sobre África nos discursos alemães dos séculos XVI e XVII', *Mare Liberum*, I (Dez. 1990), pp. 215–216.

6 João de Barros, *Década Primeira da Ásia*, Capítulo IV. 'Em que se descreve a parte da costa de África em que está situada a cidade Quíloa; à qual terra os arábios propriamente chamam Zanguebar e Ptolomeu Etiópia-sobre-Egipto'.

7 Cf. V. Magalhães Godinho, 'Índias, as Etiópias e o Nilo'. *Dicionário de História de Portugal*, Joel Serrão (dir.), 2.ª ed., vol. III. Porto: Liv. Figueirinhas, 1985, pp. 292–298.

out those where the Dominicans were well established and successful. Thus, he ends up highlighting some tiny islands in the Malay-Indonesian archipelago which the Portuguese visited regularly and where, at different points in time, they possessed fortified factories. Among the islands in this group, particular emphasis is placed on Solor, an almost unique example of Portuguese colonization that was military-missionary as well as mercantile and was run by the Dominican fathers. The only known parallel is the colonialization of what was then called the Rios de Cuama region, the Zambezi valley, where Catholic missions had been almost exclusively controlled by the Dominicans up to that point. There they actively contributed to the establishment of a Portuguese colonial system known as 'estates of the crown', since Dominican friars were among the first to receive estates of land granted by the crown in those areas.

João dos Santos was also the first European to pronounce on the uses and customs of various African ethnic groups, providing a historical and ethnographic source which remained a fundamental work of reference for a very long time. Santos's work was widely disseminated from the outset, through authors who transcribed it, summarized it or paraphrased it, such as Samuel Purchas in England and also, notably, the Jesuit Alonso de Sandoval, in his *De instauranda Aethiopum salute*, published in Seville in 1627 under the title, *Naturaleza, policía sagrada i profana, costumbres i ritos, dissiplina i catecismo evangélico de todos los etíopes*. It was then republished twenty years later under the Latin title by which it came to be best known.[8] Sandoval summarized and paraphrased entire chapters of *Etiópia Oriental*, but in contrast to Santos he reverts to the old Christian Biblical stance of problematizing and differentiating the 'negro' or the 'black man' in racial terms, whereas Santos had described an African world that was ethnically and culturally coherent.

A body of knowledge about the African interior only emerged later. Only a limited number of seventeenth-century texts about the east coast of Africa found their way into publication. The most significant narratives

8 Cf. Alonso de Sandoval, *Un tratado sobre la esclavitud* [De instauranda Aethiopum salute], E. V. Vilar (ed.). Madrid: Alianza Editorial, 1987, pp. 35 e 43.

in this category, such as *Viagem* by António Gomes[9] or the 'accounts' of Manuel Barreto,[10] and Fr António da Conceição, were not published until the nineteenth and twentieth centuries. This explains why geographers and cartographers, as well as encyclopaedists, authors of the illustrated repertories, which were much in vogue at the time, and other scholars working from the confines of their studies in Europe, continued to base their writings on the old materials they had to hand. These usually included *Etiópia Oriental*, which remained an essential reference work right up to the mid-nineteenth century, particularly the chapters dealing with Monomatapa, even though they were describing a social and political system which had long since disappeared. It was actually the travels of Livingstone which updated our scientific knowledge of these regions, since they were inaccessible to Europeans for long periods of time when the Monomatapa 'empire' was crumbling under pressure from the Roswi states. That was when the writings of João dos Santos finally became obsolete.

João dos Santos also paid fairly close attention to the Dominican presence on the Lesser Sunda Islands (Nusa Tenggara Islands), especially the islands of Solor, Lomblem and Adonara, since there were no Dominican missions, nor was there any established Portuguese presence, on Larantuca (Flores) or Timor, in this period. Dominican missionary activity on these latter islands occurred only later and was not widely disseminated until the eighteenth century by Friar Lucas de Santa Catharina, in the *Quarta Parte,* of his *História de S. Domingos*. This came out in 1733 and was based on an assortment of different accounts that had remained unpublished. However, as Hans Hägerdal has already indicated, this apologist literature, which focussed on ennobling the Portuguese empire and on the feats of the Dominican friars, distorted and minimized the role played by the Timorese

9 Eric Axelson (ed.), 'Viagem que fez o Padre Antonio Gomes, da Companhia de Jesus, ao Imperio de Monomotapa; e assistencia que fez nas ditas terras de algus anos' [1648], *Stvdia* 3 (1959), 155–242.

10 Padre Manuel Barreto, 'Informação do estado e conquista dos Rios de Cuama vulgar e verdadeiramente chamados Rios de Ouro' [1667], *Boletim da Sociedade de Geografia de Lisboa*, 4ª Série, 1 (1883), 33–59.

as protagonists in a shared history.[11] Their uses and customs receive scant attention and are invariably demonized, much as João dos Santos also did in respect of the Africans, since both author and reader shared the same prejudices in their inevitable condemnation of the beliefs and practices those texts describe.[12] João dos Santos does, however, show occasional moments of sympathy for the Africans, and certain passages provide a glimpse into their day-to-day lives.

Curiously enough, the first fairly detailed ethnographic view of the uses and customs of the Timorese was produced not by Portuguese authors, but by Dutch and French travellers and officials. In fact, in 1781 a 'Descrição da ilha de Timor até agora conhecida, com a descrição de algumas ilhas próximas' appeared written by Willem de Hogendorp[13] who became one of the first governors of the Dutch East Indies, following the demise of the VOC, the famous Dutch trading company. Hogendorp's text is actually based on a French description which remained unpublished until fairly recently, written in 1777 by Jean-Baptiste Pelon, a French engineer who worked for the VOC and lived in Timor for seven years.[14] A French translation of Hogendorp's text appeared in 1810,[15] a year before the publication of the first genuinely informational text to deal with Timor in Portuguese, the 'Instruções para o Capitão de Mar e Guerra, Vitorino Freire da Cunha Gusmão', known as 'Documento Sarzedas'. Coincidentally or otherwise, the French translation of Hogendorp figured among the sources for the 'Timor'

11 Hans Hägerdal, *Lords of the Land, Lords of the Sea: Conflict and Adaptation in Early Colonial Timor, 1600–1800*. Leiden: Brill, 2012, p. 9.

12 M. Lobato e V. Rodrigues, 'Algumas considerações sobre a circulação de informação no Império do Oriente'. In M. E. M. Santos e M. Lobato (orgs.), *O Domínio da Distância*. Lisboa: IICT, 2006, pp. 45–52.

13 W. van Hogendorp, 'Beschrijving van het eiland Timor, voor zoo verre het tot nog toe bekend is', *Verhandelingen van het Bataviaasch Genootschap der Kunsten en Weetenschappen* 2 (1781), 405–431.

14 Jean-Baptiste Pelon, *Description de Timor occidental et des îles sous domination hollandaise (1771–1778)*, texte établi, présenté et annoté por Anne Lombard-Jourdan. Paris: *Cahier d'Archipel* 34, 2002.

15 M. W. van Hogendorp, 'Description de l'île de Timor; par M. W. van Hogendorp, traduit du hollandais par M. Depping', *Annales des voyages, de la géographie et de l'histoire*, VI. Paris: F. Buisson, 1810, pp. 273–327.

entry in the *Dicionário geographico das provincias e possessões portuguesas* by Souza Monteiro, published in 1850.[16]

However, the most extraordinary piece of travel literature about Portuguese Timor probably remains the diary that resulted from Rose and Louis-Claude de Freycinet's visit to Dili in 1818. They were a French couple and they were naturalists as well as travellers. The circumstances surrounding their trip were rather colourful. As a young man, Freycinet had taken part in a scientific expedition in 1799 led by the naturalist François Peron. In 1810, he published the scientific results of that expedition. In 1818, aged 36, he decided to organize his own expedition and this initiative was sponsored by the Institut Français [French Institute] and the French Maritime and Overseas Ministry, under the personal patronage of Louis XVIII.

Louis-Claude de Freycinet had married Rose Marie Pinon when she was 18. They conspired (ostensibly together, although she seems to have come up with the idea) to smuggle Rose aboard the *Uranie* dressed as a boy, because they did not want to be separated for the three years of the expedition, and women were not, at the time, allowed onboard French navy vessels. And this is precisely what they did, to the horror of the ship's officers, although they drew more varied reactions from the newspapers of the day.[17] Rose thus became the first woman to write a diary about the circumnavigation of the globe, although the form her account ultimately took was a series of private letters addressed to her childhood friend, Caroline de Barillon, Baronness de Nanteuil. The *Uranie*'s voyage around the world was one of the most successful French scientific expeditions of its day. Inspired by the continued influence of the Enlightenment, the aim of the expedition was to deliver a huge number of observations, collections and scientific recordings. Between 1824 and 1844, L. C. de Freycinet produced a total of thirteen volumes of accounts, four of which were atlases,

16 Joze Maria de Souza Monteiro, *Diccionario geographico das províncias e possessões portuguezas no Ultramar (introdução geographico-politico-estatistico-historicada de Portugal por)*. Lisboa: Typographia Lisbonense, 1850.

17 Marc S. Rivière (ed.), *A Woman of Courage: The Journal of Rose de Freycinet on Her Voyage around the World 1817–1820*. Canberra: National Library of Australia, 2003 [1996], p. XIX.

in addition to the diaries kept by Rose Freycinet and by the expedition's artist, Jacques Arago, although these are not part of the official corpus.

The Freycinets reached Díli in November 1818. All three of the expedition diaires – Louis-Claude's,[18] Rose Freycinet's[19] and Jacques Arago's[20] – describe their arrival in Díli and their short, five-day visit to the Timorese capital. However, the sections relating to Díli in the Freycinet diaries are essentially one and the same text. Since Rose's diary was not published in its entirety until 1927, it only emerged then that her husband had in fact reproduced in 1824, with a few minor alterations and edits, the same observations and comments originally penned by Rose in her personal diary. This appropriation on the part of Louis-Claude, which may have been with Rose's consent, is readily explained by the fact that Rose did not envisage publishing it herself. Because she was not officially allowed on the expedition in the first place, she was not in a position to put her name to any accounts of it. This will also have been the reason why Louis-Claude makes no reference to Rose being part of the expedition, either in his own diary or the official account of the trip. She was also omitted from the illustrations by the artist Jacques Arago, which form part of the official account. Consequently, Rose is missing from the published engraving which depicts their arrival in Díli, an episode in which she did in fact take centre stage, as becomes evident when this engraving is compared with two water colours of the same scene, published much later, one by Jacques Arago, and the other by the *Uranie's* second in command, Alphonse Pellion.[21]

18 Louis-Claude de Freycinet. *Voyage Autour du Monde Pendant Les Années 1817, 1818, 1819 et 1820*. Paris : Pillet Aimé, 1824.

19 *Journal de Madame Rose de Saulces de Freycinet*, d'après le manuscrit original accompagné de notes par Charles duplomb. Paris: Société d'éditions Géographiques, Maritimes et Coloniales, 1927.

20 Jacques Arago, *Promenade Autour du Monde Pendant les Années 1817, 1818, 1819 et 1820 sur les Corvettes du Roi L'Uranie et la Physicienne, Commandés par M. Freycinet.* Paris: Leblanc, 1822.

21 'Rose and Louis de Freycinet in the *Uranie* an illustrated research essay for the WA Museum's Journeys of Enlightenment exhibition 2008'. Dr M. McCarthy Curator. Department of Maritime Archaeology. WA Museum. Report-Department of Maritime Archaeology No. 236.

Díli took on a lively, festive atmosphere during their visit. Arago makes it clear how much he enjoyed the warm welcome the French expeditionary party received from the governor, Azevedo de Sousa. Overwhelmed by de Sousa's hospitality, Arago appears also to have been somewhat surprised by how relaxed the Portuguese colonial system was in the matter of social relations between the Portuguese authorities and the Indian chieftaincies: 'The Rajas and their dependents are by no means cowed by the yoke of despotism as they would be in Coupang (the capital of Dutch Timor). On the contrary, they are treated with a degree of distinction. They are received in the governor's apartments at all hours and they are even allowed to eat at his table.'[22]

Enjoying a series of banquets, *soirées* and excursions to the areas around Díli, the French delegation was fêted by the most distinguished members of colonial society, including the Lieutenant Colonel and Captain-Major of the province of Belos, D. Francisco de Assis Durão. According to Rose, he introduced himself as a Portuguese person who was born in the colony, but he will undoubtedly have been a mixed-race leader, of royal blood, as evidenced by the title 'Dom', which the Portuguese administration conferred on Timor's native kings. D. Francisco was married to a lady of mixed race, named Dona Joaquina but also sometimes identified as 'Dona Joana'. The extremely old-fashioned clothes and the ostentatious jewellery worn by the elite ladies of Díli did not escape the attention of Rose Freycinet when she attended a greeting ceremony hosted by the Raja's daughters who were married to Portuguese officials. Rose noted that they all owned large numbers of female slaves, all of whom were nicely dressed in traditional Timorese style, adorned with gold chains and carrying little pouches of betel nut for their mistresses. Conversation was conducted in Malay, which Louis-Claude could speak reasonably well so he interpreted for Rose. She thought the uniforms of the Timorese soldiers looked ridiculous but she enjoyed their war dances and music. She described the gastronomy as being an inter-ethnic multicultural mix of Portuguese and Chinese, possibly rather like the cuisine in Macau.

22 Arago, cit., p. 332.

The descriptions of staying in Díli in these two texts – the diaries of Jacques Arago and Rose Freycinet – include some astute observations regarding the governor and his officials, who go out of their way to lavish hospitality on their 'wise' and 'learned' French guests. There are some positive comments from Arago regarding the governor's authority over the *liurae* or kings of the interior, as well as some rather caustic remarks about the wives of the administrative class in Díli, who are all 'Raja's daughters'.[23] Rose notes with dismay that the fashions worn by women in Díli's decent society in 1818, date back to … Louis XV, decades behind European designs. This clearly reflects the gulf that separated colonial society from European modernity, as well as the survival of a backward-looking colonial *ancien régime*, on the verge of extinction with the advent of liberalism. The governor offered the Freycinets four young Timorese slaves as a gift. Not wanting to offend their hosts, they accepted one, a boy named 'José António'. He was 6 or 7 years old at the time and died in Paris ten years later.[24]

The body of knowledge that was acquired from the *Uranie*'s expedition as well as the personal memoirs of its protagonists proved, over time, to be a rich seam of publishable material, which some of the original travellers, and even their descendants, were adept at exploiting. This was the case with the 'Nouvelle édition revue et augmentée, enrichie de notes scientifiques par M. François Arago', an 1868 re-edition of Jacques Arago's work,[25] which had already gone through several editions and had been a huge success ever since its first edition in 1822, cited above. It was followed by an English translation in 1823,[26] attesting to the commercial popularity enjoyed by this

23 Marnie Bassett (ed.), *Realms and Islands: The World Voyage of Rose de Freycinet in the Corvette Uranie 1817–1820*. Londres: Oxford University Press, 1962, pp. 104–106.

24 Bassett, cit., p. 107.

25 *Souvenirs d'un aveugle, voyage autour du monde, par Jacques Arago, … Nouvelle édition … enrichie de notes scientifiques par M. François Arago, … et précédée d'une introduction par M. Jules Janin*, Paris: 1868.

26 *Narrative of a Voyage Round the World, in the Uranie and Physicienne Corvettes, Commanded by Captain Freycinet, During the Years 1817, 1818, 1819, and 1820; on a Scientific Expedition Undertaken by Order of the French Government in a Series of Letters to a Friend, by J. Arago, Draftsman to the Expedition*, Londres: Treuttel and Wurtz, 1823.

type of travel literature. Unfortunately, the editor, who was the author's son, opted for a somewhat romanticized version, embellishing it with commentaries, dialogues and anecdotes that claimed to have been penned by the deceased Arago senior. The result was a decidedly suspicious, not to say spurious collection. It contains, for example, an erotic digression in which the supposed 'author' is to be found somewhere on the outskirts of Díli contemplating the road to Cupão, which did not exist at the time of the original text. Indeed, Cupão is two days away by car even today. This is definitely why René Pélisser decided to use the English translation of 1823 which was the first publication in English about the Freycinets' expedition.

Forty-five years earlier, in 1772, another French naval officer, François-Étienne de Rosily had spent eight months in eastern Timor. This was only three years after the Portuguese had moved their capital from Lifau to Díli, but Rosily's description of it was not published until 1982.[27] And the reader will be sorely disappointed if they are expecting a level of insight to rival the accounts of Rose Freycinet and Jacques Arago. There can be no doubt that the romantic narrative style of these two writers launched a whole new mode of travel writing about Portuguese Timor. And in so doing, it brought to an end the cycle that had begun with Friar João dos Santos, two centuries earlier.

Bibilography

Arago, J. (1822). *Promenade Autour du Monde Pendant les Années 1817, 1818, 1819 et 1820 sur les Corvettes du Roi L'Uranie et la Physicienne, Commandés par M. Freycinet*. Paris: Leblanc.

Axelson, E. (ed.). (1959). 'Viagem que fez o Padre Antonio Gomes, da Companhia de Jesus, ao Imperio de Monomotapa; e assistência que fez nas ditas terras de alguns anos' [1648]. *Stvdia*, 3, 155–242.

27 Anne Lombard-Jourdan, 'Un mémoire inédit de F.-E. de Rosily sur l'île de Timor (1772)', *Archipel* 23 (1982).

Barreto, P. M. (1883). 'Informação do estado e conquista dos Rios de Cuama vulgar e verdadeiramente chamados Rios de Ouro' [1667]. *Boletim da Sociedade de Geografia de Lisboa*, 4ª Série, 1, 33–59.

Bassett, M. (ed.). (1962). *Realms and Islands: The World Voyage of Rose de Freycinet in the Corvette Uranie 1817–1820*, pp. 104–106. London: Oxford University Press.

Feliciano, J. F. (1993). 'Os encontros e desencontros das Missões. Uma leitura sociológica sobre o Leste Africano no século XVI'. *Actas do Congresso Internacional de História. Missionação Portuguesa e Encontro de Culturas*, II, Braga, 67–76.

Freycinet, L.-C. d. (1824). *Voyage Autour du Monde Pendant Les Années 1817, 1818, 1819 et 1820*. Paris: Pillet Aimé.

Godinho, V. M. (1985). 'Índias, as Etiópias e o Nilo'. *Dicionário de História de Portugal*, Joel Serrão (dir.), 2.ª ed., vol. III. Porto: Liv. Figueirinhas.

Guerreiro, J. A. (1961). 'A acção missionária e a sua organização canónica, em Moçambique, no período filipino (1581–1640)'. *Actas do Congresso Internacional de História dos Descobrimentos*, V, Parte II. Lisbon: Comissão Executiva das Comemorações do V Centenário da Morte do Infante D. Henrique.

Hägerdal, H. (2019). *Lords of the Land, Lords of the Sea: Conflict and Adaptation in Early Colonial Timor, 1600–1800*. Leiden: Brill.

Hogendorp, W. v. (1781). 'Beschrijving van het eiland Timor, voor zoo verre het tot nog toe bekend is'. *Verhandelingen van het Bataviaasch Genootschap der Kunsten en Weetenschappen*, 2, 405–431.

——. (1810). 'Description de l'île de Timor; par M. W. van Hogendorp, traduit du hollandais par M. Depping'. *Annales des voyages, de la géographie et de l'histoire*, VI, pp. 273–327. Paris: F. Buisson.

Horta, J. d. S. (2011). *A 'Guiné do Cabo Verde': produção textual e representações (1578–1684)*. Lisbon: FCG-FCT.

Journal de Madame Rose de Saulces de Freycinet, d'après le manuscrit original accompagné de notes par Charles DUPLOMB. Paris: Société d'éditions Géographiques, Maritimes et Coloniales, 1927.

Leite, B. (1885). *D. Gonçalo da Silveira*. Lisboa: AGC, 1946; 'Relação da Viagem que fizeram os Padres da Companhia de Jesus com Francisco Barreto na conquista de Monomotapa no anno de 1569 feita pelo padre Monclaro da mesma companhia'. *Boletim da Sociedade de Geographia de Lisboa*, 4ª Série, Lisbon, 10–11: 492–508, 542–562.

Lobato, M. and Rodrigues, V. (2006). 'Algumas considerações sobre a circulação de informação no Império do Oriente'. In M. E. M. Santos and M. Lobato (eds), *O Domínio da Distância*, pp. 45–52. Lisboa: IICT.

Lombard-Jourdan, A. (1982). 'Un mémoire inédit de F.-E. de Rosily sur l'île de Timor (1772)'. *Archipel*, 23.

Lopes, M. S. (199). 'Portugal: uma fonte de novos dados. A recepção dos conhecimentos portugueses sobre África nos discursos alemães dos séculos XVI e XVII'. *Mare Liberum*, I, 215–216.

Monte, G. (1968). *Subsídios para a História da Tipografia em Évora nos séculos XVI a XVIII*, Évora.

Pelon, J. B. (2002). *Description de Timor occidental et des îles sous domination hollandaise (1771–1778)*, texto estabelecido, apresentado e anotado por Anne Lombard-Jourdan, p. 34. Paris: Cahier d'Archipel.

Rivière, M. S. (ed.). (2003 [1996]). *A Woman of Courage: The Journal of Rose de Freycinet on Her Voyage around the World 1817–1820*, p. XIX. Canberra: National Library of Australia.

Sandoval, A. (1987). *Un tratado sobre la esclavitud* [De instauranda Aethiopum salute], E. V. Vilar (ed.). Madrid: Alianza Editorial.

Souza Monteiro, J. M. d. (1850). *Diccionario geographico das províncias e possessões portuguezas no Ultramar (introdução geographico-politico-estatistico-historicada de Portugal por)*. Lisboa: Typographia Lisbonense.

JESSICA FALCONI AND ELENA BRUGIONI

4 The Indian Ocean in the commemoration of the Portuguese discoveries

Theoretical framework. Between the Atlantic and the Indian Ocean: Hegemonic perspectives in conflict

According to the ideas formulated by historian Michael Pearson, research into the Indian Ocean is characterized by a specific set of disciplinary and methodological considerations that are crucial, foundational elements in this area of studies, and at the same time enshrining some of the problems and tensions that Indian Ocean Studies must necessarily confront:

> A recent discussion in the very prestigious and widely read *American Historical Review* was guilty of a major sin of omission. A long introduction by Karen Wigen was followed by analyses of the Mediterranean, the Atlantic, and the Pacific. Curious indeed that the Indian Ocean was ignored. Could the reason for this be that for most of its history the Indian Ocean was crossed and used by people from its littorals, not by Europeans, while the three examples chosen by Wigen were all dominated by Europeans for most or all of their histories? This complaint about a Eurocentric approach applies to an extent to a very recent book, Seascapes, where again the Indian Ocean is largely absent and European-and American-controlled oceans and subjects are privileged. Indeed this Eurocentric bias goes back a long way. Braudel's study of the Mediterranean was notoriously weak on the southern, Islamic, shore of the sea. Even before this, early in the twentieth century, many European authorities considered the Indian Ocean to be only a 'half ocean' as it did not extend far into the Northern Hemisphere! Despite this neglect from the American academic mainstream, historical studies of the Indian Ocean are in fact flourishing. (Pearson, 2010: xv in Moorthy & Jamal, 2010)

The first question is related to the marginality of Indian Ocean Studies within contemporary historiographical thinking, evidencing, in the second place, the importance of these Studies in creating alternative critical and epistemological categories and paradigms for developing new perspectives for historical, social and cultural analysis. The question flagged up by Pearson regarding the absence of the Indian Ocean in the contemporary historiographical debate is posed in relation to the marginality of the European presence in this Ocean, highlighting the Eurocentric perspective as a possible reason for this apparent subalternization of the Indian Ocean. The presumption that different European nations and empires occupy marginal places within the space-time of the Indian Ocean constitutes one of the reasons why this field of study does not seem to benefit from the same dissemination that characterizes, for example, studies on the Atlantic or Pacific. In this respect, we should point out the quantitative difference between studies on the slave trade in the Atlantic Ocean and the Indian Ocean, or even the theoretical discussions concerning the identity matrix linked to the Atlantic, following on from Paul Gilroy's 1993 proposition. This aspect may also be highlighted from a more situated perspective, such as that of the historiographical, political and cultural studies pertaining to Portugal's maritime and colonial history, where the study of the Atlantic, in its different strands, stands out more than the studies of the Indian Ocean. Here the use of the category of the Atlantic is especially emblematic in relation to discourses about miscegenation and the use of the Luso-tropicalist ideology formulated by Gilberto Freyre, as well as in the theories that propose its critical and epistemological deconstruction (Almeida, 2000; Santos, 2001; Shohat & Stam, 2012). It is can therefore be stated that in postcolonial critical thinking, the Atlantic is configured as an established analytical category pointing to quantitatively significant critical production. On the other hand, it should be said that the Indian Ocean as a space of creolization, or rather, of *relation* (Glissant, 1990), signals an important set of critical and conceptual problems that have only been partially explored in the sphere of critical reflection in the field of contemporary cultural and literary studies.

Observing, for example, the theoretical production generated around what has come to be known as the South Atlantic, what stands out is a huge and heterogeneous body of historiographical, cultural and literary studies that configure the Atlantic Ocean and its southern region as an analytical category of obvious interest for a historical, political and cultural analysis, above all in the Portuguese and Brazilian context. On the other hand, taking account of the specific character of the relationship between Portugal and the Indian Ocean, both in the pre-modern period and successive eras,[1] Indian Ocean Studies offer a key critical perspective and therefore deserve a broader and more in-depth conceptual and analytical approach, especially if we bear in mind the meaning of the Indian Ocean in what we may define as the Portuguese great imperial narrative. Furthermore, with regard to Indian Ocean Studies in the contemporary Portuguese context, the pioneering work organized by Rosa Maria Perez, *Culturas do Índico* (Perez, 1998) should certainly be mentioned, as well as the *Revista Oceanos*, published by the National Commission for the Commemorations of the Portuguese Discoveries between 1989 and 2002, which will be addressed in the second part of this essay.

In addition, a broad and diverse corpus of studies – historical, anthropological, social and cultural – have characterized scientific production about the Indian Ocean, above all regarding Portuguese maritime history, between the coast of East Africa and Asia, in the imperial era and the modern period. Generally speaking, this corpus is characterized by approaches to national contexts or specific regions – Mozambique, India or, more specifically, Goa, among others. In fact, scant critical and epistemological reflection has been undertaken on what may be defined as the 'paradigm of the Indian Ocean', to be understood, on the one hand, as an analytical category, and on the other, as a fundamental premise for a transdisciplinary area of study that corresponds to the definition of Indian Ocean Studies.

Returning to Pearson's text, a second important question foregrounded by the scholar concerns the critical and epistemological potential of Indian Ocean Studies regarding the emergence of a specific conceptual and analytical lexicon (Pearson, 2010: xvi). From this perspective, the importance

1 In this regard, see some fundamental studies, such as: Alpers (1975, 2009), Bethencourt and Chaudhuri (1998–2000), Boxer (1963), Capela (2002) and Pearson (1998, 2003).

of the Indian Ocean, analysed from a literary and cultural point of view, lies not only in its uniqueness as an object of study but rather in the conceptual and epistemological renewal that can arise out of it, providing a crucial revision of the analytical categories that govern the discipline of historiography, as well as the approaches embedded in the realm of cultural studies, which constitutes the field of study that this essay sets out to address. In this regard, cultural and literary studies within the field of Indian Ocean Studies, according to Pearson, are 'unprecedented and innovative' (2011), standing out as key approaches with respect to the emergence of new conceptual and critical apparatuses. However, when we look at the essential bibliography of Indian Ocean Studies, a specific set of conflicting hegemonic disciplinary and diachronic perspectives stand out in quantitative terms, revealing what we could define as a set of hegemonic perspectives in conflict. As for the disciplinary dimension, outputs of a historiographical nature, in their political, economic, social and anthropological articulations, emerge as broader in quantitative and qualitative terms. At the same time, in a chronological perspective, approaches that examine the pre-modern era obviously predominate, while studies of the contemporary period are still relatively rare.

Our hypothesis aims to show how the initiatives and publications promoted under the aegis of the National Commission for the Commemoration of the Portuguese Discoveries are a meaningful point of departure – a beginning – for the consolidation of Indian Ocean Studies in Portuguese-speaking contexts. They are also configured as an inaugural moment for establishing a counterpoint between studies of the Atlantic and the Indian Oceans and going some way towards answering, if only partially, some of the questions that, as Michael Pearson emphasizes, underpin historiographical, anthropological and social studies of the Indian Ocean.

The National Commission for the Commemoration of the Portuguese Discoveries (CNCDP) and cultural memory

Promoting a set of initiatives in different sectors, drawing on multiple mediums and languages, the National Commission for the Commemoration of the Portuguese Discoveries[2] (henceforth CNCDP) took on an important role in the production of cultural memory in Portugal. On the one hand, the concept of 'cultural memory' as it has been formulated in the transdisciplinary field of Memory Studies, is complex and far-reaching, referring to a multiplicity of practices and objects that reveal the constant interaction between past and present, both in a given society, and on the transnational and transcultural level (Erll, 2008; Erll & Rigney, 2009). This concept relies on the interdependence between the individual and collective levels of memory, recognizing both the socially and culturally codified nature of individual memory – each individual remembers according to models that are moulded by culture and by society – and the strength of the metaphor of memory on the social and collective level – societies, when remembering and forgetting, in fact 'imitate' the functioning of the mind. Another fundamental aspect in addressing the production of cultural memory has been to explore the connection between memory and the materiality of its mediation: as Astrid Erll argues, cultural memory is comprised of a multiplicity of mediums, including religious and literary texts, audiovisual products, works of art, monuments, rituals and historiography itself, each possessing a language and a symbolic system of their own (Erll, 2008: 389). With that in mind, we intend to deal with different objects, such as the catalogue of a museum exhibition, the issue of a journal directed at the wider public and a series of television documentaries, with the aim of exploring more deeply the way that the creation of a 'pluri-media network' (Erll, 2008: 395) sought to take part in the production of a cultural memory related to the Indian Ocean, contributing simultaneously to the establishment of an area of

2 The CNCDP was created by Decree-Law n.º 391/86, of 22 November, and abolished through Law n.º 16-A/2002 of 31 May.

study relatively unknown in Portuguese-speaking contexts and which corresponds to the definition of Indian Ocean Studies.

On the other hand, speaking about the production of cultural memory in the sphere of the CNCDP's activities entails recalling the role of the institutions in the dynamic between memory and forgetfulness that Aleida Assmann (2008) codified through the concepts of canon and archive. For Assmann, remembering in social life is the exception, not the rule, and the institutions of active memory have taken charge of this exception, whose aim is to preserve the past as present. This is the construction of the canon, understood as active memory that is put into circulation by the cultural institutions; for its part, the archive represents the more passive side of memory, preserving the past as past. Though as selective as the canon, the canon produces what Assmann defines as meta-memory, a memory that preserves what was forgotten, making possible new contextualizations and, then, new lives for objects, documents, etc. In Assmann's formulation, it is in the negotiations and in the flows between canon and archive that the dynamic of cultural memory is developed, and the multifaceted activity of the CNCDP is also inscribed in this dynamic, oriented both to revive the past as present, by putting into circulation a cultural and symbolic canon to be assumed as a trace of the present, and to create and make available the archive through the edition, re-edition and digitalization of sources, to be reinterpreted and recontextualized through specialized research, artistic production, etc.

Although, in-depth analyses and overviews of the initiatives promoted and carried out by the CNCDP are still lacking both in historiography and in cultural studies,[3] different authors concur in interpreting their emergence and the main lines of their action in the light of the specific political, social and ideological context of the democratic Portugal of the 1980s (Ferreira, 2006; Cardão, 2019; Hespanha, 2019; Cardina & Fytili, 2023). This is a phase in which, after the end of the imperial cycle, the country seeks to renegotiate its position in the world system, through integration into the European Community and the construction of renewed relations with the

3 For a preliminary survey of the countless activities and initiatives promoted and sponsored by the CNCDP, see Oliveira (2003).

former imperial space (Santos, 1994). Thus, the Commemorations set out both to reinterpret the imperial and colonial legacy and to re-position it in the present of the nation, promoting a renewed evaluation of Portuguese national identity, and reconstructing an external image of Portugal, traditionally considered to be 'an enigma' by other countries (Santos, 1994: 49). For Ferreira, it was necessary to integrate 'esse legado no contexto do realinhamento internacional e europeu em que se encontravam investidas as elites dirigentes e estabelecendo ao mesmo tempo uma demarcação face ao imaginário nacionalista e imperialista que a invocação dos descobrimentos sustentou durante o Estado Novo' [that legacy into the context of the international and European realignment in which the leading elites were invested, establishing at the same time a demarcation vis-à-vis the nationalist and imperialist imaginary that the invocation of the discoveries sustained during the Estado Novo] (Ferreira, 2006: 221).

With regard to the representation of the past and relations with the former colonial space promoted by the programme of the commemorations, as Cardão states, little is known about the reception of the initiative by Portuguese society, or whether 'a diferença cultural foi efectivamente trabalhada em vez de subsumida ou superada' [cultural difference was effectively worked on instead of subsumed or surpassed] (2019: 4). This is an important question for the purposes of this chapter, which will examine a specific segment of the activities promoted by the CNCDP, namely the cycle of events and initiatives dedicated especially to the Indian Ocean and the expansion towards the 'Orient' (1997–2000), since the quincentenary of Vasco da Gama's arrival in India (1498–1998) represented the pivotal anniversary to be celebrated and commemorated in the context of this thematic cluster (Ferreira, 2006: 222).

Without intending to include all the initiatives and outputs of the Commission relating to the Indian Ocean,[4] in this chapter we will seek to answer the question posed by Cardão, dealing with academic and cultural products that strike us as significant for a contextualized reflection on the

4 It should be pointed out that Vasco da Gama's voyage and the Indian Ocean took pride of place in Expo '98, the project that emerged from the CNCDP's work, but ended up becoming autonomous (Ferreira, 2006).

representation of the Indian Ocean in the context of the Portuguese commemorations. On the other hand, we are interested in bringing to light the signs of an epistemological decentring that is important for the dialogue that has been established between the Social Sciences and the Humanities produced in Portugal, and Indian Ocean Studies, whose academic institutionalization in the international context would be consolidated from the 1990s onwards (Vink, 2007).

The prominent figure in the phase of the commemorations most focused on the Indian Ocean and the Orient was without doubt the historian António Manuel Hespanha, appointed general commissioner of the CNCDP from 1995. Hespanha wrote an important set of texts that explained the spirit of that phase of the commemorations, as well as the historiographical, epistemological and ethical perspectives at play in that new commemorative phase of the so-called 'discoveries'. Re-reading those texts in a more recent declaration, Hespanha states that the cultural policy, and consequently the commemorative practices promoted by the governments, responded to different visions and interpretations of the discoveries and the history of the so-called 'Portuguese expansion'. Inaugurated with the creation of the Commission in 1986, the Commemorations, in the historian's words, were initially oriented towards 'a more "heroic" and "sacralising" version of Portuguese history, adopting viewpoints close to Luso-tropicalism, comfortably contained in the ambiguous expression "cultural encounter"' (Hespanha, 2019: 198). The historian and former Commissioner also frames this orientation in the post-imperial discourse of *Lusofonia*, since 'Portugal, which left a colonial situation ten years before, was reframing its collective memory so as to make it shareable with the whole of the Lusophone world',[5] hence the replacement of the idea of the discovery with that of the encounter (2019: 209). In fact, as Ferreira (2006: 234) also emphasizes, the language used in the official communications of the commemorations set out to distance itself from celebrating the myth of the expansion, opting for a lexicon that was perceived as being more suitable for the new historical and political phase of postcolonial

5 For an analysis of the concept and discourse of 'Portugalidade' see Sousa (2017).

Portugal, which did not always translate into a radical change in the representation of *the others* and the self-representation of the nation.

However, the general spirit, and the strategy of Hespanha's commissariat were intended to defend the importance of cultivating 'a pluralidade de leituras e o relativismo dos olhares' [the plurality of readings and the relativism of the gazes] as the foundations of a commemorative ethic (Hespanha, 2019, 199), in a constant effort to not conceal the conflictive aspects of the 'cultural encounter', even explaining the option of that new commemorative cycle to revive the 'habit of difference', the 'spirit of curiosity' and 'self confidence' as legacies of the historical past of Portugal (Hespanha, 1997: 100). There was also insistence on the reconciliatory aspect implicit in the etymology of the verb 'commemorate': 'comemorar: lembrar em conjunto' [commemorate: to remember together] (1997: 98). This reconciliatory spirit was in part frustrated by reactions and controversies that arose in political and intellectual sectors of the old colonized spaces, namely in India and Brazil, at the time of attempts to 'remember together' Vasco da Gama's voyage to India and Pedro Álvares Cabral's voyage to Brazil. These two events were configured as controversial *lieux de mémoire* (Nora, 1984) since the narratives and the meaning attributed to them by the communities in question diverged and continue to diverge.[6]

Nevertheless, the academic and scientific initiatives driven and supported by the CNCDP in the sphere of historiography and the Social Sciences, in general, constituted a significant shift in the way of thinking about and making history and the memory of Portugal's presence not only in the colonized spaces but also in other territories connected by flows, displacements and networks, that is to say, promoting a less fragmentary and regionalized approach to the Empire.[7] In this aspect, the five-volume work

6 About the Indian reactions to the Portuguese commemorations, see Hespanha (1999) and Ferreira (2006); about the reactions to the commemorations in Brazil see Schiavinatto (2019). It should be pointed out, regarding Vasco da Gama's voyage as a place of memory, that the *Roteiro da Primeira Viagem de Vasco da Gama à Índia*, [Journal of the First Voyage of Vasco da Gama to India] attributed to Álvaro Velho was included in the UNESCO's Record of Memory of the World in 2013.

7 For a review of the historiography of the discoveries up to the 1990s, see, for example, the article by Marques (1992), in which the author backs a new impetus for Portuguese historiography to be promoted by the Commemorations.

História da Expansão Portuguesa (History of the Portuguese Expansion) coordinated by Francisco Bethencourt and Kirti Chaudhuri[8] is iconic, considered to be groundbreaking in the renewal of Portuguese historiography, because of the focus, among other aspects, on the close ties between Nation and Empire and between national history and the history of the expansion[9] (Xavier, 2000). This perspective guided a good part of the historiographical work undertaken by the CNCDP, in which a strong impact was had by the 'ventos pós-modernos' [postmodern winds] of a 'pujante nova história' [powerful new history], based methodologically on the multiplicity of tools and 'fontes de auscultação de vozes silenciadas' [sources for listening to silenced voices], in consonance with the echoes of Subaltern Studies (Hespanha, 1999: 9–10).

Culturas do Índico: Between the transnational and discontinuity

As mentioned above, the 'India programme', also called the 'India cycle', dedicated to the commemoration of the discovery of the Sea Route to India,[10] was one of the main thematic axes of Hespanha's Commissariat

8 It should be pointed out that Kirti N. Chaudhuri was a leading figure for Indian Ocean Studies, as the author of seminal volumes for the history of this ocean: *Trade and Civilisation in Indian Ocean from the Rise of Islam to 1750* (1985) and *Asia before Europe* (1990), both inspired by the work of Ferdinand de Braudel on the Mediterranean.

9 Fernando Tavares Pimenta, although recognizing the pioneering character of this work, criticizes the choice of the term 'expansion' that reveals an antiquated view of the colonial phenomenon (2010: 147).

10 The cycle was inaugurated with the commemoration of Vasco da Gama's departure in a solemn session at the Sociedade de Geografia in Lisbon on the 8th of July 1997. Among the guests were two renowned historians, Joaquim Romero de Magalhães and Sanjay Subrahmanyam. The presence of the Indian historian, author of a biography of Vasco da Gama (Subrahmanyam, 1998) caused an extremely heated controversy about the CNCDP and its view of the discoveries and Portuguese history (Hespanha, 1999; Ferreira, 2006).

and included initiatives extended to the Indian Ocean and the 'Orient'. As the former commissioner mentions, it was an area of strategic interest for Portugal, not only for reasons of a historical and cultural nature (Goa) but also because of the political and economic circumstances of the time: as an Indian Ocean territory, Mozambique demanded a special focus (Hespanha, 1999: 98). One noteworthy initiative of this cycle was the great exhibition *Culturas do Índico*, held in 1998 in the Museu Nacional de Arte Antiga de Lisboa (National Museum of Ancient Art of Lisbon) and curated by the anthropologist Rosa Maria Perez with the collaboration and consultancy of national and foreign institutions and personalities. Following the general programmatic lines of the CNCDP, it set out to valorize the gazes and reactions of other peoples to the so-called Portuguese expansion, and also to correct the distortions produced and transmitted down the centuries in representing the other. Far from constituting ephemeral and circumstantial material, the exhibition catalogue represents an enduring, autonomous product and a work of reference for Indian Ocean Studies in Portugal. In fact, the texts included in it reveal the intention and attempt to decentralize the memory of the historical past, shifting the focus from Vasco da Gama's voyage, as a controversial *lieu de mémoire*, to the space of the Western Indian Ocean and its cultural plurality. As António Hespanha argues in the opening text of the catalogue: 'Pareceu-nos [...] que, num ano em que o feito de Gama havia de ser de tantas formas relembrado, não se podia deixar de evocar esse oceano de chegada, tal como o surpreende o nosso navegador e tal como ele mesmo se surpreende com ele' [It struck us [...] that, in a year when Gama's feat should be remembered in so many ways, we couldn't fail to evoke that ocean of destination, just as our navigator surprised it, and just as he himself is surprised by it] (Hespanha, 1998: 13). Also in the opening note and in the introductory text properly speaking, the curator of the exhibition, Rosa Maria Perez defines the terms of the epistemological challenge underlying the exhibition and the organization of the catalogue, spelling out a set of topics that are important for the reflection proposed here. The exhibition project set out to 'dar inteligibilidade e, mais arrojadamente ainda, unidade às *Culturas do Índico* – assim se designando a sofisticada diversidade cultural encontrada por Vasco da

Gama na sua rota marítima para a Índia' [give intelligibility, and even more boldly, unity to the *Culturas do Índico* – thus designating the sophisticated cultural diversity met by Vasco da Gama on his sea route to India]. It also sought to 'quebrar o lusocentrismo das abordagens tradicionais' [break the lusocentrism of the traditional approaches] that is, reverse a gaze traditionally based on a lusocentric historiography of art and on an almost monolithic interest in the products of so-called Indo-Portuguese culture and later, in the presence of the Portuguese in India, more than in the Indian Ocean, understood as a plural civilizational universe (Perez, 1998a: 17–20). On the one hand, the question of the ufnity of the cultures of the Western Indian Ocean refers to the construction of the paradigm of the Indian Ocean as the analytical unit discussed above. In other words, it is still a controversial question that finds in Rosa Perez's analysis and more generally, in the academic and cultural climate designed for the India programme of the CNCDP, an expression originating in Portugal. In reality, for Rosa Perez, the *Culturas do Índico* constitute 'um sistema de contiguidades e de descontinuidades' [a system of contiguities and discontinuities] or a 'um puzzle marítimo' [maritime puzzle] within which the 'national' cultures only become intelligible on the basis of a transnational approach (1998a: 20–21). It is in this perspective, as the curator further explains, that the unity of the cultures of the Western Indian Ocean represented by the exhibition is not obvious, resulting from a construction articulated and discontinuous throughout time and space, but also from a necessarily multidisciplinary focus able to re-establish a complex dialogue between ideas, object, peoples and geographies. In fact, the texts included in the catalogue take part in this dialogue, also constructed through the thematic clusters of the exhibition, namely. (1) The ocean of departure, (2) The ocean of destination. The writings of the Indias; (3) The voyage of cultures. The body; (4) Around the Indian Ocean.

Creating a narrative chain, these subheadings suggest that both the exhibition and the catalogue that emerged from it, function as remediation of Vasco da Gama's voyage as a *lieu de mémoire* constructed by previous literary, historical, artistic, etc. representations.[11] Astrid Erll formulates the

11 In an interview given to *Expresso*, Rosa Perez declared that the guiding thread of

concept of remediation to analyse the way that cultural memory operates throughout time and in different contexts:

> With the term 'remediation' I refer to the fact that especially those events which are transformed into lieux de mémoire are usually represented again and again, over decades and centuries, in different media. What is known about an event which has turned into a site of memory, therefore, seems to refer not so much to what one might cautiously call the 'actual event', but instead to a canon of existent medial constructions, to the narratives, images and myths circulating in a memory culture. (Erll, 2009: 113)

It can therefore be stated that all the commemorative activities and initiatives of the India cycle were part of the creation of a pluri-media network to commemorate the voyage of Vasco da Gama. The exhibition took a hybrid, inter-medial and multidisciplinary form. In fact, the pieces selected and the texts that frame them in the catalogue display the multiplicity of ideologies, narratives and myths associated with the voyage and the cultural contexts in which it is framed, summoning up and deconstructing symbols and imaginaries. Therefore the 'Ocean of destination' is evoked from the knowledge, beliefs and myths about the Indian Ocean and the Orient that were developed in Europe, as the texts on the cartographical representations of the Indian Ocean and Ethiopia, Prester John and Vasco da Gama's legendary Arab pilot reveal.[12] For its part, the 'Ocean of destination' is represented through the focus on writing as record, art and symbol of composite, urbanized and cosmopolitan cultures. According to Jessica Hallett, this cluster of the exhibition set out to explicitly evoke:

> O itinerário cultural da viagem de Vasco da Gama: começa na África oriental, continua em terras islâmicas e termina na Índia. Cada porto de escala na viagem serve de ponto de embarque para uma nova viagem, em que a África oriental aparece como uma região cosmopolita, letrada, integrada no comércio transoceânico; a península da

---

the exhibition is Gama's sea route as an eminently cultural route, Gama becoming an impressive 'mediator of civilizations' <https://arquivo.pt/wayback/19991109165107/http:/www.expresso.pt/ed1335/c21.asp>

12 See the catalogue texts by A. C. Campos; M. J. Ramos; M. Barry and C. Radulet.

Arábia como a terra na qual surgiram o Islão e a civilização islâmica, e o subcontinente da Índia como berço de uma civilização antiga que deu origem a três das grandes religiões mundiais – hinduísmo, budismo e janismo – e a um complexo mosaico de práticas sociais e religiosas. (Hallett, 1998: 106).

[The cultural itinerary of Vasco da Gama's voyage: it begins in East Africa, continues in Islamic lands and ends in India. Each port of call on the voyage serves as the point of embarcation for a new voyage, in which East Africa appears as a cosmopolitan, literate region integrated into transoceanic commerce; the Arab peninsula as a land in which Islam and the Islamic civilization emerge, and the subcontinent of India as the cradle of an ancient civilization that gave rise to three of the great world religion – Hinduism, Buddhism and Jainism – and a complex mosaic of social and religious practices.]

The focus on literate cultures, on the writings and languages of the Western Indian Ocean becomes especially relevant if viewed in the light of the political, cultural and ideological context in which the commemorations of the discoveries arose and which was briefly discussed in the introduction to this chapter. In fact, after the golden years of the discourse of *Lusofonia* and the celebration of the 'expansion' of the Portuguese language assumed as a gift from Portugal to the colonized peoples, the mid-1990 onwards saw broad a criticism of this discourse and the actual concept of *Lusofonia* as the stamping ground of privileged relations between the old metropolis and ex-colonies (Lourenço, 1999; Almeida, 2000). Thus, the work of the cultural memory set in motion by the exhibition looks to the past from a present that needs other positionings, able to rethink postcolonial relations in which the languages and writing carry out a crucial role.

Regarding the third thematic cluster of the exhibition, Rosa Perez reflects on the concept of culture in the realm of anthropology and broaches the topic of the voyage of cultures; she goes on to explain the narrative construction underpinning the exhibition, alluding both the mutual flows of cultural translation between Europe and the Indian Ocean, and the dynamic of cultural memory, between past and present, in which the commemorations are framed:

> O núcleo agora apresentado completa uma fiada diacrónica que, ao longo da Exposição, da Europa conduz às Índias pela via das suas culturas e das suas histórias recíprocas. Ele tenta traçar o encontro de homens e de ideias, de crenças e de credos no palco do Índico, das formas como as suas culturas, cruzadas entre si de margem a margem no esteio da peregrinação religiosa e das trocas comerciais, se atraíram apaixonadamente ou se repeliram nas suas singularidades – ao mesmo tempo que influenciavam uma certa construção de Europa e que ela própria as marcava a elas. Este núcleo completa ainda um itinerário lógico e temporal desde sempre anunciado: privilegiadamente situado no presente, ele fecha o anel dos percursos e recursos posteriores ao encontro entre a Europa e o Índico, das suas conjunções e disjunções mútuas. (Perez, 1998b: 228–229).

> [The cluster now presented completes a diachronic thread which, throughout the Exhibition, leads from Europe to the Indias through their cultures and their reciprocal histories. It tries to trace the meeting of men and ideas, of beliefs and credos on the stage of the Indian Ocean, the ways their cultures criss-crossed from shore to shore in the wake of the religious pilgrimage and the commercial exchanges, they were passionately attracted or repelled in their singularities at the same time as they influenced a certain construction of Europe and Europe itself marked them.
>
> This cluster also completes a long announced logical and temporal itinerary: privilegedly situated in the present, it closes the ring of the journeys and recourses that came after the meeting between Europe and the Indian Ocean, its mutual conjunctions and disjunctions.]

Both the Ocean of departure and the Ocean of destination arise as universes of cultural memories intersected and resignified by the voyage of cultures, from perspectives that claim not to be lusocentric and which rely on a complex construction of the unity of the cultures of the Western Indian Ocean, founded on exchanges and multidirectional transits between material cultures and spiritual and human universes. In respect of this cultural unity, through pieces and texts related to Ethiopia, Mozambique, Madagascar and the Swahili civilization,[13] the exhibition highlights the centrality of East Africa in the universe of the Western Indian Ocean, translating a stance in the debate about the inclusion of the East coast of Africa in the sphere of influence of this ocean, that is,

13 See the catalogue texts by A. Sheriff, J. C. Senna-Martinez; R. Pankhurst; J. Mack; J. C. Penrad and A. C. Roque.

an aspect that is controversial and at the same time crucial for the construction of a historiographical paradigm of the Indian Ocean (Alpers, 2009; Bose, 2006; Pearson, 2003). This positioning is consistent with the general prominence of the African Indian Ocean in the sphere of the commemorations, through other academic and cultural initiatives supported by the CNCDP, for example, the retrospective exhibition of José Henrique Silva *Pescadores Macua* (Macua Fishermen) curated by Joana Pereira Leite, and the resulting catalogue (Silva, 1998); and also, issue number 25 of the review *Oceanos* (January/March 1995), dedicated to the Island of Mozambique.

If the topos of the 'cultural encounter', introduced by the commemorations instead of the idea of the 'discoveries', played a central role in the conception of the exhibition *Culturas do Índico*, it was used less to emphasize the universalist vocation of the Portuguese people, than to work the difference – taking up the question posed by Cardão – giving prominence to the diversity and cultural sophistication of the peoples of the Western Indian Ocean, and underscoring the interactions and transits in an ocean which, as has already been mentioned, even today remains underexplored in comparison with the Atlantic. Thus, the *remediation* of Vasco da Gama's voyage, constituted by the articulation of the various thematic clusters of the exhibition, contributed significantly to a renewed approach to the Indian Ocean and the relationship of Portugal with its cultures and peoples, opening the way to an epistemological decentralization still relevant today in the realm of historiographical, anthropological, cultural and literary studies (Leite, Brugioni and Falconi, 2019).

Issue number 24 of the review *Oceanos* also contributed to this decentralization, bringing together the interventions from the colloquium preparatory to the exhibition held in 1997 and entitled *The Portuguese Expansion and the Cultures of the Indian Ocean*. The review *Oceanos*, published by the CNCDP,[14] began in June 1989 and ended in 2002 after a total of 49 numbers published every three months. With a strong graphic impact, due also to the size, 37.5 × 27.5 cm, the review set out to reach a

14 In addition to *Oceanos*, the Commission also published the reviews *Mare Liberum*, scientific in nature, and *Na Crista da Onda*, aimed at children and juveniles.

wide audience. It launched with miscellaneous numbers then published monographical numbers intended for scientific dissemination in the various knowledge areas related to the discoveries. More specifically the double number 19/20 of 1994 was dedicated to the Indian Ocean, *Indo-portuguesmente*; number 25, mentioned above, focused on the Island of Mozambique, number 33 dedicated to Vasco da Gama and number 34, entitled *Culturas do Índico*. The latter bears witness to the presence of Indian Ocean Studies in Portugal, presenting essential texts produced in this field. Among them, Sugata Bose's seminal article "Estado, economia e cultura na orla do Índico: teoria e história" (State, economy and culture on the shore of the Indian Ocean: theory and history)[15] stands out in particular. The historian formulates the definition of the Indian Ocean 'enquanto arena inter-regional de interação econômica, política e cultural' [as an inter-regional arena of economic, political and cultural interaction] (Bose, 2019: 32). On the other hand, he points to the most important limitations of the approaches to this ocean rehearsed up till then and which are still current, principally in literary and cultural studies. As Bose asserts, in fact '[a]s fronteiras coloniais constituíram um obstáculo ao estudo de comparações e laços entre regiões e deixaram, como um legado duradouro, o confinamento generalizado do foco académico ao âmbito estrito dos estudos regionais' [[t]he colonial frontiers were an obstacle to the study of comparisons and bonds between regions and left, as a lasting legacy, the general confinement of the academic focus to the strict sphere of regional studies] (Bose, 2019: 33).

Mirroring the multiple topics, periods and territories addressed in the 1997 colloquium, the volume established a dialogue between different disciplinary fields and subjects of research, with the aim of both reconstituting the plurality of the Ocean which the Portuguese reached and analysing the impact of that arrival on the societies involved, illuminating political, religious, commercial and cultural dynamics of great complexity. Just like the exhibition, the volume also focused on the sophistication of

15 The article was included in an anthology of theoretical texts on the Indian Ocean (Leite, Brugioni & Falconi, 2019).

those cultures, addressing not only material artefacts but also systems of thought and organization.

This perspective was then expressed in the editorial written by António Hespanha, which also gave an account of the immediate reception of the CNCDP's activities by Portuguese society. Alluding to the controversies generated by the not very celebratory and heroic way of dealing with the theme of the voyages and the figure of Vasco da Gama, Hespanha presented one of the objectives of the *Culturas do Índico* project, namely to correct the distortions and biases that characterize the representation of the Other in Portugal and in particular, the Orient:

> Fomos nós, há cerca de 500 anos, que mais contribuímos para modelar aqui a imagem do Oriente. Comerciantes, missionários, viajantes, capitães, homens de letras e homens de ciência, oriundos de Portugal ou levados pelos portugueses, descreveram o Oriente. Como sempre acontece, essa descrição foi unilateral. Foi-o pelo enviesamento próprio de quem vem e vê de fora. Mas foi-o também pelo carácter auto-apologético, 'imperial' e 'de cruzada' que caracterizou, fundamentalmente, a expansão portuguesa. (Hespanha, 1998: 7)

> [We were the ones who around 500 years ago contributed most to modelling here the image of the Orient. Traders, missionaries, travellers, captains, men of letters and men of science, proceeding from Portugal or taken by the Portuguese, described the Orient. As always happens, that description was unilateral. It was unilateral because of the actual bias of someone who arrives and looks in from the outside. But it was also unilateral because of the auto-apologetic, 'imperial' and 'crusading' character that characterised, fundamentally, the Portuguese expansion.]

This editorial, just like the other texts produced by Hespanha, explains the confrontations and rifts that the 'remediations' driven by the CNCDP caused in the construction of a public memory of the discoveries. Paolo Jedlowski defines public memory as the set of discourses about the past that circulate in the public sphere, 'ou seja, naquele espaço que não coincide nem com as instituições do Estado nem com os círculos de cidadãos privados, mas se configura como lugar de encontro dos indivíduos e da crítica (...). Em outras palavras, é a imagem do passado discutida publicamente' [or rather, in that space that does not coincide either with the institutions of the State or the circles of private citizens, but is configured as a meeting place of individuals and criticism (...). In other words, it is the image of the past discussed in public] (Jedlowski, 2007: 14) that includes not only monuments and ceremonies but also debates associated with them or even the forms of fruition and the conflicts they generate.

Mar das Índias: Between ruins and (post)imperial melancholy

Other activities, complementary to the Exhibition and the academic publications, were pursued in the context of the India cycle, allowing us to consider the construction of the cultural memory of the Indian Ocean in the domain of popular culture as well.[16] We are referring in particular to the field of audiovisual production through the television series *Mar das Índias*, coordinated by the commissariat of the exhibition and co-produced by RTP and the CNCDP.[17] The sponsorship of television series and programmes on the subject of the discoveries was another important initiative of the CNCDP, where television was an unsurpassable source of historical information for millions of people and a powerful means of moulding the collective memory (Edgerton & Rollins, 2001). As Cardão

16 The complementary programme *Culturas do Índico* included commercial iniciatives such as the Mercado das Índias [Market of the Indies], for the sale of products, and the Loja dos Artistas [Artists' Shop], where different Portuguese artists created pieces inspired by the objects and topics of the exhibition. It also included the Sons e Sabores [Sounds and Tastes] programme, a co-production between the CNCDP, the Ethnomusicological Institute of the Universidade Nova de Lisboa, the Oikos Space and the National Museum of Ancient Art, bringing together musical groups from the migrant communities residing in the Greater Lisbon area. Source: *Culturas do Índico* – page preserved at arquivo.pt, <https://arquivo.pt/wayback/20000229230710/http://www.cncdp.pt:80/expo/indico/int1.html>.

17 Credits:– Original conception: Camilo Azevedo, Paulo Varela Gomes and Rosa Maria Perez; Script: António Cabrita (with Luís Carlos Patraquim for Mozambique and Rosa Maria Perez for India). Production: Camilo Azevedo.

emphasizes, the role of television products in constructing history and memory in Portugal is a little explored subject, to which the present chapter intends to contribute, by examining, albeit succinctly, the series.

With the scientific coordination of Rosa Perez, the series took up the general perspective of the exhibition, flagging up the cultural complexity of the Indian Ocean at the time of Vasco da Gama's expedition and drawing on the thematic clusters of the Ocean of destination and the voyage of cultures, as could be read in the introductory note on the web page dedicated to the *Culturas do Índico* project:

> Quando Vasco da Gama penetrou pela primeira vez no Oceano Índico, em 1498, este Oceano não era uma simples via de acesso à Índia. Como o Gama e os portugueses descobriram, era ele próprio um mundo.Em volta das margens do Índico, várias culturas e civilizações haviam estabelecido redes de relações que se sobrepunham e interligavam: relações comerciais, é verdade, mas também cultos religiosos que, da Arábia, tinham vindo até à Índia e ao norte de Moçambique, dialectos que tinham viajado da Pérsia para o arquipélago das Comores, formas artísticas muçulmanas ou hindus que apareciam, transformadas, em sítios inesperados a centenas de milhas de distância.[18]
>
> [When Vasco da Gama entered the Indian Ocean for the first time, in 1498, this Ocean was not a simple pathway to India. As Gama and the Portuguese discovered, it was a world in itself. Around the shores of the Indian Ocean, various cultures and civilizations had establish networks of relations that overlaid and interlinked trade relations, it is true, but also religious cults that, in Arabia, had came to India and the north of Mozambique, dialects that had travelled from Persia to the Comoros archipelago, Muslim or Hindu artistic forms that appeared, transformed, in unexpected places hundreds of miles away.]

On the other hand, the introductory text would suggest the *remediation* that characterized the documentary series in relation to Vasco da Gama's voyage, inviting the public to establish a bridge between the historical past and the present of the narration, through the figure of the 'traveller-presenter' – the journalist and politician Miguel Portas – a contemporary

18 Source: <https://arquivo.pt/wayback/19990528125951/http://www.cncdp.pt/expo/indico/mar_indias.html>

Gama entrusted with the task of reconstituting, on the basis of the gaze of the present, the connections between times and spaces closed by the Portuguese:

> A série *O Mar das Índias* conta a história de um viajante culto que, nos nossos dias, procura reconstituir este mundo. Através de portos e aeroportos modernos, pelas ruas das cidades de hoje, em aldeias paradas no tempo e noutras que o tempo transformou, aventurando-se no interior à procura dos vestígios monumentais de grandes culturas, este viajante apresentador comenta aquilo que no passado unia e dividia o mar da Índia e aquilo que, no presente, é vestígio desse passado.
>
> Nas aldeias de Moçambique ou no forte de Jesus em Mombaça, buscando os antigos templos e ritos do lendário Prestes João na actual Etiópia, visitando as cidades cor-de-terra do Iémen ou, na Índia, a capital monumental dos Mogóis, as aldeias do Gujarat, estranhos mercados onde confluem hábitos muito diversos, seguindo as rotas da fé que cruzam pessoas e lugares de peregrinação, passando por estes sítios e unindo-os entre si pelo comentário ou as imagens, o viajante procurará reconstituir o olhar espantado do Gama e completá-lo com um olhar de hoje, informado e curioso.
>
> [The series *O Mar das Índias* tells the story of an educated traveller who, in our times, seeks to reconstitute this world. Through modern ports and airports, through the streets of the cities of today, in villages frozen in time and others that time has transformed, venturing into the interior in search of monumental traces of great cultures, this presenter traveller comments on what in the past united and divided the sea of India and what, in the present, is the trace of that past.
>
> In the villages of Mozambique or in the Fort of Jesus in Mombasa, seeking the ancient temples and rites of the legendary Prester John in present-day Ethiopia, visiting the earth-coloured cities of Yemen or, in India, the monumental capital of the Moghuls, the villages of Gujarat, strange markets where very different habits converge, following the faith routes that cut across people and places of pilgrimage, passing through these sites and uniting them through the commentary and the images, the traveller will seek to reconstitute Gama's shocked gaze and complete it with today's gaze, informed and curious.]

The first episode, entitled 'A costa dos três ouros' (The coast of three gold coins) presents images of the Island of Mozambique, Ibo Island and Zanzibar. In the opening shots, the inscriptions in Arabic and other non-Western languages, as well as the figures and landscapes shrouded in mist immediately raise the question of accessibility, both of the past and other cultures, of which rituals and fragments are shown throughout the episodes, in most cases without any particular contextualization. In the opening sequence, walking along beside the sea in the Island of Mozambique, Miguel Portas introduces the space dealt with in the first episode, stating that this is the African Indian Ocean, different from the Atlantic, made up of cities whose names are known to the Portuguese from their school days. Portas's allusion, then, suggests a relationship of complementarity between institutionalized historical knowledge and the information made available by television, evoking the complex debate around the legitimacy of 'popular history' in the face of academic historiography, an especially pertinent aspect in the case of *Mar das Índias*, since, although under the scientific coordination of anthropologist Rosa Perez, neither of the series authors was a professional historian. As Miguel Portas declared in an interview at the time of the première: 'Não há aqui verdade histórica e pisámos um terreno em que ainda há muito por estudar.'[19] [There is no historical truth here, and we are treading ground on which there is still much to be studied.] Instead of seeking historical truth, the series uses the evocation of the past to mould the collective memory which, according to Gary Edgerton, is configured as 'the site of mediation where professional history must ultimately share space with popular history' (Edgerton, 2001: 5).

As well as focusing on the geographical and cultural space of the African Indian Ocean, the first episode also makes evident the treatment of time given to the verbal content of the series: instead of a chronologically ordered narrative, the historical reconstruction presented by Portas moves between different eras, linking the eighteenth, nineteenth and twentieth centuries with the present of the images, to return again to the sixteenth century. Thus, the relationship between past and present is given through the articulation between sound and the visual, where the past is narrated through the voice-over and evoked by the soundtrack, while the present unfolds in the images and is intersected by the past in the shots of ancient monuments and ruins. This treatment of time produces a representation

19 'Um belo poema sobre a tristeza' by Carlos Câmara Leme, *Público*, 2 April 2000.

of the world of the Indian Ocean, founded on multiple temporalities, decentralizing the primacy of Portuguese time, that is, of the hegemonic narrative of the voyages and the discoveries. On the other hand, it is the actual space addressed that contributes to this decentralization, since most of the episodes focus on places that do not belong to the 'world that the Portuguese created', in other words the world of *Lusofonia*, territories where there were only passages and presence, such as Zanzibar, Ethiopia and the Yemen. These options correspond to the non-lusocentric attitude adopted by the curator of the exhibition and, more generally, by the cycle of the Commemorations presided over by Hespanha, where the Indian Ocean is in fact the best space to rehearse this decentralization. Even the topos of the 'cultural encounter' also seems mitigated when evoking episodes of violence and conflict as in the case of the sacking of Quiloa (Kilwa) reconstructed in the reading of an extract from Gaspar Corrêa beside the Fortress of Jareza.

However, as a whole, the series communicates an exclusively Portuguese perspective and narrative about the past and present of the Indian Ocean. The presenter's choice of a single perspective and the voice-off commentaries guarantee the coherence of the focus, leaving no space for dissonant voices or gazes. While the contemporary 'voyage' moves forward, the historical reconstruction also becomes marginal, leaving room for ethnographical description, in line with widely consumed documentary products. In fact, both the voice-over and the images that depict landscapes and communities of the Mozambican islands, of Quiloa, Zanzibar and Ethiopia, offer a representation of the African Indian Ocean as a world of negligence, abandonment and underdevelopment, in which the presenter-traveller seeks both vestiges of ancient splendour and traces of modernization. The last sequence of the second episode, filmed in Ethiopia, is especially revealing of this attitude, with Miguel Portas declaring that he has found in the legendary 'lands of Prester John' a 'fourth world', alluding to the excess of underdevelopment. The perspective is also confirmed in the above-mentioned interview with Miguel Portas and Camilo Azevedo:

CAMILO AZEVEDO Há uma característica comum aos quatro episódios: são terras tristes, tudo aquilo é triste: as pessoas, o ritmo em que vivem. Vivem num estado de inacção do qual nós podemos ser culpados ou não. […]

MIGUEL PORTAS Olhando para aquelas pedras e aquelas terras, que os portugueses lá deixaram há cem anos, não tenho dúvidas: podíamos ter escolhido o desenvolvimento, optámos pelo subdesenvolvimento

[CAMILO AZEVEDO There is a feature common to the four episodes: they are sad lands, all of that is sad: the people, the rhythm at which they live. They live in a state of inaction for which we may be to blame or not […]

MIGUEL PORTAS Looking at those stones and those lands, which the Portuguese left a hundred years ago, I have no doubts: we could have chosen development, we opted for underdevelopment.]

We should ask whether the sadness of the land is not, in reality, the melancholy of a post-imperial gaze that lingers in a romantic and nostalgic contemplation of the ruins, in the construction of a cultural memory scarcely inhabited by the other and in the myth of western development. To sum up, the ruins and remains of the Portuguese empire are not presented as strategic points of view for rethinking the past (Stoler, 2013) and yet the present, nor are they constituted as aesthetic and conceptual devices able *to ruin the ruin* (cf. *to ruin*, *ruinations*, Stoler, 2013). That is, as a critical paradigm that reveals and focuses on the vital and present representation of the inert remains of past empires (Stoler, 2016).

In the third episode of the series, Africa is left behind, and we transit to Yemen, and in the fourth and last episode, to India. In these two episodes, the historical information is reduced to a minimum, primacy being given to descriptions of religious and architectural traditions and present-day lifestyles. The difference in the treatment of spaces and the communities that inhabit them in comparison with the African territories is flagrant; the bustle and vitality of the urban centres stands out, which risks transmitting a somehow hierarchical view of the cultures addressed in the series which, opening on the beach in the Island of Mozambique, tends with a shot of commercial signage lighting up the night in Colaba, Bombay. If the historical reconstruction rehearsed in the first episode of the series suggests a narrative of multiple temporalities of the Indian Ocean, the treatment in the last episodes ends up reintroducing a linear view of time, based on the myth of progress and modernization.

Conclusions

Inserted into the broader programme of the CNCDP during the three-year period 1997–2000, the activities and subsequent products addressed in this chapter allow us to draw some provisional conclusions about the way that different mediums, and therefore different languages and disciplines took part in the representation of the Indian Ocean in the sphere of the commemorations and simultaneously contributed to a possible beginning – in the sense that Edward Said attributed to the concept *beginning* – to what might be defined as the field of studies of the Indian Ocean.

As we have seen, the project entitled *Culturas do Índico* produced, among other results, the catalogue of the great exhibition held in 1998, and the number of the review *Oceanos*, and another product – less direct – the documentary series *Mar das Índias* arising out of that event. As we have tried to demonstrate, the project as a whole set out to provide an alternative way to commemorate the discovery of the sea route to India, that is, an unprecedented 'remediation' aimed at shifting the focus of the voyage and the figure of Vasco da Gama to the multifaceted and plural cultural universe of the Indian Ocean. In our opinion, this 'remediation' sought to illuminate alternative routes and narratives of the so-called 'discoveries', opting to downplay the Eurocentric and hegemonic perspectives in various disciplines. On the one hand, we consider that today the catalogue and the review number are works of reference for rethinking the reception – still underway – of a multidisciplinary paradigm based in the Indian Ocean able to transmit a change in the Portuguese gaze on this ocean. On the other hand, we consider that the 'remediation' constituted by the RTP documentary series, although operating the same shift in focus, did not achieve the same objectives, but ended up reproducing a nostalgic, monolithic and sometimes stereotyped gaze on the Indian Ocean in general, and its African shores in particular. If the non-lusocentric posture summoned up by the exhibition *Culturas do Índico* managed to find some fertile ground in the audiovisual translation, the same did not happen in relation to the winds of a 'powerful new history' (Hespanha, infra), marked albeit incipiently by the epistemological rupture in postcolonial studies.

It should also be pointed out that in light of the recent controversies in Portuguese society on the construction of a Museum dedicated to the discoveries,[20] it becomes evident that the memory of the colonial past is far from being consensual, still comprising an area of ideological, political and intellectual disputes. The terminological question – Expansion, Discoveries, etc. – reanimated the debate, highlighted the difficulties of naming this past, but above all 'remediating it', both in the sense of *remediation*, understood as a dynamic of cultural memory, and in the sense of *reparation*, a key concept in the last decade's debates about the decolonization of museums and heritages which has been the object of different interventions in the international sphere.[21] The reaction of different sectors of Portuguese society against the concept of the Museum bears witness to the importance of re-examining the most recent commemorative cycles with the aim of

20 The controversy broke out between 2018 and 2019, following the proposal to create the Museum, supported by the President of Lisbon Town Council, Fernando Medina. Various publications of the Portuguese press gave space and coverage to the matter, in which personalities from the political, cultural and academic world intervened. Among the first interventions to liven the debate, the letter signed by Portuguese and foreign historians and social scientists opposed to the denomination 'Museum of the Discoveries' stands out: <https://expresso.pt/cultura/2018-04-12-A-controversia-sobre-um-Museu-que-ainda-nao-existe.-Descobertas-ou-Expansao->.

21 See, for example, the intervention by Achille Mbembe at the international conference *Recognition, Reparation, Reconciliation*, Stellenbosch University, 5–9 December 2018. On this matter see also the report *Rapport sur la restitution du patrimoine culturel africain. Vers une nouvelle éthique relationnelle* by Felwine Sarr and Bénédicte Savoy commissioned by President Macron: <http://restitutionreport2018.com/.sec>

measuring distances, identifying ruptures and continuities, and imagining future developments.

Finally, here it should be stressed that notwithstanding some impasses and ideological and perhaps conceptual and aesthetic limitations, the different initiatives, publications and productions promoted within the CNCDP's sphere of action constitute, in our opinion, an important point of departure for establishing the field of Indian Ocean Studies in Portuguese-speaking contexts, posing and answering some of the questions and problems that govern this field of study in national and academic contexts where Indian Ocean Studies benefit from a longer and more established tradition.

Bibliography

Reviews in Print

Oceanos n. 25, *Ilhas de Todos*, 1995.
Oceanos n. 34, *Culturas do Índico*, 1998.

On-line Newspapers

Expresso, 12.04.2018, <https://expresso.pt/cultura/2018-04-12-A-controversia-sobre-um-Museu-que-ainda-nao-existe.-Descobertas-ou-Expansao->

RTP Archive

Series *Mar das Índias:* 1. *A Costa dos Três Ouros*; 2. *Os Terraçőes de Prestes João*; 3. *O Jardim da Rainha de Sabá*; 4. *A Ilha da Boa Vida*.

Pages Preserved by Arquivo.pt

<https://arquivo.pt/wayback/19990826030700/http:/www.expresso.pt/ed1334/c362.asp>

<https://arquivo.pt/wayback/20000229230710/http://www.cncdp.pt:80/expo/indico/int1.html>

<https://arquivo.pt/wayback/20000414203009/http://www.cncdp.pt:80/expo/indico/mar_indias.html>

Almeida, M. Vale de (2000). *Um Mar da Cor da Terra: Raça, Cultura e Política da Identidade*. Oeiras: Celta.

Alpers, E. (2008). *East Africa and the Indian Ocean*. Princeton: Marcus Wiener Publishing.

Assmann, A. (2008). 'Canon and Archive'. In Cultural Memory Studies*: An International and Interdisciplinary Handbook*, pp. 97–107. Berlin: De Gruyter.

Bethencourt, F. and Chaudhuri, K. (eds). (1998). *História da Expansão Portuguesa*, 5 vols. Lisbon: Temas e Debates.

Bose, S. (2019 [1998]). 'Estado, economia e cultura na orla do Índico: teoria e história'. In A. M. Leite, E. Brugioni and J. Falconi (orgs), *Estudos sobre o Oceano Índico. Antologia de textos teóricos*, pp. 29–52. Lisbon: Colibri.

——. (2006). *A Hundred Horizons: The Indian Ocean in the Age of Global Empire*. Harvard: Harvard University Press.

Boxer, C. (1963). *Race Relations in Portuguese Colonial Empire – 1415–1825*. Oxford: Oxford University Press.

Capela, J. (2002). *O tráfico de escravos nos portos de Moçambique*. Oporto: Afrontamento.

Cardão, M. (2019). 'A grande aventura. Televisão, nacionalismo e as comemorações dos Descobrimentos portugueses'. *Práticas da História*, 8, 17–47.

Cardina, M. and Fytili, M. (2023). 'From History to Memory: Representations of Regime Change in Portugal, Spain and Greece'. *Mélanges de la Casa de Velázquez*, 53 (1), 151–176.

Edgerton, G. R. and Rollins, P. C. (eds). (2001). *Television Histories: Shaping Collective Memories in the Media Age*. Lexington: University Press of Kentucky.

Erll, A. (2008). 'Literature, Film, and the Mediality of Cultural Memory'. In A. Erll and A. Nunning (eds), *Cultural Memory Studies: An International and Interdisciplinary Handbook*, pp. 389–398. Berlin: De Gruyter.

——. (2009). 'Remembering across Time, Space, and Cultures: Premediation, Remediation and the "Indian Mutiny"'. In A. Erll and A. Rigney (eds), *Mediation, Remediation, and the Dynamics of Cultural Memory*, pp. 109–138. Berlin: De Gruyter.

Erll, A. and Rigney, A. (eds). (2009). *Mediation, Remediation, and the Dynamics of Cultural Memory*. Berlin: De Gruyter.

Ferreira, C. C. (2006). *A EXPO'98 e os imaginários do Portugal contemporâneo. Cultura, celebração e políticas de representação*. Tese da Faculdade de Economia da Universidade de Coimbra.

Gilroy, P. (1995). *The Black Atlantic, Modernity and Double Consciousness*. London: Verso.

Glissant, E. (1990). *Poétique de la relation*. Paris: Gallimard.

Hallett, J. (1998). 'Introdução II'. In AA.VV. *Culturas do Índico*, pp. 105–116. Lisboa: Comissão Nacional para as Comemorações dos Descobrimentos Portugueses.

Hespanha, A. M. (1997). 'Comissão Nacional para as Comemorações dos Descobrimentos Portugueses'. *Acervo*, 10 (1), 97–102.

——. (1998). 'Apresentação'. In AA.VV. *Culturas do Índico*, pp. 13–14. Lisbon: Comissão Nacional para as Comemorações dos Descobrimentos Portugueses.

——. (coord.) (1999). *Há 500 anos. Balanço de três anos de Comemorações dos Descobrimentos Portugueses 1996–1998*. Lisbon: Comissão Nacional para as Comemorações dos Descobrimentos Portugueses.

——. (2019). 'Comemorar como política pública. A comemoração dos Descobrimentos Portugueses, ciclo 1997–2000'. *Práticas da História*, (8), 198–220.

Jedlowski, P. (2007). 'La Memoria pubblica: che cos'è?'. In Rampazi, M. and Tota, A. L. (a cura di), *La memoria pubblica: trauma culturale, nuovi confini e identità nazionali*, pp. XIII–XVIII. Milano: Utet.

Leite, A. M., Brugioni, E. and Falconi, J. (eds). (2019). *Estudos sobre o Oceano Índico. Antologia de textos teóricos*. Lisbon: Colibri.

Lourenço, E. (1999). *A Nau de Ícaro seguido de Imagem e Miragem da Lusofonia*. Lisbon: Gradiva.

Marques, A. Pinheiro (1992). 'A historiografia dos descobrimentos e expansão ultramarina portuguesa'. *Revista de História das Ideias* (14), 439–458.

Moorthy, S. and Jamal, A. (eds). (2010). *Indian Ocean Studies: Cultural, Social, and Political Perspectives*. London: Routledge.

Nora, P. (1984). *Les lieux de mémoire*. I vol. Paris: Gallimard.

Oliveira, A. de (2003). 'The Activities of the CNCDP: A Preliminary Assessment'. *E-Journal of Portuguese History*, 1 (1), 1–12.

Pearson, M. N. (1988). *Port Cities and Intruders: The Swahili Coast, India and Portugal in the Early Modern Era*. Baltimore: Johns Hopkins University Press.

——. (2003). *The Indian Ocean*. London: Routledge.

——. (2011). '*History of the Indian Ocean:* A Review Essay'. *Wasafiri*, 26 (2), 78–99.

Pimenta, F. Tavares (2010). 'Perspectivas da Historiografia Colonial Portuguesa (século XX)'. In M. M. Tavares Ribeiro (ed.), *Outros Combates pela História*. Coimbra: Imprensa da Universidade de Coimbra.

Santos, B. de Sousa (1994). *Pela Mão de Alice. O Social e o Político na Pós-modernidade*. Oporto: Afrontamento.

——. (2001). 'Entre Próspero e Caliban: colonialismo, pós-colonialismo e inter-identidade'. In M. I. Ramalho and A. Sousa Ribeiro (eds), *Entre ser e estar. Raízes, percursos e discursos da identidade*, pp. 23–85. Oporto: Afrontamento.

Schiavinatto, I. L. (2019). 'Entre celebrações e exposições: algumas visibilidades em jogo nos Descobrimentos (1990–2000)'. *Práticas da História* 8, 49–83.

Shohat, E. and Stam, R. (2012). *Race in Translation: Culture Wars around the Postcolonial Atlantic*. New York: New York University Press.

Silva, J. Henriques (1998). *Pescadores Macua* (Coord. Joana Pereira Leite). Lisbon: Comissão Nacional para as Comemorações dos Descobrimentos Portugueses.

Sousa, V. (2017). *Da 'portugalidade' à lusofonia*. Famalicão: Húmus.

Stoler, A. L. (2013). *Imperial Debris: On Ruins and Ruination*. Durham: Duke University Press.

——. (2016). *Duress: Imperial Durabilities in Our Time*. Durham: Duke University Press.

Subrahmanyam, S. (1998). *The Career and Legend of Vasco da Gama*. Cambridge: Cambridge University Press.

Vink, M. (2007). 'Indian Ocean Studies and the "new thalassology"'. *Journal of Global History*, 2, 41–62.

Xavier, Â. Barreto (2000). 'Tendências na Historiografia da Expansão Portuguesa'. *Penélope*, 22, 141–179.

PART II

Networks, Narratives, Imaginaries: Literary and Artistic Dialogues

CARMEN LUCIA TINDÓ SECCO

5 Religiosity and universalism(s) in the Indian Ocean

How far can a text imbued with memories give texture, perspectives, significance and meaning to a place? 'Indian Ocean traces' (Borges Coelho, 2005): trails, tenuous threads between memories and forgetfulness. Itineraries through the Indian Ocean: crossings, diversity, mysterious relations between heaven and earth. In these transits, some questions in particular stand out, such as the complexity of histories and philosophies; the great interest in this plural ocean, permeated with inter-relations between peoples, cultures, religions and imaginaries of East Africa and the Orient.

Studies on the Indian Ocean must necessarily be interdisciplinary, demonstrating that, from the medieval period onwards, there was already an ancient, complex, commercial, religious and cultural network in it. In the opinion of historian Sugata Bose:

> A rápida expansão do Islão pelo Índico entre os séculos XIII e XV teceu um novo padrão de unidade económica e cultural em toda esta vasta arena inter-regional. No século XV, mercadores árabes e indianos, na sua maior parte muçulmanos, mas também alguns hindus e jainistas, estavam na vanguarda da actividade económica marítima desde a costa de Moçambique, a oeste, até às ilhas de Maluco, a leste. Ao mesmo tempo, pregadores sufis, espalhando-se das cidades portuárias para os *hinterlands* agrícolas, teciam um universo comum de atmosfera e sensibilidade religioso-cultural. (Bose, *in* Leite et al., 2019: 41–42)

> [The rapid expansion of Islam through the Indian Ocean between thc thirteenth and fifeenth centuries wove a new pattern of economic and cultural unity in all this vast inter-regional arena. In the fifteenth century, Arab and Indian merchants, mainly Muslims, but also some Hindus and Jainists, were in the vanguard of the maritime economy from the coast of Mozambique to the west, as far as the Molucca Islands

> to the east. At the same time, Sufi preachers, spreading through the port cities to the rural hinterlands, created a common universe with a religious- cultural atmosphere and sensibility]

Currently, various other researchers are focusing on the Indian Ocean, exploring not only its geography and history but also investigating different material and immaterial knowledge attached to its waters and coasts, both insular and continental. Like Sugata Bose, Isabel Hofmeyr, Emeritus Professor of African Literatures in South Africa, emphasizes the multiplicity of cultural circulations around this ocean, calling attention to the strong interaction of religions, mainly those of oriental origins, like Hinduism and Buddhism among others. Citing Bose, Professor Hofmeyr underlines the important role of the diasporas when they crossed the Indian Ocean:

> As Bose points out, while there is a rich tradition of work on the distant past of the Indian Ocean world , there is comparatively less on the nineteenth and twentieth centuries. One notable exception has been the work of Mark Ravinder Frost (2002), which has started to outline a distinctive Indian Ocean public sphere that flourished from the 1880s to the 1920s. Based in the port cities of the Indian Ocean and sustained by the intelligentsias of intersecting diasporas, this public sphere was rooted in pan-religious movements, be these Buddhist, Muslim or Hindu. (Hofmeyr, 2007: 1)

Hofmeyr proposes research that goes beyond the mere tracing of the maritime movement of merchants, people and ideas. She encourages research around the ecological and material qualities of the Indian Ocean that make possible the decoloniality of the countries of the Indian Ocean coast. According to her, transnationality is part of the historical development of that ocean. For this reason, new gazes on the Indian Ocean may uncover other ocean sensibilities, other pacts and exchanges between the religiosities present there.

Sílvia M. Pizzetti, an Italian historian who also works on Indian Ocean Studies, observes that, like the Mediterranean, the Indian Ocean is plural. There are various Indian Oceans: the Indian, European, Arab, African, etc. Like the researchers Isabel Hofmeyr and Françoise Vergés, she points out the factors that, historically, gave a certain unity of comprehension to studies on the Indian Ocean space: the colonialism, struggles, migration and anticolonial resistance; the religions, among other things. However, the researcher warns, in the last twenty years, research has turned to new methodologies and to comparative studies that deal not only with historical similarities but with differences, particularities and transversal universalisms that exist in these cultures.

According to Isabel Hofmeyr, during the colonial period, the Indian Ocean was a 'world of universalisms' (2007: 7), for 'it was linked to pan-Islamism, Hindu reformism, pan-Buddhism, the ideas of Greater India, acting as a source of transnational forms of imagination' (*idem, ibidem*). This theorist conceives of the Indian Ocean as a source of universalisms, given that the commercial and imperial diasporas favoured exchanges, tensions, desires, powers and religions that had wider ambitions, and, for this reason, cannot be understood, merely, on the basis of reductive dichotomies. The religions that came with the diasporas mingled among themselves and with some of the local beliefs. For Hofmeyr, the Arab, Hindu and Catholic religions were not limited uniquely to the sphere of the sacred; they had political projects and allied themselves to the dominant powers. However, they must be analysed in a problematizing way, bearing in mind that, according to Meg Samuelson, they were formed through a 'uma densa teia de trocos' [dense web of exchanges] (Samuelson, *in:* Leite et al., 2019: 145).

Around the Indian Ocean, multiple hybrid perspectives of African, Arab, Hindu and Catholic sacrality have intertwined, giving rise to a rich religious and cultural polyphony. The Indian Ocean may then be understood as a 'arquivo líquido' [liquid archive] (Leite et al., 2019: 12) not only of 'narrativas e histórias subalternas' [subaltern narratives and stories] (*idem, ibidem*), but of oppressed religions that, with the unfolding of the historical process, were silenced by credos brought and imposed by the colonizing peoples. The friction between oppressing religions and different subalternized religions resulted in religious syncretisms that were forms of resistance and persistence of specific beliefs that did not let themselves die out altogether. Thus, according to the Mauritian poet and essayist Khal Torabully, 'casitas, credos, línguas, autênticos mood's de vida que se avia mantido apartados passam a confundir-se uns nos outros' [castes, credos, languages, authentic ways of life that had kept themselves apart now begin to mingle with each other] (Torabully, *in* Leite et al., 2019: 231).

Indian Ocean religions: Representations in Mozambican words and canvases

Mozambique, a country on the Indian Ocean coast, was not only home to indigenous African peoples, but experienced colonization by the Arabs and Portuguese, as well as migrations of Indians, Goan and Chinese, all of which had a profound impact on its hybrid cultural imaginary. Complex networks of religions were thus formed in its territory, making it possible for animist African beliefs, Catholicism, Islamism and Hinduism, among others, to exist side by side.

Both in literature and painting and in the cinema in Mozambique, we find works in which various of these intricate religions are presented, allegorically or metaphorically, represented and discussed. In order to observe different artistic treatments given to that religious diversity, we have chosen the short story 'O Pano Encantado' (The Enchanted Cloth), written by João Paulo Borges Coelho (see Brugioni, Grossegesse & Medeiros, 2020); some poems by Rui Knopfli and Sónia Sultuane, in which there are references to Islamic and Hindu divinities; the story 'Entrada no Céu' [Entry into Heaven] and the novel *O Outro Pé da Sereia* [The Other Foot of the Mermaid] by Mia Couto which broach, reflexively, the Catholic religion imposed by Portuguese colonialism; the films *Ngwenya, o Crocodilo*, by Isabel Noronha, and *O Dia em que Explodiu Mabata-Bata* [The Day when Mabata-Bata Exploded] by Sol de Carvalho, in which local African cults are the object of scrutiny; some paintings by Malangatana Valente, in which are found, embodied on canvas, mythical beings from the Ronga imaginary and Christian symbols from Roman Catholicism. In those words and canvases, we will try to point to gods, beliefs, objects and places of worship, producing a reading of Mozambican cultural-religious hybridity, constituted by material and immaterial knowledge that has transited, for centuries, along the shores of the Indian Ocean.

The story 'O Pano Encantado', that opens *Setentrião*, the first of the two volumes that comprise *Índicos Indícios* [Indian Ocean Traces] by author João Paulo Borges Coelho, is set in the Island of Mozambique. This is sung from the memory of what it once was – full of life, trade, exchanges and productivity. The story describes the space, capturing historical fragments in each alley, in each monument. However, the description is shot through with imagination and subjectivity, intertwining the real and the fictional, amid landscapes that are laden with hybrid and abstruse cultural, economic and religious webs.

Many stories are spun in 'O Pano Encantado', with the sea as witness and a bridge linking memories, places, beliefs, temples and cultures. An uneasiness, or rather a desire to belong to the continent on the other side and, at the same time, the isolation, the insularity, in other words, the feeling of being an island, are present in the story. Cloths and threads, motifs of memory and forgetfulness, disputes and confraternities are knotted together. The Island, first capital of Mozambique, gathers in itself a plurality of cultures and religions. Its history is criss-crossed by numerous intersections: of Arabs, Black Macuas, Indians, Asians and Portuguese, with their respective beliefs. On arriving at the Island, the Portuguese caravels met with intense cultural interchanges between Africans and people of oriental heritage:

> Entre casamentos, trocas comerciais e o Islão, veio a nascer e a desenvolver-se na região uma civilização original de matriz banto – a 'civilização suaíli'. Fruto de misturas ao longo dos séculos, Africanos, Árabes e Persas (e mais tarde Indianos muçulmanos) cruzaram-se entre si, originando, [em consequência desse] mestiçamento, o aparecimento de um novo tipo étnico. A par do cruzamento de sangue, influenciaram-se reciprocamente religião e culturas, nascendo dessa forma os suaílis, a quem os cronistas e navegadores portugueses chamaram 'mouros'. (Rocha, 2006: 29)

> [Between marriages, commercial exchanges and Islam, an original civilisation of Bantu origins – the 'Swahili civilisation' came to be born and developed in the region. The fruit of mixtures down the centuries, African, Arabs and Persian (and later Muslim Indians) commingled and as a consequence of this miscegenation, there emereged a new ethnic type. Along with the mixing of blood, religion and culture were reciprocally influential and in this way gave rise to the Swahili, whom the Portuguese chroniclers and navigators called 'Moors']

Borges Coelho's story takes place, largely, in Alfaiataria 2000, the property of Mr Rashid, in which the tailor Jamal is employed. The story problematizes conflicts between coexisting religions on the Island of

Mozambique. Jamal tries to mend his frayed identity. His needlework traces the path of his house, in the neighbourhood of Macaripe, on the Island, as far as Makkah, the holy city of Mecca. It passes through traditionally Islamic African spaces, like the Quirimbas Islands and Zanzibar. On the other side, 'é normalmente aí que se acham as raízes, o segredo do enigma que é cada bordado' [it is normally there that the roots are found, the secret of the enigma that is each embroidery] (Coelho, 2005: 38). It is where Jamal begins a new work, no longer heading towards the geography of the Island, but plunging into the past, in search of his beliefs, the history of his people: 'a história da sua dikiri, a sua confraria' [the history of his dikiri, his confraternity] (Borges Coelho, 2005: 38). According to Ana Mafalda Leite, 'mito e história entrelaçam-se assim num outro itinerário, secreto e sagrado, que se desenha/borda em 'O Pano Encantado', e que refaz uma pertença não apenas moçambicana, mas islâmica e índica' [myth and history are thus intertwined in another itinerary, secret and sacred, that is drawn/embroidered in 'O Pano Encantado', and which refashions a belonging that is not only Mozambican, but Islamic and of the Indian Ocean] (Leite, 2018: 115).

Through the commentaries of the narrator, we note that Jamal belonged to an Islamic confraternity more conservative in its customs than that of Mr Rashid: 'Ma'ruf pretendeu sempre (era esta a sua luta) que o Islão fosse purificado e limpo de costumes infiéis, e o alfaiate crê profundamente que é assim que deve ser' [Ma'ruf always claimed (this was his struggle) that Islam was purified and clean of infidel customs, and the tailor believes profoundly that this is how it should be] (Borges Coelho, 2005: 40). According to José Luís Cabaço:

> Os islâmicos de Moçambique pertencem a duas confrarias principais: Qadiriya e Chadhiliya; a primeira originária do Iraque e a segunda das Comores. Predomina a influência da escola *shafita* da ortodoxia suni. As escolas corânicas instaladas no território, ensinando em árabe ou em *swuahili*, tinham um conteúdo meramente religioso. (Cabaço, 2009: 214)

> [The Islamic citizens of Mozambique belong to two main confraternities: Qadiriya and Chadhiliya; the first originates from Iraq and the second from the Comoros

Islands. The influence of the Shafita school of the Sunni orthodoxy predominates.[1] The Koranic schools established in the territory, teaching in Arabic or Swahili, just had a religious content.]

In Borges Coelho's story, the reverse of Jamal's embroidery expresses religious divisions that exist not only in the Island of Mozambique but within Islam itself: its traditions are diverse, some more orthodox than others. According to Nazir Can:

> o imaginário que se borda no pano lembra-nos, pois, 'o choque entre o que o Alcorão revela, ou o que a tradição sunita (isto é, a ortodoxia) crê que revela, e o que os homens que se sentem muçulmanos creem na prática, se faz cada dia mais inevitável'. É evidente uma fratura interna com base na religião, que anuncia distintos focos de tensão, apresentados nesta narrativa por uma memória 'possível' e seletiva (...). Assim, para além de estetizar a existência de uma geografia santa, o pano de Jamal projeta, na contramão das representações artísticas sobre o hibridismo índico, um Islã fraturado. (Can, 2016: 9)

> [the imaginary that is embroidered on the cloth reminds us then, that 'the collision between what the Koran reveals, or what the Sunni tradition (that is, orthodoxy) believes it reveals, and what men who call themselves Muslims actually believe, is becoming more and more inevitable' (Geertz, 1971: 15, cited by Macagno, 2006: 230). An internal fracture based on religion is evident, announcing difference focuses of tension, presented in this narrative by a 'possible' and selective memory (...). Thus, as well as aestheticising the existence of a holy geography, going against artistic representations on Indian Ocean hybridism, Jamal's cloth projects a fractured Islam.]

A transnational space of different religions, the Indian Ocean has lived with many of these religions, and various of its poets have exalted them, sometimes challenging and questioning them. The poet Rui Knopfli, for example, demonstrates the Muslim presence in the Island of Mozambique alongside Catholicism. In the poem 'Mesquita Grande' (Great Mosque), there is a criticism of the Catholic religion and a celebration of the God Allah and his prophet Mohammed. The Koran, holy book of Islam; the mosques with their minarets; the skullcaps, Muslim caps. The churches,

1 South of the Zambezi, many Muslims who originate from Pakistan and India belong to the Hanafita school, also of the Sunni orthodoxy.

the crucifixes on top of the chapels, the palaces dedicated to saints attest to Catholic religiosity. Thus, material knowledge and immaterial beliefs that prove the coexistence of different religions in the Island of Mozambique circulate both in the architecture and in objects of worship. Knopfli is one of the first Mozambican poets to reflect on these questions, representing the Island as a 'espaço que partilha diferentes geografias e culturas, em que predomina a presença do Oriente, configurado e construído citacionalmente por via da referência de textos sagrados do hinduísmo e do islamismo, de línguas como o árabe, o urdu, o sânscrito e da geografia de origens das diásporas indianas' [space that shares different geographies and cultures, in which the presence of the Orient predominates, configured and constructed citationally by reference to sacred texts of Hinduism and Islamism, of languages like Arabic, Urdu, Sanskrit and to the geography of origins of the Indian diasporas] (Leite, 2018: 203).

Another poet who refers to the Islamic religion in Mozambique is José Craveirinha; in 1955, he published in the newpaper *O Brado Africano* the chronicle 'Maulide Rifai na Mafalala' which, in 1999, was inserted into the book *Contacto e outras crônicas*, by José Craveirinha, a publication that was organized with the assistance of António Sopa and Fátima Ribeiro.

Mafalala was a peripheral Mozambican neighbourhood that continued to exist after the independence of Mozambique, when the capital Lourenço Marques became known as Maputo. In the first decades of the twentieth century, many Muslim Macuas came from the north to live in Mafalala. Even today there are mosques in this neighbourhood, where some followers of confraternities originating in the Island of Mozambique are to be found. José Craveirinha, who lived in Mafalala, was present at the Maulide rituals, that is, of the Rifaiyya Confraternity[2]:

2 'A Rifaiyya é uma das mais proeminentes ordens sufistas. Seu fundador foi Ahmad al-Rifai, nascido no início do século XII, no Iraque. A chegada da Rifaiyya à Ilha de Moçambique pode ter ocorrido entre o final do século XIX e o início do XX. Supõe-se que Comores ou Zanzibar possam ter sido os locais de sua origem, já que mantinham, nesse período, um intenso contato com o norte de Moçambique' (Capela & Medeiros, 1987).

[The Rifaiyya is one of the most prominent Sufi orders. Its founder was Ahmad al-Rifai, born in the early twelfth century in Iraq. The arrival of the Rifaiyya

> A noite de Mafalala enche-se de sons para guiar os homens. As estrelas parecem outras mais brilhantes, quase humanamente comunicativas apesar da lonjura espacial. Seguimos condensados na noite morna, nossos pés afeitos aos brilhos e labirintos dos quintais de caniço. A voz do Maulide chama-nos. Estacas, barrotes, sacas e encerados e os homens lá dentro acocorados em largas esteiras. Num dos lados um lençol amarelo como pano de fundo. Um amarelo carregado de açafrão. Velhos patriarcais lêem o livro sagrado e entoam monótonas cantilenas. As pernas flectidas, os olhos baixos, o tronco balanceando ao ritmo dos cânticos. (Craveirinha, 1999: 11)

> [The Mafalala night is filled with sounds to guide the men. The stars seem like other, more brilliant ones, almost humanly communicative, despite their distance in space. We follow, condensed in the tepid night, our feet attached to the flashes and labyrinths of the yards with reeds. The voice of Maulide calls us.
>
> Stakes, bars, sacks and tarpaulins and the men inside squatting on wide mats. On one of the sides a yellow sheet as a backdrop. A yellow laden with saffron. Old patriarchs read the holy book and sing monotonous chants. The bent legs, lowered eyes, the trunk swaying to the rhythm of the songs.]

Craveirinha describes the Islamic cult sinaesthetically, with emphasis on the musicality of the songs, in the chromatism and taste of the Arab condiment of saffron. Vision, smell, taste, hearing and the touch of the devoted bent over the mats express, in a plastic and very connotative way, the mystical atmosphere of the Muslim ceremony.

Among younger Mozambican poets, belonging to other generations, religions of oriental origins are also mentioned. In the book *Minarete de medos e outros poemas (*Minaret of fears and other poems), authored by Mbate Pedro, although the poems do not stick to religious matters, Islam is present in the metaphor of the title: the Minaret is the highest part of a mosque, from which the faithful are called to prayer. However, in Mbate Pedro's work, the minarets do not summon them to pray; metaphorically, they warn of different kinds of fear. The lyrical subject reflects on the destiny of the poet, his relationship with his fears and sorrows, the geography

to the Island of Mozambique may have occurred between the end of the nineteenth century and the beginning of the twentieth. It is assumed that Comoros or Zanzibar may have been the places of their origin, since theywere in close contact with the north of Mozambique at that time.]

of other terrors in the body of the people and the city, the anguish of other poets, including, for example, the young Sangare Okape.

The poet Melita Matsinhe, in *Ignição dos sonhos* (Igniting dreams), presents poems that sing the age-old waters of the Indian Ocean. Others speak of hands, winds, words, sacred blankets in painful destinies. References to beliefs, to spirituality, may be inferred from some of her metaphors, as we observe in the following verses: 'Sou/de imortais almas/infinitas chamas/vidas que tocam além/do pão e da dor/Reencarnações' [I am/of mortal souls/infinite flames/lives that touch beyond/the bread and the pain/Reincarnations] (Matsinhe, 2017: 37).

Another feminine voice in present-day Mozambican poetry that also tackles the themes of other lives after death is Sónia Sultuane. In *Roda das encarnações* [*Wheel of incarnations*], the poems lead the reader through a sensorial, mystical universe in which spiritual and earthly experiences mingle in a voyage through Mozambique, India and the poet's own interior. The body and the movement of the words transit through the Indian Ocean, through its religious legacies, giving meaning to the life that is perpetuated in the wheel of incarnations:

> Por todos os lugares agrestes e sagrados que piso,
> pelas savanas, florestas e montanhas que me povoam,
> caminho com as palavras impressas em meus pés
> e viajo no mundo a qualquer hora do dia e da noite
> falo em qualquer língua
> rezo as palavras das mais diversas religiões
> sem amarras ou falsas convicções
> (Sultuane, 2017: 14)
>
> [Through all the wild and sacred places I tread,
> through the savannahs, forests, mountains that populate me,
> I walk with the words printed on my feet
> and I travel in the world at any hour of day or night
> I speak in any language
> I pray in words from the most diverse religions
> With no ties or false convictions]

The poems of *Roda das encarnações* operate with faint and delicate frontiers between the emotional and the rational, between life and death, between the sensorial and the spiritual, between body and soul, between the local and the cosmopolitan, between chaos and the cosmos – but not in a dichotomous but rather a plural way, intertwining knowledge, smells, tastes, beliefs, feelings, vibrations, prayers, thoughts. Following senses essential both for her life and for her poetry, the lyrical I, in constant motion, travels through many religions, expressing its gratitude: 'namaste' (Sultuane, 2017: 19). The Indian 'thank you' coexists with the Arab call to prayer of the mosque: '*Adhan Mashallah*!' (Sultuane, 2017: 34), with the Muslim greeting '*Masha 'Allah*', which means 'God so wished'. Permeated by all these identities, the voyage of the lyrical subject is not outwards; it is inwards, towards self-knowledge and comprehension of its own existence.

The metaphorical voyage is, simultaneously, a writing about self and writing about the world. It is configured as a transit of knowledge. 'Durante a Idade Média, a viagem fazia parte da vida dos letrados do mundo islâmico [...], era uma expressão religiosa e, ao mesmo tempo, uma forma de construção do saber' [During the Middle Ages, the voyage was part of the lives of the scholars of the Islamic world (...), it was a religious expression and, at the same time, a way of constructing knowledge] (Bissio, 2012: 145). In *Roda das encarnações* it was constituted as a means of the lyrical I incarnating its origins and previous lives, shot through with mixed cultural heritages – Indian, Islamic, Mozambican.

Christianity, Buddhism and Hinduism[3] connote a supreme sense, that of the sacred word of the actual poetry. The search for vital plenitude, cosmic sacrality, the universe as creative energy. Taking on this search, the poems of *Roda das Encarnações* are realized not only as a cosmopoetics of the Indian Ocean but also as a profound metaphysical and existential crossing.

For Pamila Gupta, the 'Oceano Índico é mais que uma mera geografia estimulante. É uma ideia porque nos proporciona um panorama elucidativo

3 'Para o hinduísmo, a alma humana passa por várias encarnações, até atingir um estado de consciência (*moksha*); busca religar a essência sagrada ao ego, isto é, possibilita o encontro entre o eu e a centelha divina que há em todos os seres humanos.' [For Hinduism, the human soul goes through various incarnations, until it reaches a state of awareness (*moksha*); it seeks to reconnect the sacred essence to the ego, that is, it makes possible the meeting between the I and the divine spark there is in all human beings] (Blanc, 2015: 8).

do Islão, (...) porque nos permite ver o mundo como um todo, num enquadramento simultaneamente muito novo e muito antigo, completado pelas próprias tradições e características' [(the) Indian Ocean is more than a mere geographical stimulant. It is an *idea* because it furnishes us with an elucidatory panorama of Islam (...) because it allows us to see the world as a whole, in a framework that is simultaneously very young and very old, completed by its own traditions and characteristics.] (Gupta, *in* Leite et al., 2019: 168–169).

The Indian Ocean has always been a space of cosmopolitanization and Afropolitanism, insofar as it has made possible historical, commercial, social, cultural and religious exchanges. According to Achille Mbembe, African identities cannot be discussed just from the starting point of colonialism, the anticolonial struggles, Negritude, Pan-Africanism. He himself proposes a 'relativização das raízes e pertencimentos' [relativization of the roots and affiliations] (Mbembe, 2015: 70). He warns that many African countries were formed by exchanges and crossbreeding prior to colonialism. However, for him, 'a história cultural do continente africano praticamente não pode ser compreendida fora do paradigma da itinerância, da mobilidade e do deslocamento' [the cultural history of the African continent cannot practically be understood outside the paradigm of iterancy, mobility and displacement] (2015: 69); African identities involve the reinvention of their cultural traditions in 'contextos de cosmopolitização que ele conceitua como afropolitanismo, ou seja, como uma sensibilidade histórica, cultural, estética' [contexts of cosmopolitanization that he conceives as *Afropolitanism*, or rather, as a historical, cultural, aesthetic *sensibility*] (2015: 70, our emphasis) which is woven through constant interactions and transhumances.

Plural, religious asymmetrical relations, expressed in works by various writers and poets, among whom we may number José Craveirinha, also represent, critically, the animism of the ancestral African culture mixed with symbols of Christianity, re-agented in tense miscegenation:

> Efígies de Cristo suspendem ao meu pescoço
> em rodelas de latão em vez dos meus autênticos
> mutovonas da chuva e da fecundidade das virgens.
> (Craveirinha, 1980: 16)

> [Effigies of Christ hang around my neck
> in slices of tin instead of my authentic
> mutovonas of rain and the fertility of virgins.]

This cultural and religious hybridization is revealed both by the verses of the poet José Craveirinha and by the canvases of the Mozambican painter Malangatana Valente, among which we remember *Adão e Eva em frente da Catedral de Lourenço Marques* (Adam and Eve in front of Lourenço Marques Cathedral) (1960)[4] and *Nu com Crucifixo* (Nude with Crucifix) (1960),[5] in which icons of Christianity are found immersed in African cults and customs, attesting to the strong presence of Catholicism imposed by the Portuguese colonization in Mozambique.

In another phase of Malangatana's painting,[6] the canvases express allegorically the African ancestral imaginary, in which animals and men, *xicuembos* (spirits of ancestors) and *shetanis* (magical and phantasmagorical figures), lumpfanas (lizards) which according to the Mozambican legends and traditions recorded by Henri Junod, in *Usos e costumes dos bantu*, 1996, vol. II: 297, were responsible for the transformation of men into beings condemned to death, no longer able to be resurrected, and *ngwenhas* (alligators with sharp teeth) and hybrid beings and mythical birds like the *ndlati* (known as 'the cockerel of the heavens, the bird of thunder and lightning', also according to Junot, 1996, vol. II: 264) are interwoven in metamorphoses, sometimes monstrous, revealing profound fears, rooted in the soul of the people reduced by so many aggressions suffered under colonialism and during the war of liberation, violence materialized by sharp and feral claws.

The African religious imaginary is also brought into play in the documentary of the Mozambican film-maker Isabel Noronha, *Ngwenya, o crocodilo* (2007), which deals with the life of the painter Malangatana Valente. This enters the mythical past of Master Malangatana who, as well as a painter, was a poet and healer.

4 Cf. Figure 5.1.
5 Cf. Figure 5.2.
6 Cf. Figure 5.3.

The camera, capturing the onyrical and magical climate of the traditional local cults and habits, accompanies Malangatana's journey, which takes him back to the religious and social customs and practices of the Ronga people of southern Mozambique. For example: the image of the birth of a boy child is depicted on film and the Master celebrates the child's arrival, throwing a stick burned with mother's milk, a Ronga ritual practised there, the *quenguelequezê*, whose purpose is to make the child grow up correctly, following tradition, in good health. The child's mother shows him to the moon and his grandmother gives thanks for her grandson's arrival: '*kanimambo! kanimambo! kanimambo!*'

Malangatana sings to his mother and his ancestors who have already died. He teaches painting to the children,[7] showing them how to grind the mulala, the plant with which he makes the dyes to colour their pictures. The cinematographic gaze captures and superimposes scenes: the painter sees a trunk, imagines a woman then a crocodile, and there he has an erotic relationship with the wood before beginning to carve it. The documentary enters the traditional region of Matalana, where the painter lived. He describes the nearby forest that was inhabited by sorcerers and spirits, by owls and nightbirds. Hens' feathers and ashes invade the film space, symbolizing spells. Legends and stories from the traditions intersect: the figure of the Bunhale emerges, the friend of the hippopotami and sorcerers, which, according to local beliefs, flew at night, *xinguilavam* (they shook their body and arms) in order to incorporate spirits and they cast the *tintholos*[8] (little magic bones) in order to foretell the future. The film therefore reinvents rites, songs and the magic of the sorcerers, owners of the sacred Mozambican forest.

Another film that also deals with traditional Mozambican beliefs is *O Dia em que Explodiu Mabata-Bata* (The Day when Mabata-Bata Exploded)

7 Cf. Figure 5.4.

8 'Os ossículos divinatórios constituem um sistema admirável de adivinhação nas tradições moçambicanas e o papel que desempenham é enorme.' [The little divining bones are an admirable system of divination in Mozambican tradition and they play an enormous role.] (Junod, 1996, vol. 2: 463).

(2018), the adaptation of the homonymous story by Mia Couto, directed by Sol de Carvalho, winner of various prizes.

O Dia em que Explodiu Mabata-Bata by Sol de Carvalho is not a documentary. It is a fictional film. It reinvents Mia Couto's narrative, maintaining some faithfulness in relation to the literary plot, but moves away from it at different moments, mainly when he broaches African religiosity, regarding ancestor worship. Mia Couto's tale narrates the story of Azarias, the little orphaned herdsman who wanted to learn to read at school, but whose uncle Raul stopped him, obliging him to take care, every day, of his oxen who were intended for the lobolo ceremony, the dowry that he would pay to his fiancé's family. In the story, at the beginning, the ox steps on a landmine and is blown up; at the end, the same thing happens to the child who, deceived by his uncle, had decided to return to his grandmother Carolina's house. The little herdsman suffers the same fate as Mataba-Bata: he steps on another mine and is splintered in the air. Here literary and cinematographic language come together. The film, resignifying customs from local African traditions, focuses on rituals of consultation of the family with the ancestors so that uncle Raul's marriage can take place. Azarias's spirit is summoned. But how would he judge those who caused his death? Starting from this question, crucial in the film plot, Sol's work innovates Mia Couto's story, taking a fairly African point of view in respect of the representation of the Bantu religious imaginary through cinematographic language. Azarias, having died tragically, had no right to the traditional funeral ceremonies.[9] For this reason, in the film, he walks about like a 'xipoco', that is, a lost soul seeking its home. The camera accompanies him, as well as the rites practised to calm his spirit. The cinematographic gaze thus manages to capture knowledge of the Mozambican material and immaterial culture, for it expresses and calls into question representations of African animism.

9 According to the explanation in the book *Usos e costumes dos bantu*, 'o cadáver que não foi enterrado com os ritos e cerimônias de rigor, ou seja, o defunto que não foi cuidado como devia ser, torna-se um espírito inquieto e cheio de amargura'. [the body that was not interred with the correct rites and ceremonies, or rather,the dead man who was duly cared for, becomes an unquiet spirit full of bitterness]. (Junod, 1996, vol. 2: 320).

Likewise, the novel *As duas sombras do rio* (The Two Shadows of the River) by João Paulo Borges Coelho, reveals, through fiction, the ideological and religious shock between traditionalists and materialists. The novelistic enunciation points critically to these questions, as, for example, when the character Sigaúke, the administrator, finds himself divided between the myths and religiosity of foundational Mozambican traditions and the materialism imposed by the government of the first phase of the Independence: 'Na escola de administradores me ensinaram que era preciso acabar com os obscurantistas, fazendo valer o materialismo e a lei.' [In the school of administrators they taught me that it was necessary to get rid of the obscurantists, making materialism and the law prevail] (Borges Coelho, 2003: 47).

However, there are two characters who have dealings with spirits that they conjure in or order to help the apparent madness of the fisherman Leónidas: Harkiriwa, a rural woman refugee, linked to the earth, who comes from the North, and Gomanhundo, the town's healer, whose spirit is a Portuguese Dominican friar from the beginning of the nineteenth century. This fusion of an African healer incorporating the spirit of a European friar is highly symptomatic of the religious hybridism that exists there, arising out of the mixing of ancestral African religions with the beliefs imposed by the colonizers.

According to the author Mia Couto in an interview, one cannot:

> compreender a África se não compreender uma coisa que nem tem nome, que é a religião africana, que chamam às vezes de animista. Os próprios africanos também não entendem que têm de procurar esse entendimento do que eles são, das suas dinâmicas atuais, a partir deste entendimento do que é a sua ligação com os deuses. (Couto, 2002)

> [understand Africa if one does not understand a thing that does not have a name, which is African religion, that they sometimes call animist. The Africans themselves do not understand that they have to seek this understanding of what they are, of their current dynamics, on the basis of this understanding of what their connection to the gods is.]

Nor can we help observing how Catholicism is staged in fiction. The story, 'Entrada no Céu' (Entry into Heaven), from the book *O fio das missangas*

(The String of Beads), and the novel *O Outro Pé da Sereia* (The Mermaid's Other Foot), both written by Mia Couto, question the Catholic religion imposed by the Portuguese colonizers.

The story tells of a black man who in the final moments of his life, asks himself if he is worthy to enter heaven: 'Se faça-me o favor, senhor padre, me diga: cuja essa entrada no Paraíso é à moda da raça, ou das cláusulas de sermos um zé-alguém? Os pretos como eu, salvo sou, apanham licença? Ou precisam pagar umas facilidades, encomendar um abre boca nalgum mandante?' [If you please, Father, tell me: that entry into Paradise, does it have to do with race, or the clauses of a nobody's sermons? Blacks like me, am I saved, do they get permission? Or do they have to pay some instalments, order an appetizer for someone in charge?] (Couto, 2003: 77). The irony of Mia Couto's language places Padre Bento's catechism under suspicion, revealing that this was placed at the service of colonialism:

> O imaginário colonial português, intermediado pela religião católica, confronta-se com a tradição africana. O narrador quer entrar no céu, mas não sabe como, porque o padre fabrica incertezas que funcionam como sistema de exclusão. A angústia do narrador, seu mal-estar, é a condição que o sistema de representação colonial impõe a esse indivíduo. O colonialismo coisifica o africano, aniquilando-o como cidadão, como sujeito. (Araújo, 2013: 419)

> [The Portuguese colonial imaginary, mediated by the Catholic religion, is confronted with African tradition. The narrator wants to enter heaven, but does not know how, because the priest creates uncertainties that function as a system of exclusion. The narrator's anguish, his malaise, is the condition that the system of colonial representation imposes on this individual. Colonialism reifies the African, destroying him as a citizen, a subject.]

In the novel *O outro pé da sereia* by Mia Couto, which takes place in 2002, ten years after the peace treaty in Mozambique, a shepherd and his wife, Mwadia Malunga, find an image of Our Lady buried on the banks of the river that cuts through the little village named Antigamente [Long Time Ago]. The healer warns that they have tarnished the spirit of the river and for that reason they may be punished. Mwadia then decides to return to Vila Longe to protect the saint.

In the course of the narrative, it becomes known that the image had come to Mozambique in 1560 with the Jesuit Gonçalo da Silveira, whose expedition was intended to convert the emperor of the kingdom of Monomotapa to Christianity. It is symptomatic that the Catholic saint is called Kianda and Nzuzu, African water divinities, by the slaves who accompanied Dom Gonçalo. Religious syncretism and culture shock between Portuguese, Indians and Africans are present the whole time in the novel.

The Catholic saint without the foot, seen by the slave Nimi Nsundi as an African aquatic goddess, functions as an allegory that reveals the ethnic, religious, cultural and historical mixing that pervades Mozambican society.

Such hybrid religions, however, do not achieve complete adherence. To the present day, there are radical religious groups in Mozambique whose practices are harmful to society, since they exercise intolerance and cause deaths. This is the case, for example, of albinos who are discriminated against by animist religions who see them as damned beings who have to die. Also children who are born with defects, or some syndrome, are taken to be sorcerers or witches and are excluded from social life. The same occurs with twins, in some ethnicities, whose beliefs preach the sacrifice of one of them at the moment of birth.

These discriminations are not only found in animist religions. In the North of Mozambique, in recent years, Muslim beliefs have been shaken by radical Islamic groups, whose violence and prejudices have provoked attacks and deaths in Mozambican cities in the province of Cabo Delgado:

> Numa altura em que Moçambique ainda estava num longo processo negocial com vista a pôr termo ao conflito decorrente dos resultados eleitorais das eleições gerais de 2014, o país foi surpreendido, a 5 de Outubro de 2017, com notícias de um ataque armado às instituições do Estado na vila-sede de Mocímboa da Praia, província de Cabo Delgado. Perpetrado por um grupo desconhecido, com reivindicações da prática de um Islão radical, este ataque armado era um fenómeno novo no processo político moçambicano e trazia uma série de questões não só do ponto de vista da natureza do grupo e suas motivações, como também no que se refere às implicações políticas, sociais e económicas do próprio fenómeno para o país. (Habibe, Forquilha & Pereira, 2019: 6)

> [At a time when Mozambique was still in a long negotiating process with a view to putting an end to the conflict resulting from the electoral results of the general

> elections of 2014, the country was surprised, on the 5th of October 2017, with news of an armed attack on the institutions of the State in the town-headquarters of Mocímboa da Praia, province of Cabo Delgado. Perpetrated by an unknown group claiming to practise radical Islam, this armed attack was a new phenomenon in the Mozambican political process and raised a series of questions not only from the point of view of the nature of the group and its motivations, but also in respect of the political, social and economic implications of the actual phenomenon for the country.]

The religions of Mozambique and other countries bathed by the Indian Ocean are complex: This ocean extends between Africa and the Orient, but was and still is sailed by ships coming from the West. For this reason, it bears a plural historiography, pervaded with different perspectives, beliefs and knowledge:

> O oceano Índico foi um caminho, um cruzamento de culturas. Por suas águas, chegaram navegantes de outros continentes, de outras raças, de outras religiões. Na costa moçambicana os navios eram a agulha que costurava esse imenso pano em que ainda hoje se estampam diversidades. (Couto, 2010: 67)

> [The Indian Ocean was a pathway, a crossing of cultures. Through its waters, sailors arrived from other continents, other races, other religions. On the Mozambican coast the ships were the needle that sewed that immense cloth on which diversities are still printed today.]

To close, we remember that the Atlantic Ocean was a route by which Catholicism came to Africa. The Indian Ocean brought Islamism, Hinduism and other religions from the Orient. In the course of history, the Black Atlantic and the Indian Ocean touched on one another, insofar as both of them trafficked slaves, and carried out commerce and exchanges. The two oceans ended up forging innumerable, different modes and models of eating, dressing, thinking, feeling and praying. Asymmetrical, transnational relations were constructed, they crossed over compasses and they crossed frontiers; they went beyond norms and frameworks of eurocentric paradigms, placing in dialogue economies, politics, cultures, knowledge, religions and tastes both from Europe and Africa, Asia, the Americas, Oceania.

Appendices

Figure 5.1: *Adão e Eva em frente da Catedral de Lourenço Marques* (Adam and Eve in front of Lourenço Marques Cathedral) 1960

Figure 5.2: *Nu com Crucifixo* (Nude with Crucifix) 1960

Figure 5.3: *O Poço Sagrado* (The Sacred Well), 1985

Figure 5.4: Scene from the film *Ngwenya, o Crocodilo*

Bibliography

Bissio, B. (2012). *O mundo falava árabe. A civilização árabe-islâmica clássica através da obra de Ibn Khaldun e Ibn Battuta*. Rio de Janeiro: Civilização Brasileira.

Blanc, C. (2015). *Guia das religiões*. São Paulo: *On-line* Editora.

Borges Coelho, J. P. (2003). *As duas sombras do rio*. Lisbon: Caminho.

——. (2005). *Índicos indícios – Setentrião I*. Lisbon: Caminho.

Bose, S. (2019). 'Estado, Economia e Cultura na Orla do Índico: Teoria e História'. *Oceanos*, 34, April/June 1998, 25–36. Translation by José Vieira de Lima. Reprinted in Leite et al., pp. 29–52.

Brugioni, E., Grossegesse, O. and de Medeiros, P. (2020). *A Companion to João Paulo Borges Coelho. Rewriting the (Post)Colonial Remains*. Oxford: Peter Lang.

Cabaço, J. L. (2009). *Moçambique: identidade, colonialismo e libertação*. São Paulo: UNESP.

Can, N. A. (2016). 'A Ilha de Moçambique na ficção de João Paulo Borges Coelho: tensões islâmicas, um pano encantado e os índicos indícios'. *Todas as Musas*, 8(1). <https://www.todasasmusas.com.br/15Nazir_Ahmed.pdf>. Accessed: 10 September 2019.

Capela, J. and Medeiros, E. (1987). *O tráfico de escravos de Moçambique para as Ilhas do Índico – 1720–1902*. Maputo: Núcleo Editorial da Universidade Eduardo Mondlane.

Couto, M. (2002). Mia Couto e o exercício da humildade. Entrevista à Marilene Felinto publicada em SP, *Folha de São Paulo*, 21/07/2002. *In site: Acervo Racismo Ambiental*. <https://acervo.racismoambiental.net.br/2012/02/17/mia-couto-e-o-exercicio-da-humildade/> Accessed: 11/09/2024.

——. (2006). *O outro pé da sereia*. São Paulo: Companhia das Letras.

——. (2009) *O fio das missangas*. São Paulo: Companhia das Letras.

——. (2010). 'Um mar de trocas, um oceano de mitos'. In *Pensageiro frequente*. Lisbon: Caminho.

Couto, M. (2016). Intervenção no lançamento de *Roda das encarnações* de Sónia Sultuane. *Nósdiario*. Maputo, 16/11/2016. <https://www.nosdiario.gal/articulo/cultura/sonia-sultuane-caminhar-com-palavras-impressas-nos-pes/20161122151349052995.html> Accessed: 11/09/2024.

Craveirinha, J. (1980). *Xigubo*. Lisbon: Edições 70.

Craveirinha, J., Sopa, A. and Ribeiro, F. (eds). (1999). *Contacto e outras crônicas*. Maputo: Centro Cultural Português.

Geertz, C. (1971). *Islam Observed: Religious Development in Morocca and Indonesia*. Chicago: Chicago University Press.

Habibe, S., Forquilha, S. and Pereira, J. (2019). *Radicalização Islâmica no Norte de Moçambique. O Caso de Mocímboa da Praia. Cadernos IESE*. N.º 17. Maputo: Instituto de Estudos Sociais e Económicos (IESE).

Hofmeyr, I. (2007). 'The Black Atlantic Meets the Indian Ocean: Forging New Paradigms of Transnationalism for the Global South – Literary and Cultural Perspectives'. *Social Dynamics*, 33 (2), 3–32.

——. (2018). 'O Atlântico negro encontra o Oceano Índico: forjando novos paradigmas do transnacionalismo para o Sul Global – perspectivas literárias e culturais'. Trans. Ana Mafalda Leite, Elena Brugioni and Jéssica Falconi. In *Remate de Males. Revista do Instituto da Linguagem*, IEL, UNICAMP. Campinas-SP, 38:1, Dossiê: *Espaços transnacionais: narrativas do Oceano Índico* (Jan./June), pp. 10–41.

Junod, H. (1996). *Usos e costumes dos bantu*. Colaboração dos linguistas Bento Sitoe, Félix Khosa, José Fialho. Maputo: Arquivo Histórico and II. Coleção Documentos, 3 de Moçambique, 1996, vol. I.

Knopfli, R. (1982). 'A Ilha de Próspero. Roteiro privado da Ilha de Moçambique'. In *Memória consentida*. Biblioteca de Autores Portgueses. Lisbon: Imprensa Nacional-Casa da Moeda.

Leite, A. M. (2018). *Cenografias pós-coloniais & estudos sobre literatura moçambicana.* Lisbon: Colibri.

Leite, A. M., Brugioni, E. and Falconi, J., (eds). (2019). *Estudos sobre o Oceano Índico. Antologia de textos teóricos.* Lisbon: Edições Colibri.

Macagno, L. (2006). *Outros muçulmanos: Islão e narrativas coloniais.* Lisboa: Imprensa de Ciências Sociais.

——. (2007). 'Islã, transe e liminaridade'. *Revista de Antropologia*, 50 (1), 85–123, São Paulo, USP. Available at: <http://dx.doi.org/10.1590/S0034-77012007000100003>. Accessed: 8 June 2020.

Malangatana. (1998). *Álbum das pinturas* (org. de Júlio Navarro). Lisbon: Caminho.

Mata, I. and Padilha, L. (2007). *Mulher em África. Vozes de uma margem sempre presente.* Lisbon: Colibri.

Matsinhe, M. (2017). *Ignição dos sonhos.* Maputo: Fundação Fernando Leite Couto.

Mbate, P. (2009). *Minarete de medos e outros poemas.* Maputo: Índico Editores.

Mbembe, A. (2015). 'Afropolitanismo'. *Áskesis*, 4 (2), 68–71. Available at: <http://www.revistaaskesis.ufscar.br/index.php/askesis/article/view/74>. Accessed: 14 October 2018.

Mendonça, F. and Saúte, N. (1993). *Antologia da nova poesia moçambicana.* Maputo: AEMO.

Morais, M. P. A. (November 2013). *Religião e colonização em Mia Couto. Anais do I Simpósio de Linguística, Literatura e Ensino do Tocantins*, pp. 416–424. Araguaína (TO): UFT.

Noronha, I. (2007). Ngwenya, o crocodilo. Documentário africano de 90 min. Maputo: Ébano Multimídia. Available at: <https://www.youtube.com/watch?v=sAyyPhljCY0>. Accessed: 20 July 2015.

Pizzetti, S. M. (2014). 'Mare Indicum Temps et espaces d'une autre modernité'. *Annales historiques de la Révolution française*, 375. <http://ahrf.revues.org/13046>. Accessed: 15 November 2018.

Rocha, A. (2006). *Moçambique história e cultura.* Maputo: Texto Editores, LDA.

Said, E. (2013). *Orientalismo: o Oriente como invenção do Ocidente.* São Paulo: Ed. Companhia das Letras.

Secco, C. L. T. (2003). *A magia das letras africanas.* Rio de Janeiro: Quartet.

——. (2016). As índicas águas da (na) poesia moçambicana. *Diadorim.* Revista Científica do Programa de Pós-Graduação em Letras Vernáculas da F. Letras/UFRJ. V. 18. número especial. pp. 61–82. Available at: <https://revistas.ufrj.br/index.php/diadorim/article/view/4047>. Accessed: 21 September 2019.

Sultuane, S. (2017). *Roda das encarnações.* São Paulo: Kapulana.

Vergés, F. (2002). 'L'océan Indien, un territoire de recherche multiculturelle'. *Hermès*, 32–33, 447–456. Available at: <https://shs.cairn.info/revue-hermes-la-revue-2002-1-page-447?lang=fr> Accessed: 11/09/2024.

JOANA PASSOS, KAMILA KRAKOWSKA RODRIGUES AND MARTA BANASIAK

6 Writing on the Indian Ocean rim: Literary representations of migration and transit

Africa in the Indian Ocean

In 2009, Moradewun Adejunmobi wrote an article about the absence of an internationally recognized notion of regional identity regarding the cultures of the Indian Ocean. Other regions such as the Caribbean, which is geographically and demographically much smaller, have been comparatively successful in establishing themselves as points of reference. In the article in question,[1] Adejunmobi reflected specifically on the lack of visibility of African elements in any self-defining discourse for the creole society of Madagascar, a gap which has done much to ensure that the island remains a case in isolation from the African continent, consigned to the margins of international debates that address it. Madagascar is not the subject of this paper, but her ideas on self-representation and the silencing of African genealogies in Malgache identity were inspiring for our work. The point is that there is a connection between a society's self-representation and the area of field studies where it claims membership.

For example, if the focus of a specific academic debate highlights Africa or 'black cultures' ('culturas negras') say, in island societies, then, Madagascar is left in a kind of limbo, because it is not 'sufficiently' African,

1 M. Adejunmobi, 'Claiming the Field: Africa and the Space of Indian Ocean Literature', *Callaloo* 32.4 (2009), 1247–1261.

nor is it one of the most representative examples of a black culture. In part, this international perception follows the representation that the Malgache society itself projects to the world.

If, on the other hand, we turn from self-representation to the fields traditionally covered by, say, Asian Studies, the key focus of the debates will probably fall on India, China and Japan, clearly established national units, with a powerful international image situated in that field of research. In other words, we are also saying that in established area studies there are geographical areas, peoples and societies, which are overshadowed by other, more obvious case studies, which tend to be considered the most relevant by communities of researchers. The unintended consequence is that certain subjects are bypassed, becoming less visible than they should be. For example, in European postcolonial studies, the Atlantic coast of Africa has been more studied than the Eastern, Arab one, and currently, the star of Asian studies is China, because of its economic relevance.

What we find interesting in Adejunmobi is that besides engaging in an identification of the blind spots or shadow zones which emerge from either imagined self-representations or established conceptual frameworks, she proposes that Indian Ocean Studies can become a new frame of regional reference, underlining the connections and exchanges of influence which exist between the coastal areas of the region and the islands of the Indian Ocean. Indirectly, through the case of Madagascar, she is suggesting that the coastal area of Eastern Africa has to gain enhanced relevance in Indian Ocean Studies, and that is precisely the point of our paper, by addressing literary representations of the area that highlight contact, travelling and exchange between India and East Africa, be it in Portuguese or in the English language.

Another point worth is that the underrepresentation of African presence (and circulation) throughout the Indian Ocean has had epistemological consequences, misleading researchers to miss/dismiss the relevance of the African continent and of African subjects in Indian Ocean Studies.

Adejunmobi also offered an explanation for the tendency to belittle African influence/presence on Indian Ocean Studies: this legacy is minimized, or suppressed, since any reflection on one's collective relation to Africa would require a creole society to come to terms with its African ancestral past and recognize the role played by slavery in the constitution of its own local identity. And this is our second point – one for which we are also indebted to Adejunmobi: it is time to underline the importance of Africa and of the history of slavery for Indian Ocean Studies *per se,* suggesting that even to this day, the concerned academic debate has rather focused on the history of port cities (Pearson, 2003, 2015), the lived experiences of diaspora communities (Gupta, 2019), the possibilities opened up by new paradigms of comparative literary history (Hofmeyr, 2012), and comparative history (Vink, 2007). The thing that gets forgotten in all of these valid (and foundational) contributions, is a more far-reaching perception of the influence that the African continent has had on Arab and Indian societies, via the African diasporas (forced or otherwise), that were scattered across the whole of the Indian Ocean. African kingdoms may not have possessed navies or fleets, but huge numbers of Africans were involved in all those enterprises of production and trade which required crossing the sea and engaging in business at port cities. Secondly, they were courageous soldiers and sailors, belonging to a host of different armies and ship crews across the Indian Ocean.

Picking up on Adejunmobi's ideas on the relevance of 'repressed Africa', and mindful of the potential Indian Ocean Studies affords us, we decided that the first section of this chapter should explore the power of literature to evoke the past, bringing the memory of slavery and its specific history into contemporary debate. Minimizing the importance of this history effectively impedes the construction of an appropriate self-image for various, different societies of the Indian Ocean. It is understandably uncomfortable, and even painful, recalling how our own past was complicit with slavery – reading the number of people who were displaced, realizing how many families were separated and confronting the sadistic practices that were used to break the will of those who resisted – but facing up to this history is the only way to understand on a collective (national) level, those deeper layers of identity which are still manifest in present-day tensions and conflicts. And what is more, denying the past can never produce a functional mode of engagement with the present, and there is no denying that the history of slavery is woven into the very fabric of the Indian Ocean.

Isabel Hofmeyr (2012) correctly observes that practices of slavery in the Indian Ocean region differed from those which European Expansion would later go on to adopt, but she does not deny that they existed nonetheless:

> While plantation slavery was evident in the Indian Ocean world (especially the Mascarenes), other forms of slavery were evident in the Indian Ocean arena. In the Atlantic world, slavery was starkly racialized and the boundary between slave and free unequivocally drawn. In the Indian Ocean world, slavery was not racialized and there was some mobility between the states of being slave and free. (Hofmeyr, 2012: 588)

With this in mind, we would like to point out that, although the practices associated with slavery were different in these two ocean contexts, in neither case can we consider the history of Indian Ocean slavery independently from transnational trade routes, nor can we ignore the fact that the agricultural produce transported on these ships, relied on slave labour. We realize that the argument Hofmeyr wants to make (see, in particular, her article of 2007) has less to do with the history of slavery per se, and more with endorsing the need to study the history of 'the ancient Indian Ocean', as a model of globalization from the south, prior to, and therefore independently of, European Expansion. In recuperating this history of the ancient Indian Ocean, the old colonial north/south axis is replaced by a transnational south-south dynamic, as the basis for organizing commercial exchange: this alternative paradigm for comparative analysis is one of the ways in which Indian Ocean Studies is an innovative frame to conceive research.[2]

In any case, whether we are discussing the 'ancient Indian Ocean' period or the era of European imperialisms, retrieving the history of slavery in present-day research is important to legitimize pride in this ancestry, granting it a form of epic memory as the history of survivors, of those who managed to endure under conditions of the greatest adversity. This

2 Other relevant aspects that testify to the innovative potential of IOS are its transnational focus in relation to traditional approaches centred on (land) territory and national culture, or, the research perspective attuned to journeys and migratory movements, highlighting the relevance of the move of people and goods to understand coastal societies and their inland contacts.

is precisely what Vergès and Marimoutou (2012) mean by their poetic formulation that 'islands always remember their continents' (15). What this effectively means is that the different societies of the Indian Ocean must recognize and reclaim the implications, and material consequences, of the crossing of maritime routes that transported goods and people. This is why all the creole societies[3] of the Indian Ocean possess their own, specific African roots, particularly as this relates to the history of slavery which, as Vergès and Marimoutou point out, is not a simple tale of victimization: it is a history which also encompasses a struggle for justice, collective courage and an extraordinary level of resilience.

The silenced history of slavery

In relation to the issues discussed above, we have selected for discussion a short story by Adelaide de Souza,[4] 'The House of Wailing' (2010: 169–174), published by Peter Nazareth in a recent anthology of Goan literature. This narrative is a clear example of the power of literature to rekindle debate and raise awareness, combining creative art with civic education and intellectual training. In more specific terms, this narrative explores the haunting presence of slave memories in the contemporary collective consciousness, highlighting how impossible it is to silence and suppress these memories.

3 By 'creole' I mean a 'miscegenated' society, with various influences coming together to create something that is considered 'native' to the islands, a product of different cultural fusions.

4 I could not find much information on Adelaide de Souza. According to Peter Nazareth, the editor of *Pivoting on the point of Return, Modern Goan Literature* (2010), the anthology that includes her story, de Souza brought out two other stories, 'Tricky of Uganda', published in *The Sunday Standard* of the *Indian Express*, and 'Uganda Adieu!', published in *Eve's Weekly* (Nazareth, 2010: xiii). The editor does not give any dates for these two publications. The short story covered in my research here had already appeared in 1983 in *Journal of South Asian Literature*, vol. 18, nº 1, Winter, Spring, 1983.

'The House of Wailing' is a short story which takes place in Zanzibar, the port Arab traders have used since the first centuries AD to establish trade links between the east coast of Africa, the Arab/ian Peninsula (Yemen and Oman) and Persia (Iran) – across the Persian Gulf, and the west coast of India, especially along the coastline that runs from Gujarat to Goa. Consequently, writing a story set in Zanzibar inevitably brings to mind one of the busiest sea routes of the Indian Ocean, where the principal trade was in ivory and slaves.

During the sixteenth century, Zanzibar was controlled by the Portuguese, but in 1650, the Sultan of Oman expelled them and switched his own capital from Oman to the trading port of Zanzibar. Economic activity flourished under the control of these Arab traders and Zanzibar became one of the main ports for slave traffic in the centuries that followed, right up to the eighteenth and nineteenth centuries. This historical context is essential if we are to understand the significance of this narrative being set in a place redolent with slave history. The villa where the story takes place is known locally as 'the House of Wailing', and it is commonly believed that it was built on the ruins of the former local slave market. Consequently, people think there is a curse on the place, and according to local superstition, a terrible, heart-rending wailing can be heard there at night (Souza, 2010: 170).

At the beginning of the story, this local belief provides a conversation topic for a dinner party attended by a group of people who have taken advantage of a passenger liner stopover to meet friends on shore. The various dinner guests include the female narrator; her cabinmate, Mrs Sharma; her good friend, Marcia (the woman they are visiting); Marcia's husband, a British anthropologist; his friend Mr Macfarlane, a Scottish archaeologist; and a local couple Mr and Mrs Suleman who own the villa, and have rented half of it to Marcia and her husband. They form a chance gathering, as the narrator has simply decided to use the stopover in Zanzibar to visit Marcia. During that entertaining evening, the guests start discussing the possibility of supernatural experiences, and this conversation sets the atmosphere for the plot of a horror story, further emphasized by the setting in a cursed place.

During the night, when the dinner is over, the female narrator becomes aware of the sorrowful wailing sound which people say they have heard echoing through the house. All of a sudden, she feels as if she is being transported back in time, and, apparently, she incarnates in another body, in another epoch. When she looks around her, she realizes that she has turned into a young African woman who is being brought to the slave market in Zanzibar, by Arab traders. Through this dream/trance, the narrator lives her own experience of slavery:

> Our arms ached and heads spun with the heavy tusks that were carried on our heads. Black ivory carrying white ivory to the coastal markets, hundreds of miles away from our native homes in the great lakes. (…) Prowling hyenas now followed on our trail, for more than two thirds of our number were claimed by death through disease and sheer starvation. A deep moan and a child's frightened cry rent the stillness of the night, as a mother stumbled on a rock and her baby was flung into the bush. (Souza, 2010: 171–172)

The thing that this young African woman (avatar of our narrator) fears the most on her way to the slave market in Zanzibar is the possibility of being separated from her husband, who is in the same column of prisoners. During their forced walk, her husband fights the slavers holding them captive in defence of another woman in the group. As punishment, they are both subjected to torture:

> A bamboo slit half-way down the centre was the halter round our necks – the cane made deep cruel gashes. We dared not move our heads, for every movement meant agonizing pain for the other. At last, we reached the coast, from where we were shipped with cattle to the slave markets of Zanzibar. The *dhows*[5] were small and there was barely standing room in the dark, damp holds with bilge water up to our ankles. The welts on our necks were flaming ulcers. The scars would be indelible on our skins and on our souls. (Souza, 2010: 172)

When they reach their destination on slave market day in Zanzibar, the protagonist's husband is soon sold to a plantation owner, and they are separated for good. On her turn, the young woman is sold to a repulsive

5 A sail boat commonly used in the Indian Ocean.

old man who leaves no doubt as to what her future fate will be: '[Y]ou will have no work but to make me happy' (Souza, 2010: 173). Although she is inwardly appalled, her instinct for survival forces her to smile:

> Resentment grew deeper and deeper, till it was a burning volcano, [...] with the duplicity of the eternal Eve, I smiled at my master. (Souza, 2010: 173)

Suddenly, the narrator wakes up from her dream/trance. It is dawn, and she feels relieved that she will soon be leaving this haunted place behind her. But the final twist in this curious tale is yet to come. When Mrs. Suleman joins her guests that morning, bringing them some fruit for breakfast, she is not wearing her *buibui*, the veil that is worn by Muslim women. In that moment, the startled narrator notices the scars on Mrs Suleman's neck, which she recognizes as the scars of the punishment inflicted to the young woman and her husband, in that other life she had just incarnated during the night:

> Marcia noticed the direction of my gaze and said, half apologetically, 'Oh, you are looking at Salwa's birthmarks.' [...] But Mrs. Suleman returned my gaze unflinchingly, 'It's not just a birthmark – it's the branding of a slave. You know that,' she said in an almost inaudible whisper. (Souza, 2010: 174)

This surprising end indicates that the narrator's supernatural experience was, in some way, deliberately induced by Mrs Suleman, for she knows the narrator has recognized the story behind those scars on her neck. Salwa (Mrs Suleman) – or rather, as we read it, the ghost who personifies the grief and suffering associated with the Zanzibar slave market and the tragedies lived there for centuries, – wants what happened in the slave market to be remembered. And this ghost will continue to haunt her guests, making them witnesses to the past, so that people living in the present remain aware of what happened, in all its shameful, dehumanizing detail.

The plot of this short story is a powerful, postcolonial metaphor, in so far as it deals with a cursed and tragic history that needs to be remembered, if not for retribution (Salwa is not a character seeking revenge) at the very least as a matter of acknowledgement and respect. The supernatural element of this narrative takes on the role of a compensatory alternative to other official forms of silence, and through the character of Salwa, a suppressed history makes itself known.

Faced with a history that has been silenced, or at least inadequately represented in the dominant historical archives, the writer's use of the supernatural seems appropriate, by way of a provocatory gesture that confronts established historical narratives. In our view, she draws on the supernatural to symbolize alternative ways of seeking (and transmitting) silenced knowledge about the history of slavery. This story goes far beyond mere entertainment to provide an intelligent reflection on the need to maintain historical memory in all its complexity. It is also suggested that the gaps in official archives can be engaged with, and although some information may be lost for good, it might still be possible to retrieve elements of collective memory through artefact, myth and conjecture. It is no coincidence that the local people believe this place to be haunted. They have informally kept alive the memory that this used to be a slave market. To the point, the two academics at the dinner party are, precisely, an archaeologist (Marcia's husband) and an anthropologist, Mr. Macfarlane, both of them working in fields of research which dig up valuable evidence for reconstructing the historical past and the histories of peoples. In any case, the memory of what actually happened in Zanzibar has yet to find its proper place, starting with acknowledgement, as the story discussed here suggests. We believe this is precisely what Adejunmobi meant when she talked about the need to put out the suppressed African memory in Madagascan society, acknowledging the contribution and the legacy of slavery to the formation of present-day society.

The Bandung Conference on a beach in Goa

On 18 and 19 April 1961, a group of activists committed to the liberation struggle of Portugal's colonies in Africa met in Casablanca, Morocco, in what would become the first Conferência das Organizações Nacionais das Colónias Portuguesas (CONCP – Conference of the National Organizations of the Portuguese Colonies). The aim of this conference was to coordinate the national liberation struggles of all the Portuguese colonies, at an international level. Its president was Mário Pinto de Andrade, who was also one of the leaders of the MPLA, an Angolan party which was engaged in the War of Liberation, and was set to gain power after a long civil war against UNITA (1975–2002).

Between the 20th and the 23rd of October of that year, Mário Pinto de Andrade visited India, attending a seminar in New Delhi entitled 'The Problems of the Portuguese Colonies'. Mário Pinto de Andrade had been invited by the 'Indian Council for Africa' and his participation in this event attests to the political affinity which existed between India and the liberation struggles of the Portuguese colonies in Africa, in 1961 (the year when Goa was annexed by India and also the year when the War of Liberation began in Angola).

CONCP ran an official bulletin,[6] written in French, and published by the 'gabinete de coordenação [coordinating office] ('Secrétariat Permanent'). The articles in it were not individually signed, but the office was coordinated by Mário Pinto de Andrade. The section from this bulletin which interests us here[7] is the report on the CONCP representative's trip to the seminar in New Delhi:

> [...] This visit to India has permitted us to:
>
> Inform the government of India, and the people of India about the key principles of the 1st Conference of the National Organizations of the Portuguese Colonies which was held in Casablanca.
>
> Learn what the Government of India thinks about colonialism in general and the case of GOA in particular.
>
> Prepare the organization of a United Front of Nationalist Organizations in Goa by means of bilateral discussion.
>
> [...] Furthermore, the liberation of Goa by the Indian army would, at the present time, make a concrete contribution to the liberation of Portugal's African colonies.][8]
>
> ('Secrétariat Permanent', CONCP, emphasis in the original)

6 This bulletin can be accessed in *Casa Comum*, the digital archive of Fundação Mário Soares.

7 Cf. <http://casacomum.org/cc/visualizador?pasta=07059.021.008#!10>

8 [...] Cette visite aux Indes nous a permis:

What you can read in this document is, in fact, a declaration of mutual support in the struggle against colonialism, which was important for all the involved colonies, either African or Asian. Secondly, that co-operation between Asia and Africa, on a south-south axis, seemed a good way of strengthening coordinated resistance against western colonialism (not only European colonialism but also American influence, which was seeking to polarize the world in the context of the Cold War). In this sense, the 1961 CONCP document seems to follow the ideas of the earlier Bandung Conference (1955, Bandung, Indonesia) which tried to promote cooperation between Asia and the newly independent African countries, in the quest for an international peace which would be independent from the spheres of influence of both the Soviet Union and the United States. This third way became known as the Non-Aligned Nations movement.

Mário Pinto de Andrade's notes (in the form of the quoted report) actually take up ideas of the Bandung Conference, in respect to the specific territories held by Portugal. As is evident in the CONCP document referred to, the destiny of Goa seems to be linked to events in other Portuguese colonies, in Africa, and, indeed, it must have been very encouraging for all activists to witness the fall of Goa, the first colony to split off from an empire that had existed for centuries.

Mário Pinto de Andrade's Indian trip is significant because it documents the moment when bonds of political solidarity were forged across the Indian Ocean, linking the destiny of Goa with that of other Portuguese colonies, especially Mozambique, which needed to find regional allies as it

---

De faire connaître au gouvernement Indien, au peuple de l'Inde, les grandes thèses de la 1ére conférence des Organisations Nationalistes des Colonies Portugaises, qui s'est tenue à Casablanca ;

De connaître le point de vue du gouvernement Indien sur le colonialisme en général et sur les cas de GOA en particulier.

De contribuer par de discussions bilatérales à la réalisation d'un Front Uni des Organisations Nationalistes de GOA.

[...] Par ailleurs, la libération de Goa par l'armée Indienne serait, à l'heure actuelle, une contribution concrète à la libération des colonies africaines du Portugal.

made ready for the war of liberation which began in 1964. It is no coincidence that the second CONCP conference took place in Dar-es-Salaam, on 5 October 1965, which made it one of the first political summit meetings held in Tanzania, shortly after that country gained independence in 1964. The Indian Ocean region was thus effectively organizing itself, in the process of transition to decolonization.

This historical framework affords essential background for understanding the explosive political potential of writing a short story set in Goa, prior to liberation, with a storyline which depicts the unexpected friendship between a political prisoner and the Mozambican soldier tasked with guarding him. The story in question was written by Laxmanrao Sardessai (1904–1986), a Goan author whom I discovered in the pages of a Goan literary anthology edited by Vimala Devi and Manuel de Sebara[9] (1971, *A Literatura Indo-Portuguesa).* This anthology includes a section on L. Sardessai (1971: 351–359), where we found a selection of his poems and the short story 'The Boat from Africa', which we will next discuss. Laxmanrao Sardessai was a famous author in Goa, but he wrote primarily in Marathi, although he was also fluent in Konkani and English. He only began to write in Portuguese in the sixties, mainly producing poetry which he published in two Portuguese language newspapers in Goa, namely *A Vida* and *O Heraldo.*

Although Laxmanrao Sardessai is recognized as being a politically committed writer, what he writes is not mere propaganda. In both his poems and his short stories, he goes far beyond the instrumentalization of literature for political ends. This caveat is an important one, in light of the general literary climate of the period, the 1950s and 1960s, when militant writing, in the service of national resistance and consolidation, was highly valued, especially in territories that were still under colonial rule. However, in the case of Laxmanrao Sardessai, his social and political convictions did not diminish his commitment to stylistic elegance, replete with images that were finely drawn. Vimala Devi and Manuel de Seabra

9 Devi and Seabra's study consists of two volumes. L. Sardessai is referred to on p. 331 of volume I which contains a collection of critical essays on Goan literature. The second volume is an anthology of texts. The work of L. Sardessai is on pp. 351–359.

evidently shared this view. Nonetheless, looking beyond questions of aesthetics and form to matters of impact and content, a strong political commitment underpins the writings of L. Sardessai. As D. A. Smith observes, (2017) it was precisely a political matter which caused Sardessai to turn to the Portuguese language somewhat belatedly in life, after the liberation of Goa, when this territory was no longer a Portuguese colony. Sardessai knew that if he wanted to gain the support of the Catholic community of Goa in opposing Goan integration with the neighbouring state of Maharashtra, he would have to write in Portuguese, and publish his writings in a newspaper which that community read, in order to win their sympathy to the cause (Sardessai founded the Anti-Integration league in 1967).[10] L. Sardessai is mostly recognized in contemporary Goa for his writings in Marathi, but the community of researchers that studies Goan literature in Portuguese is currently addressing with a renewed interest his works written in Portuguese (and he also has published work in Konkani).

In the story 'The Boat from Africa', Francisco, the African guard at Aguada Fort (a Goan prison for political dissidents[11]) becomes friends with the Goan activist that he is guarding, to the point where he secretly releases him for a while, so that they can go for a walk along the beach together. He also takes him to visit his own humble household and introduces him to his newborn son. Francisco wants the prisoner to meet his family before his tour of duty ends the following week. At that point, Francisco will have leave from the Portuguese authorities to return to his native land in Africa. The two men talk to each other, sharing their hopes and plans.

10 'Goa Opinion Poll': this was the referendum conducted among the population of Goa on 16 January 1967. The Goans had to decide if they wanted to remain their own independent state within India or become part of the neighbouring state of Maharashtra. The Catholic community of Goa did not want to join a Hindu majority Indian state. Various sectors of the population voted against fusion and Goa remained independent, achieving full statehood in 1987. The will of the Goan people as expressed in the referendum was respected by the government in Delhi.

11 During the Dictatorship of Estado Novo (1933–1974), Aguada Fort, which had long been a military facility to protect ships in the harbour of Goa, became a prison for political dissidents. T. B. Cunha, the most prominent political prisoner of Goa was held at Aguada Fort until his transference to the Peniche prison, in Portugal.

This simple, moving plot represents, in symbolic terms, a bond of friendship between the two anticolonial struggles in the region, united by their opposition to a common enemy, the Portuguese colonizer. Within this line of interpretation, the birth of Francisco's son also evokes the future of an Afro-Indian union which, like the newborn baby in the story, would be the child of an Indian/Goan and African/Mozambican partnership.

Beyond its political frame of reference, the narrative also represents an example of dignity. Neither does the prisoner try to escape nor does the guard doubt that he can be trusted. But the guard's gesture of kindness does help him find the strength to face his destiny when he goes back to his cell at Aguada Fort.

This story is a clear example of politically committed writing, encouraging political literacy in relation to the international debates of the day. It is the context, meaning the timing and the political scene of Goa and Portuguese colonialism in the 1960s that provide the frame of reference to understand this text. One might even claim that this story reproduces the cooperation ideals of the Bandung Conference (1955), on a smaller scale, reduced to the dimensions of the beach in Goa where the two protagonists come to embody the wider international objectives that were discussed at the political summit. Nowadays, for the contemporary reader, the relevance of such a story is related to the ability of literature to illuminate possible, unsuspected exchange and collaboration among people from different places around the Indian Ocean, which can then be transformed into valid pathways of inquiry for different areas of research in the humanities. It should also be noted that L. Sardessai was a politically committed writer, who was seriously involved in the social and political issues related to Goan integration in India, and very much aware of the waves of decolonization that were sweeping the mid-twentieth century. It is no coincidence that this Goan author decided to include a character from another Portuguese colony in his narrative. 'The Boat from Africa' must have been written a few years before the date when it was published in the anthology (1971) by Devi and Seabra. It was around this time that he began to write in Portuguese. The Goa Opinion Poll had been conducted five years earlier, in 1967, and that would have been one the reasons for Sardessai's decision to write in Portuguese. However, the setting of story refers to a pre-1961 period, when Goa still remained a Portuguese colony.

The reference to an African prison guard in Goa should not be surprising. Portuguese army troops were moved against their will and posted thousands of kilometres away from their families. The forced nature of this exile becomes evident when the guard Francisco joyfully announces the arrival of the boat from Africa, as it approaches the coast of Goa:

> 'Boat!' cried the black corporal on the veranda suddenly. (…) That dark silhouette awakened memories of his life in Africa, that were so joyous and yet so painful. His elderly parents eagerly awaiting him, the girlfriends who had made him happy, vast lakes of clear water (…). (Sardessai, 1971: 355–356)

The corporal Francisco is guarding the narrator, the political prisoner in Fort Aguada. Francisco tells the prisoner that he has been in Goa for four years, and that his tour of duty will be ending soon. However, during this period of service, Francisco has fallen in love with an Indian woman, and the couple are expecting their first child. This information emerges from confidences exchanged between the two characters, prisoner and guard, drawn together by an unusual level of mutual esteem.[12] Indeed, the narrator remarks that during the period of this friendship, he has observed a change in Francisco:

> When I came to Fort Aguada as a political prisoner two years ago, Francisco was a different man. He was rough and brutal; he treated the prisoners badly and insulted them the whole time. One day he even attacked a fellow inmate with his bayonet. (Sardessai, 1971: 357)

The personal love story with the Indian woman is the implied reason for the character's transformation. But the affinity the two men share as they confide in each other has also had some bearing on this. The source of this affinity between them is never referred to directly, but both of them, prisoner and guard, are forced to endure circumstances imposed upon

12 However, note that there are some racist stereotypes that permeate Sardessai's writing (e.g. 'o seu espírito ingénuo de selvagem'/his wild, naïve spirit 1971: 355; 'a sua índole de fera africana'/his character like an African beast 1971: 357).

them by the Portuguese. If Sardessai were simply a writer of propaganda, he would have highlighted the reason why the prisoner is in jail, and the rebellion of the soldier on his tour of duty would be the character's main personality trait. But all the political references in the story are kept in the background, as if they were minor, passing comments, such as those which describe conditions in the prison:

> Two long years had gone by since I came to this fortress and I was still a prisoner awaiting trial. We prisoners had nothing to pass the time except reading the few novels that had got past the censors. We could only leave the confines of our cells twice a day. (1971: 358)

The reference to Aguada Fort, for political dissidents, and to the two years in prison without a proper trial should immediately be recognized as a veiled reference to the fearful standard practices of the Portuguese dictatorship of Estado Novo (1933–1974). In the same vein, there is a brief reference to 'black soldiers (…) who were not happy with their forced service' (1971: 358). However, the main narrative focus when you read this story is the way it exemplifies humanity and solidarity, far beyond what might normally be expected from the respective functions and positions of the characters in the prison system. As a narrative of Afro-Asian solidarity, united by shared opposition to Portuguese colonialism, this story has an international, political dimension which justifies its inclusion in a putative canon of Indian Ocean literature. One of our aims, with the research we present here, is to contribute to creating such a canon.

A further important point in establishing the relevance of this story, and which relates it to the previous analysis of the narrative set in Zanzibar, is that there is a need to remember the role of African diasporas in the Indian Ocean region insofar as they form an integral part of the dynamics of its history. This is precisely the situation that is portrayed in the story 'The Boat from Africa' (1971: 355–359), which, through the African character, recalls the plight of Goa's Mozambican soldiers. These soldiers are an example of forced displacement, which, although it was not slavery as such, certainly did not recognize the troops' freedom of choice. In his article 'African Troops in the Colonial Army, 1961–1974', João Paulo Borges Coelho has collected information on the process known as 'Africanizing the war effort'

(2002: 130), highlighting changes in Portuguese army policy, particularly in the 1960s during the Wars of Independence in Angola, Guinea-Bissau and Mozambique. However, even prior to this war context, it was common practice to enlist Mozambicans in the colonial army. In fact, according to Maria Eugénia Rodrigues, Macua and Swahili soldiers were first recruited as far back as 1784.[13] Note that, in articulation with the setting of the short story discussed in this section, Borges Coelho stated explicitly in his article that 'units from Mozambique and Angola were also posted to distant colonies such as Goa, Macau and Timor', (2002: 133) since the tendency was to use African troops in the colonial army, either as soldiers or as 'mobile companies', who would then move 'en bloc' (2002: 131).

Like 'The House of Wailing' which we discussed in the previous section, this story also reveals something of the complex social, political and historical jigsaw of the Indian Ocean region, allowing us to make connections among different pieces in this puzzle. If it is imperative that we recuperate and acknowledge the history of slavery, which relates the West Coast of Africa to its neighbours, we must also, by the same token, explore the role that Africans played in the military history of the region, and what forms of collaboration and support these individuals might have fostered to mutually encourage decolonization across different shores of the Indian Ocean.

Narrating the transnational Indian Ocean: The interwar period in historical novels

Transnationalism, according to Vertovec, 'describes a condition in which, despite great distances and notwithstanding the presence of international borders (…), certain kinds of relationships have been globally intensified

13 See Maria Eugénia Rodrigues (2006), 'Sepoys from India or African Soldiers? Dilemmas in the naturalization of the Portuguese Colonial Army in the 18th century'.

and now take place paradoxically in a planet-spanning yet common – however virtual – arena of activity' (Vertovec, 1999: 447). However, this intensification of interrelationship, which has been so evident in the past few decades with evolving communications technology, and greater accessibility for modes of transport, does not mean that transnationalism is a contemporary phenomenon nor even an intrinsically modern one. Transnationalism, understood as a phenomenon of interaction between persons and/or institutions across frontiers (of nation-states), predates the legal and physical existence of the nation-state in its modern form, just as it also predates the national sovereignty of various countries which have experienced the continuous ebb and flow of peoples, goods and ideas in transit. The Indian Ocean is 'by far the oldest of the seas in history, in terms of it being used and traversed by humans' (Pearson, 2003: 3), and it acquired an (inter)regional dynamic all of its own, known as the 'Indian Ocean world' (Pearson, 2010: 32), right at the beginning of the modern age. For this reason, it offers a paradigmatic case study for exploring the emergence and formation of national identities in the context of regional and global movements of transit and migration.

As Isabel Hofmeyr states, the Indian Ocean invites us to rethink the very concept of the nation-state. Analysing a collection of academic writings from the last few decades in Indian Ocean Studies, alongside some fictional narratives set on the Ocean's shores, Hofmeyr argues that the presence of so-called 'ancient' diasporas, undermines any supposedly intrinsic or normative connection between the sovereignty of the nation-state and the emergence of a transregional trade network: '[T]he nation-state in the Indian Ocean is subordinated to older diasporic aspirations' (Hofmeyr, 2010: 724). The Indian Ocean thus has to be conceptualized and studied as a matrix of 'alternative modernities' – with the idea of nationhood itself being one of its axes – where 'formations of modernity have taken shape in an archive of deep and layered existing social and intellectual traditions' (Hofmeyr, 2007: 13).

As we will argue in this section, literature and the historical novel in particular, offer valuable sources for understanding not so much historical reality per se, but the importance of remembering, narrating and imagining the various routes and lines of transit that crossed, circumnavigated and intersected with the nations of the Indian Ocean, at the moment when they were first becoming nations, as the independence movements were formed. Those narratives that are located in the spaces that are washed by the tides of the Indian Ocean invite us to rethink the conventional paradigms that structure the study of identity and identity narratives, namely the national paradigm, and the linguistic paradigm, as discussed by Elena Brugioni (2013), and to explore the intersection of cultural, political and economic routes of transit, contextualized in terms of their spatial and material dimensions. How does literature go about problematizing that moment of (proto-)national awakening in the context of political events occurring on a global, international scale, such as world wars? How do economic lines of transit, meaning here the import and export of goods as well as labour migration, impact the local cultural identity and imaginary of a place? Finally, how does the infrastructure which both enables and impedes travel, develop out of certain dominant social structures and influence the process of identity construction?

Positioning the study of national identity and its narratives in the transnational context requires thus engagement with the theoretical and methodological principles of the new mobility paradigm (Sheller & Urry, 2006). This new paradigm proposes, on the one hand, a systematic interrogation of the 'sedentary' perspective that predominates in certain areas of the social sciences, which 'treats as normal stability, meaning, and place, and treats as abnormal distance, change, and placelessness' (Sheller & Urry, 2006: 208). On the other hand, this paradigm rejects the 'grand narrative' of mobility celebrated by theories of 'nomadism', and criticized by feminist studies, because it projects a romantic image of mobility while failing to recognize the underlying structures of power which can, in fact, either promote movement or constrain it (Sheller & Urry, 2006: 211). This new axis of differentiation accounts for both structures of power and the relationship between mobility and materiality, since '[m]obilities cannot be described without attention to the necessary spatial, infrastructural and institutional moorings that configure and enable mobilities' (Hannam, Sheller & Urry, 2006: 3).

These issues will be discussed here with reference to João Paulo Borges Coelho's *O Olho de Hertzog,* and *Tivolem* by Victor Rangel-Ribeiro, as narratives which refuse to adopt a telluric perspective, fixed upon and grounded in national identity, in order to focus on the different levels of mobility that the formation of this identity implies. *O Olho de Hertzog* tells the story of Hans Mahrenholz – a German soldier who adopts the fictitious identity of Harry Miller, an English businessman – travelling to Lourenço Marques, present-day Maputo, on a secret mission in the closing stages of the First World War. The novel intercalates his memories of a heroic journey to the German front in a zeppelin, and fighting alongside General Lettow-Vorbeck in Zambezia, with the story of his search for the mysterious 'eye of Hertzog', a much sought-after diamond. During his sojourn in the capital of colonial Mozambique, the protagonist encounters various (real) historical figures such as the social activist and journalist João Albasini who founded the *O Africano* and *O Brado Africano* newspapers, as well as numerous fictitious characters (although many of them are based on real, historical events), such as the Austrian painter, Wally; Florence, the daughter of a South African hero, and Natalie, the lover of a notorious robber, who have sailed to the capital on the same ship as Hans/Harry; and also Rapsides, a mixed race man with a dark past who might hold the key to solving a mystery. His wanderings are set in the urban matrix of Lourenço Marques, with its various cafés, shops and establishments full of imported goods. The theme of labour migration between Mozambique and South Africa is also relevant to the context. As Elena Brugioni argues, the writing of João Paulo Borges Coelho, 'proposes an original outline of the relation between space and time, thus redefining the so-called national cultural context through an emblematic repositioning of facts and subjects within a transnational space and time framework' (Brugioni, 2013: 95).

Tivolem, on the other hand, is set in a rural community, but the movement of ideas, goods and peoples also stands out clearly as the narrative's *leitmotif*. This Goan narrative opens with the return of Marie-Santana to her native village after more than two decades as a migrant in Quelimane, Mozambique, where her parents had set up a successful business. This subsequently went bankrupt following the death of Marie-Santana's father and the fraudulent practices of her suitor, a mysterious man named John Frenshaw. The arrival of the protagonist also coincides with the return to Goa of two other inhabitants of the village: Simon Fernandes, a gifted musician, who has spent a large part of his life in Kuala Lumpur where he emigrated with his family as a child; and Eusébio Pinto, who made a fortune in the Persian Gulf, in the aftermath of the Great War. The novel follows the complex social dynamics of an apparently closed community – where nearly all the inhabitants have known each other since childhood – which is also in a sense, very open, characterized by travel and the exchange of ideas. Given the diasporic circumstances of the author's own life, this is a work that could be classified as 'migration literature' as Joana Passos argues, a text 'written outside of the immediate dialogue with local Indian literatures, and it is at least partially free from their influence, in terms of literary tradition and trends, since migrant literature is in contact with other literary universes as well, as a consequence of the author's everyday immersion in the host nation's culture' (2018: 218).

On journeys and moorings

'Não fosse o calor nem esta vegetação tão particular, não trouxesse o ar este vago cheiro a queimado, e Hans Mahrenholz dir-se-ia de volta a uma sua muito antiga vida. Hamburgo talvez …' [If it were not for the heat, the very specific vegetation, and the slightly burnt smell that was floating on the air, Han Mahrenholz would have said he was back in some part of what was now very much his former life. Hamburg perhaps …] (Borges Coelho, 2010: 13). The opening passage of Borges Coelho's novel immediately draws the reader's attention to the idea of movement – between spaces, temporalities and identities. The arrival of Hans/Harry in Lourenço Marques is tinged with a degree of nostalgia, triggered by a range of different sensations (the smell, the light, the humidity, the rocking of the ship's deck). Neither the bay nor the port is observed or described directly. Rather they are experienced in sensory terms, which demonstrates the relevance of 'sensory aspects of affect, event and encounter', which, according to Nayak, implies 'more open understandings of [how] people's 'sense of place' and how this registers with ideas of nation, region, home or locality as geographically located and emotionally experienced' (Nayak, 2010: 2389). The experience of walking through the city streets and squares invades Hans's senses, creating a complex dynamic of being *in and out of place.* The protagonist seems, on the one hand, to be drinking in the city and its atmosphere, for example, by noticing the signs and posters – 'Está já escuro, até os anúncios quase se apagaram.' [It has gone dark and even the adverts are almost invisible] (Borges Coelho, 2010: 103) – the day to day routines – 'À entrada dos bares, criados com compridos aventais [que se afadigam] a varrer com vassouras de corda' [In the doorways of bars, waiters in long aprons [...] are wearily sweeping the floor with rope-handle brooms] (Borge Coelho, 2010: 93) –, or observing another aspect of urban life – Por vezes saía de noite [...], ao fazê-lo, descobria uma outra cidade, contígua no tempo, inimiga do sol, brilhando à sua maneira no escuro' [Sometimes he went out at night [...], and when he did, he would discover another city entirely, adjacent in time to the first one, but entirely alien to the sun, casting its own light into the darkness] (Borges Coelho, 2010: 150). His wanderings through the city seem to reflect a kind of 'mobile ethnography' which entails 'participation in patterns of movement while conducting ethnographic research' (Sheller & Urry, 2006: 217). At the same time, these encounters have both a familiar and an alien feel for Hans. Hearing a French song in a bar makes him feel as if 'O mato toma conta da rua, silenciosa e quieta' [the bush were invading the street which is calm and silent] (Borges Coelho, 2010: 95). And Hans travels from Lourenço Marques back to the world of the German front and Colonel Glück's hut; the noises and lights of the Gato Preto [Black Cat] Bar (and the glasses of absinthe) transport the soldier back to the battle front ('Os askaris ébrios da vitória, libertos enfim do medo, progrediam sem tino à caça das suas vítimas e dos seus fantasmas [...] A pequena orquestra soluça, atordoada pelo agudo som dos vidros' [The *askaris* drunk on victory, and free of fear at last, were pursuing their victims and their ghosts without mercy [...] And the little orchestra plays on as if dazed and weeping amidst the piercing sound of the glass]) (Borges Coelho, 2010: 106–107).

In fact, the repeated scenes of Hans alone or in company, crossing the city of Lourenço Marques shows that the city is not simply represented as a backdrop or a stage upon which people experience events, but rather as a space that is being actively produced, as the seminal work of Lefebvre (1991 [1974]) would have it, by the different experiences of its inhabitants and the relations of power that govern them. The question of agency, of who is able to go where, and how, is particularly evident in the scene when Hans and Albasini visit the municipal market. When they get there, these two potential customers wander around the various stalls while the produce, and the people selling it, stay still.

> Percorrem os húmidos caminhos de pedra entre as bancas de produtos que João Albasini vai inspeccionando atentamente, sem no entanto abrandar o passo. É como se os vendedores, organizados por raças, tomassem parte no empreendimento colectivo de revelar as cores que existem na paleta da natureza. Os europeus propondo uma vasta gama desde os verdes dos legumes aos vermelhos das carnes, os indianos ripostando com a gradação completa dos temperos, do amarelo ao castanho passando pelo laranja vivo e por um vermelho de fogo. [...] Estão já nas traseiras do mercado, às quais se segue um baldio pejado de vendedeiras negras sentadas no chão enlameado, expondo os seus produtos sobre esteiras de palha semiapodrecida ou pequenas caixas de madeira, desta feita sem nenhuns anúncios que os nomeiem.... (Borges Coelho, 2010: 31–32)

> [They walk through the damp stone alleyways among the stalls selling produce which João Albasini inspects carefully but without slowing his pace. It is as if the stall holders, organised according to race, were engaged in some collective endeavour to display the full pallet of nature's colours. The Europeans offer a huge range, from the green of the vegetables to the red of their meat products, and the Indians, not to be outdone, respond with every possible hue of spices – from yellow all the way to chestnut brown, through bright orange and fiery red [...] And now they are at the very back of the market, which leads out onto a scrap of wasteland, teeming with black women sitting on the muddy ground displaying their produce on mats of half rotten straw or little wooden boxes, with nothing to indicate what they actually are ...]

This passage highlights how spatial organization is already implied in existing power relations, creating hierarchies between different social groups. The space of the market comes to metonymize colonial society itself, indicating not only the social barriers that exist (i.e. racial categorization) but also the way these barriers affect 'potential for mobility', or 'motility' (Hannam, Sheller & Urry, 2006: 3). The narrator's focus on who moves where in the market, and on the black women taking great care, at the end of the day, not to pile their produce too high 'em grandes cestas que levarão à cabeça (às crianças, levá-las-ão às costas)' [in big baskets they carry on their heads (while carrying their babies on their backs)] (Borges Coelho, 2010: 33), highlights the idea of anchored or 'moored' mobilities, advanced by Hannam, Sheller and Urry (2006). The concept of mooring, the infrastructure and institutional support which make mobility possible, or rather, the absence of that support for certain layers of society is brought to life here in the image of the woman with the basket on her head and the baby on her back. This image evokes a much wider tradition in the representation of African womanhood, as (a racialized depiction of) iconic maternity which was 'the main subject of a large quantity of photographic postcards of African women which circulated over a long period of time from the late nineteenth century to the 1970s, and even later' (Vicente, 2017: 30). In this novel, however, the image seems precisely to deconstruct this discourse, inviting the reader to question how the intersection of race and gender actually serves to constrain mobility in this context.

In fact, the plot of the novel draws our attention to the entire infrastructure which enables or impedes travel. From the minutely detailed project of the Zeppelin, built to send supplies to General Lettow's beleaguered forces, through recurring references to the city's various modes of transport: taxis, rickshaws, trams, even Hans's trip in Albasini's car to lunch with his family: all these scenes seem to suggest that the action can only move forward, and the mystery surrounding 'the eye of Hertzog' can only be resolved, through movement and travel. Indeed, this idea is already sown at the beginning of the novel with the description of the various ships anchored in the bay:

> Por ora a *Ferreira* apalpa atentamente o canal que leva ao porto, passando não muito longe de um punhado de navios quietos, fundeados. De amurada, o comandante Vieira de Sousa, há quatro anos no governo da galera, vai explicando aos cinco passageiros o historial de cada um desses navios: o *Niassa*, que foi *Bulow* antes de ser há dois anos capturado pelas autoridades portuguesas; o velho *Admiral*, glória da German East Africa Line, também ele obrigado a ser *Lourenço Marques*; e finalmente o *Beira*, nome pintado de fresco por cima de outro nome que ainda se consegue ler, por estar marcado em relevo leve na chapa negra do casco. *Hertzog* é o nome. (Borges Coelho, 2010: 14)

> [At this precise moment, the *Ferreira* is feeling its way carefully along the canal which leads to the harbour, passing close to a cluster of ships lying calmly at anchor. Standing at the rail, Captain Vieira de Sousa, who has had command of the galley for four years, is providing the five passengers with some historical background on each of the ships: the *Niassa*, which was previously the *Bulow*, before it was captured by the Portuguese authorities two years ago; the old *Admiral*, the pride of the German East Africa Line, which was also made to change its name and become the *Lourenço Marques*; and finally the *Beira* with its name freshly painted over the top of another, previous name which remained legible underneath because it had been lightly embossed on the black metal plating of the ship's hull. The name was *Hertzog*.]

Tivolem also starts with the arrival of a ship: The *Lilavati* approaches the mouth of the River Mandovi, bringing Marie-Santana back to her homeland. The image of the ship sailing around Fort Aguada entering the mouth of the Mandovi evokes the centuries-old history of travel, contact and conflict in the Indian Ocean: 'Three centuries earlier, from atop that same stark promontory in Goa, that venerable muzzleloader, guarding the jewel of Portugal's Indian empire, twice kept an entire Dutch fleet at bay' (Rangel-Ribeiro, 1998: 5). This architectural heritage offers, however, only a partial view of the network of maritime routes that was woven between the shores of this most ancient of oceans (Pearson, 2003: 3), because it represents Goa from the perspective of Portuguese expansion history. However, as the narrator accompanies the Lilivati's journey up the mouth of the river, he effectively contextualizes the narrative of Goa as the pearl of the Portuguese empire, in terms of other voices and other crossings. Thus 'sleek Arab dhows still furling their sails' (Rangel Ribeiro, 1998: 7) connote the transregional circulation of people and goods that preceded the primacy of nation-state and empire, precisely one of the issues singled out by Isabel Hofmeyr in her conceptualization of the Indian Ocean as the space of 'alternative modernities' (Hofmeyr, 2007: 13). These two aspects of Goan history – which are both complementary and intertwined – are also represented in the language spoken on board: 'Around her Marie-Santana heard the familiar sounds of English and Portuguese interspersed with the babble of more ancient tongues – Konkani, Marathi, even the elegant Hindi of the north' (Rangel-Ribeiro, 1998: 7). In this sense, the opening scene of the novel draws an intrinsically transnational map of Goa as a part of what may be termed the 'Indian Ocean World'.

At the same time, the ship's passage evokes the personal stories of Marie-Santana and her family, weaving a narrative that combines elements of memory, stories and History. As the journey progresses, the happy memories of beach holidays with her parents and grandmother, fishing for crabs, climbing cashew nut trees, and sheltering from the sun in the shade of palm trees, give way to sadder memories about the death of her parents in Mozambique ('Coming home like this, even coming home to Granny, was itself a wrench; it was she who was leaving them behind, abandoning them' Rangel-Ribeiro, 1998: 7). While she is in transit, onboard the ship, the protagonist manages to reflect on her life, past and present ('Despite the growing excitement around her, Marie-Santana's mood remained pensive. She became concerned at the prospect of facing her grandmother' Rangel-Ribeiro, 1998: 8). This becomes a recurrent motif throughout the novel: in moments of tension and anxiety, Marie-Santana goes to the river and travels across it in a quest for self-knowledge. While the memories of Hans in *O Olho de Hertzog* surface when he is strolling through the city, Marie-Santana climbs into a small rowing boat to get to the other side of the river, the Panaji/Panjim side. Her thoughts during the crossing help her to make sense of her return to Tivolem, her social relationships and her (re)integration into the community, even though people suspect that Marie-Santana possesses the power of the evil eye, a suspicion put about by the local gossip, Josephine Aunty, and in spite of her closely guarded secret about the role which her suitor's brother, Simon, played in the collapse of her parent's business. As she explains to Shankar, a boatman who knew her father, the river crossing allowed her to get physical and mental distance from her problems: 'It calms me down. And it gives me time to think' (Rangel-Ribeiro, 1998: 328). Indeed, throughout the novel as a whole, the river seems to function as a symbol of fluid, transformative identity, formed in the constant movement between departure and return, between 'being' and 'becoming', as described in the seminal work of Stuart Hall (1990: 225). This idea is particularly salient in the dialogue between Marie-Santana and Shankar in the closing chapters of the book.

> 'I like your notion of life as crossing and recrossing a river,' she said, 'but I have always thought of my life as being itself a river. Where river melts into the ocean, that ocean itself could be the All-Pervasive One; the crossing would then be over.' He laughed.

> 'Perhaps. But think carefully: that blending of sea and river is not an end, but part of a cycle.' (Rangel-Ribeiro, 1998: 329)

While the symbolism of the river seems to possess a more metaphorical significance relating to journeys and crossings, allowing us to explore notions of home and belonging in the context of migration and return, *Tivolem* also highlights the relevance of *moorings* (Hannam, Sheller & Urry, 2006) in stimulating and constraining mobility. The various modes of transport are a source of income for some: the bicycle is crucial for the work of the postman Mottu; Ramu earns a living transporting rich folk to the city in his oxcart (and they are well-trained, as Dona Esmeralda finds out; Rangel-Ribeiro, 1998: 77); while the boatmen take passengers to the other side of the Mandovi. And it is there, in the river port that Marie-Santana notices the changes which are gradually transforming the type of work that is required to keep passenger transport going: 'Marie-Santana strolled along the square overlooking the quay, watching the launches and gasolinas come and disgorge their human cargo and take on another boatful, while the canoes still pulled slowly across the river with their load of passengers' (Rangel-Ribeiro, 1998: 112). The introduction of motor-boats is presented as inevitable, while the 'old' mode of transport is described with a certain nostalgia. However, it becomes evident throughout the narrative that although the power of mobility seems to be increasing, it is not yet equally accessible to all, as becomes evident in Marie-Santana's conversation with her grandmother. The old matriarch explains to her granddaughter why she only visits Panjim once a year, for the festival of Nossa Senhora da Imaculada Conceição: 'It's a three-hour walk over the hills just to get to the river; can I do that at my age? So now I must share a taxi to the landing, and you know how much that costs' (Rangel-Ribeiro, 1998: 96). As is also the case with *O Olho de Hertzog,* Rangel-Ribeiro's novel challenges the romantic image of travel, disseminated by the narratives and critical approaches associated with 'nomadism' (Sheller & Urry, 2006: 211), choosing instead to focus on the socio-economic and technological factors that come into play.

Economic lines of transit in the Indian Ocean

As he crosses the frontier, Hans introduces himself to the Portuguese colonial authorities "súbdito inglês embarcado na cidade de Durban, viajando para Lourenço Marques em sondagem de oportunidades de negócio.' [an English subject who boarded in Durban, and travelled to Lourenço Marques to explore business opportunities] (Borges Coelho, 2010: 17). Indeed, the city of Lourenço Marques is presented in the novel as a hub of international trade, supplied with a wide variety of products from across southern Africa, the other side of the Indian Ocean and also Europe. Restaurants, kiosks and shops pull their customers in with posters that highlight the variety on offer, creating a cosmopolitan urban atmosphere. 'Bebidas geladas nacionais e importadas da Metrópole, do Transval e das principais capitais europeias' [Cold drinks both home-produced and imported from the metropolis, the Transvaal and the major capitals of Europe] (Borges Coelho, 2010: 19); 'Importação directa de todos os géneros nacionais e estrangeiros' [direct imports of all kinds, from home and abroad] (Borges Coelho, 2010: 129); 'Compra e venda de toda a espécie de cereais, produtos da África e outras mercadorias' [Buying and selling of all types of grain, from Africa and other markets] (Borges Coelho, 2010: 129). Others engage in marketing that compares local products with foreign brands: 'melhor não se fabrica na Inglaterra' [they don't make them better in England] (Borges Coelho, 2010: 147). The kind of adverts quoted create an aura of urbanity, portraying the city as a cultural contact zone that is also intrinsically material. The adverts and signboards for the banks and overseas travel companies also highlight a constant ebb and flow of capital and people. 'Banco Nacional Ultramarino, Sede em Lisboa na Rua do Comércio n°74, Filiais e agências em Londres, Paris, Continente e ilhas adjacentes, colónias e Brasil' [National Overseas Bank, Headquarters in Lisboa, Rua do Comércio, no 74, Branches and agencies in London, Paris, Continent and adjacent islands, the colonies and Brazil] (Borges Coelho, 2010: 128).

As already noted earlier in this chapter, the evocation of the linguistic landscape, which is so striking throughout this novel, creates a strongly localized sense of presence, confined to the urban matrix of historical Lourenço Marques. As Brugioni argues, 'ao trazer para a página literária este género de testemunhos, o texto de Borges Coelho vai ganhando a fisionomia de um 'arquivo menor', tornando acessível e evidente – isto é, público – um aspecto do passado moçambicano quase totalmente rasurado do espaço urbano contemporâneo' [by bringing this kind of testimony to the pages of literature, Borges Coelho's text takes on the form of a 'minor archive', making accessible and available – i.e. bringing into the public sphere – an aspect of Mozambique's past which has been almost totally erased from contemporary urban space] (2012: 395). It is through these different signs that Hans – and via Hans, the reader of the novel – gets to know the city with its flows of movement and its boundaries. As Stroub and Jegels observe, '[t]he complexity of place is reflected in the complexities of linguistic or semiotic landscapes' (2014: 180). In the context of studying Manenberg, a multilingual township of Cape Town, linguists have adopted a performative approach to exploring how 'signage and the material semiotic landscape play an important role in organizing place, and how place in turn determines the reading of signage'. They argue that 'a central aspect of place-making is in fact the way affect and movement through space is organized, narrated and interactively accomplished by means of – direct or indirect – engagement with situated material semiotic artefacts' (Stroud & Jegels, 2014: 180). Their approach seems particularly applicable for a close reading of how signs and billboards are deployed in Borges Coelho's novel. The inclusion of signs in Portuguese, English and Ronga not only leaves a textual trail which reflects (albeit partially) the local multilingualism and its associated linguistic ideologies, it also enables us to explore how the narrative of cosmopolitanism – implied in the images of goods and capital circulating – is witnessed and experienced according to the socio-economic, cultural and economic positioning of the inhabitants of (post-)colonial cities. These different signs ultimately function as an additional voice, intercalated between the various narrative layers woven by the characters, to constitute an act of rewriting which Calafate Ribeiro would consider 'a political movement of conquering the territory and its history by narrating versions of history told from other voices and other places, making the history of the city from its beginning textually more polyphonic' (2016: 104). It is this other voice, the voice of the signboards and adverts, that locates Lourenço Marques specifically as a transnational, transregional and cosmopolitan city which cannot simply be reduced to the confines of a Portuguese colonial dominion; at the same time, however, it also seems to capture a series of lacunae and implicit silences.

It is the motif of labour migration, however, that highlights most explicitly the structures of power and control which hold sway in this colonial society, in the world of migrant workers. The very fact that Hans comes to the city disguised as a British commercial traveller, raises the issue of borders, as physical and legal boundaries which control movement between sovereign territories. At that point, the trip which Hans and Albasini make to visit the recruitment camp for the Transvaal mines, is also distinguished by the motif of (im)mobility. The two men '[a]tendem a longa fila que se estende, tortuosa, a perder de vista' [look at the long queue which stretches further and further back, twisting out of sight] (Borges Coelho, 2010: 82); the crowd of men seems to form one single, huge living organism assembled from a series of disjointed parts, like a 'a gigantesca cobra chinesa de papel' [giant Chinese paper dragon] (Borges Coelho, 210: 82), its waves of motion driven alternately by the tiny movements of the hopeful recruits waiting, and by the sepoys [Muslim colonial policemen] giving orders and exerting their authority with truncheons. Both of these two scenes focus on an act of passage: the importance of 'papers', the different levels of hierarchy and authority, the long wait for authorization to go ahead with plans and aspirations - whether that means carrying out a secret mission as is the case with Hans, or getting a highly exploitative labour contract and climbing into the 'vagões que ronronam pacientes, prontos a partir' [trucks of the train purring, getting ready to leave] (Borges Coelho, 2010: 84). Focussing on these turning points, of either authorization or refusal, the novel reminds us that the question of labour migration and mobility, which has become so politicized in recent decades, is not an intrinsically contemporary phenomenon. It also reminds us that regional dynamics such as those represented by Borges Coelho, from the African shores of the Indian Ocean, can make a valuable contribution to how we think about the connection between the movement of capital and the movement of people, and how mobility, which is now often conceptualized and debated in terms of human rights (Pécoud and Gutcheneire, 2007: 10–11) has been, and continues to be, implicated in relations of power.

In contrast with Borges Coelho's literary Lourenço Marques, the setting of *Tivolem* does not initially appear to be marked by the passage of capital or commerce. While the Mozambican capital represented in *O Olho de Hertzog* is a lively contact zone made up of the coming and going of workers, goods and traders, Ti*volem* symbolizes the idea of home, as a point of departure and return rather than transit. The migrants who return to the village after decades away do observe various changes but they have not changed the familiar atmosphere they still remember. 'The village has changed but not much' (Rangel-Ribeiro, 1998:22) observes Simon. The only small aspect of urban life which draws attention to both the passage of time and the impact of transnationally circulating capital is the new house Eusébio is having built.

> Senhor Eusébio knew just the kind of house he wanted. Not for him the slopping roofs with interlocking rows of curved mud-brown tile found in every village in Goa; those tiles tended to break if any hard object fell on them; but even if they didn't, they had to be removed and reset each year, since they shifted with the action of the wind and the sun and the rain. Besides, he wanted a house that would be unique – made just out of red laterite rock but with outer walls made of poured cement concrete, a flat roof to match, and a terrace on top that would be open to the skies. (Rangel-Ribeiro, 1998: 40)

This construction does not go down at all well with the other inhabitants of the village. Dona Esmeralda sums up her view succinctly: 'An ugly monstrosity – it does not belong. The man has much money but absolutely no taste' (Rangel-Ribeiro, 1998: 41). While taste per se is a personal matter, Dona Esmeralda's conversation with her friend Dona Elena draws the reader's attention to some important local dynamics. Not belonging and standing out is the main reason for her criticism. While the arrival of some capital is clearly regarded by the women as a blessing, ('Let's thank God for the money and blame the bad taste on the devil.') as Dona Elena says (Rangel-Ribeiro, 1998: 41), the villagers seem to like the traditional character of their village and they want to preserve its architectural features which symbolize certain kinds of social relationship and interaction.

However, in spite, or precisely because, of the rural, parochial nature of the village, the novel portrays a rich web of local dynamics, seen through the optic of migration in the context of trade and labour. All the characters who have left the village have done so for financial reasons. Marie-Santana and Simon with their parents, Eusébio as an adult. Obviously, these three characters represent different routes and different kinds of occupation, thus creating a clearly differentiated picture of the Goan diaspora: traders in Africa, colonial administrators in South East Asia and oil company workers in the Middle East. However, it should be emphasized that these are all 'passengers', travellers who have been able to pay their own fares, in contrast to the contract workers whose travel has been included as part of their contract (Hansen, 2010: 110). In this novel, migration seems to be more a question of opportunity than of strict necessity. This view is suggested by the relatively untroubled and successful if not ideal lives led by Eusébio and Simon ('Now back in Tivolem and embarked on a life of leisure, Simon found loneliness pressing on him' Rangel-Ribeiro, 1998: 106), or by the passages which tell us how the men were able to retire: 'Eusebio's diligence earned him promotions, though not large salary increases; however as he had found no place in which to spend his earnings, and he had absolutely no vices, his savings mounted. When his parents died, he transferred to his now substantial bank balance both his attention and his affections. He retired and returned to his native village thirteen years after he had left it' (Rangel-Ribeiro, 1998: 35). These success stories contrast with Marie-Santana's, as the village gossip Josephine Aunty points out: ' "What bothers me," Josephine Aunty said, "is not why she has never married, but why she has come back poor from Africa." ' (Rangel-Ribeiro, 1998: 58). Weaving these different stories of migration together, the narrative of the novel explores the various meanings of the Goan diaspora, which cannot be categorized simply in terms of labour migration. As Margret Frenz argues, Goans clearly stand out from other migratory groups of South Asia, owing to their high levels of education, their perceived Westernized lifestyle and their right to citizenship. This has led Frenz to classify Goan migration according to a different model altogether, calling it 'subaltern elite migration' (Frenz, 2014: 18). Focussing on narratives of return, *Tivolem* effectively offers a literary perspective on the processes of identity and alterity pertaining to local social hierarchies, which are themselves simultaneously created and interrogated in the context of capital circulation across the shores of the Indian Ocean.

Transnational (proto-)nationalism

The two novels discussed here also incorporate into their storylines the issues that initially sparked political consciousness in these two Portuguese colonies, taking into account the wider context of international politics. Thus, the two fictional narratives draw attention to the local dynamics which are developing beyond the colonial/metropolitan axis, extending out to the entire network of regional and global relations. In *O Olho de Hertzog,* Hans's memoirs of his time on the German Front in the Great War recall 'o grande Lettow de Tanga e de Mahiwa' [the great Lettow of Tanga and Mahiwa] (Borges Coelho, 2010: 44), and the contingents of Askaris 'gente experimentada, aguerrida, diria mesmo que tresloucada, como se só na destruição e no aniquilamento dos outros pudessem encontrar um fim para a humilhação que era terem sido obrigados a fugir da sua terra' [battle-hardened men, one might even say, fierce to the point of crazy, as if they needed to destroy and humiliate to assuage the humiliation they had suffered themselves when they were driven out of their homeland] (Borges Coelho, 2010: 76). These memories pave the way for a narrative exploration of Mozambican history as the setting for a world war, focusing on what motivated the various alliances, loyalties and rivalries. As Nazir Can (2020: 70) notes, João Paulo Borges Coelho's novels play with the 'grand' narratives of History, placing them alongside everyday personal stories, and thus exploring 'the ties between the poetics of memory and the poetics of amnesia to identify some continuities between the colonial and post-colonial times, hence refuting the official narratives of both periods'. A similar narrative strategy is discernible in *Tivolem,* where Marie-Santana's memories of her life in Mozambique also provide the context for the German [military] intervention, which her father visualizes using a map on the wall: 'With tiny green flags he marked the Portuguese positions on the south bank of the Rovuma River; '"I'll use red for the Germans," he explained' (Rangel-Ribeiro, 1998: 52). However, in both novels, the principal motif used to develop the idea of nationhood, in the 'inter' and transnational context, seems to be the circulation of news, the periodical press in particular. In fact the circulation of newspapers throughout the twentieth century played an important role in awakening various, different notions of national identity and nationalism. As Benedict Anderson observes (1991), in Europe, the press enabled the creation of 'imagined communities' because it sustained a sense of simultaneity – all the members of a given community shared the ritual of reading the same news, on the same subjects, at the same time. In the colonized territories, the press became an important locus of ideological struggle, where the dominant discourse was disseminated and, as far as possible, resisted. This was the case in Mozambique at the time of João Albasini, as depicted in Borges Coelho's novel. As Fátima Mendonça (2012: 17) argues, 'a Imprensa constituiu a forma mais eficaz de consolidação das ideias base do discurso legitimador da empresa colonial e – em simultâneo – da irrupção de um contra-discurso com diversas nuances e com uma evolução sintonizada com as fracturas produzidas no interior da sociedade colonial' [the Press offered the most efficient means of consolidating the basic, core ideas of the colonial enterprise's legitimizing discourse – and also, at the same time – of fostering the emergence of a counter-discourse with various shades of opinion, which started to develop along the very fault-lines that were fracturing colonial society itself from the inside] (2012: 17). In the case of Portuguese Goa, Sandra Lobo (2013: 22) notes the tension between the transcontinental and the transnational nature of the Goan intellectual elite, stuck in the provincial confines of this economically deadlocked enclave, as evidenced by the local periodical press, in the period between Portugal's Liberal Revolution in the 1820s and the founding of the New State in 1933, the year when the action of *Tivolem* is set.

Hans's wanderings in *O Olho de Hertzog* are also punctuated by his reading of the news. When he arrives in Lourenço Marques, he leafs through the paper, reading about the surrender of General Lettow-Vorbeck, official reports of the military campaign, the assassination of the Portuguese president, Sidónio Pais, and an advert for the 'milagrosos produtos Melcin' [miraculous Melcin products] imported from Johannesburg, which treat venereal diseases and skin conditions (Borges Coelho, 2010: 21). His departure from the city is similarly marked by news items: the 'incansável Albasini' [indefatigable Albasini] and his criticism of how the city organizes native burials (Borges Coelho, 2010: 435), Albasini's own trip to Lisbon for health treatments, and also the departure of the *Beira-Hertzog*, the ship which Hans had spotted when he first sailed into the bay. This news item seems to function, on one level, as a moment of symbolic closure for the narrative axis, making Hans's whole mission a kind of maritime stopover between two instances of arrival and departure. At the same time, there is an 'anúncio discreto, ao fundo de uma página anterior' [inconspicuous announcement, at the bottom of a previous page] (Borges Coelho, 2010: 435) stating that 'partiu para a Europa o Navio Beira-Herzog, Comandante Van Zyl, com escalas na Beira, Porto Amália, Zanzibar, Maqtishu, Suez, Alexandria, Atenas-Pireu, Rimini e Trieste' [the *Beira-Hertzog* has set sail for Europe, under Captain Van Zyl, with stopovers in Beira, Porto Amália, Zanzibar, Mogadishu, Suez, Alexandria, Athens-Pireus, Rimini and Trieste] (Borges Coelho, 2010: 435–436). This indicates the ongoing relevance of transit of both people and goods and also the importance of contextual information explaining the rise of political consciousness and proto-nationalism on the shores of the Indian Ocean.

In *Tivolem*, the news becomes the subject of regular debate among the local male elite, men being the gender explicitly singled out for this activity by the narrator (Rangel-Ribeiro, 1998: 10). While Gupta in the decolonial reading of *Tivolem* explores how the local 'elite stayees and returnees' need to reclaim the past' functions 'not as personal nostalgic *re–membering* but as a challenge to colonial power in their capacity as discursive agents of anti–colonialism and indigenous modernity' (2014: 254), here we would like to explore the acts of *re-telling* the present as a means of political mobility. The men – the priest, the governor, the secondary school headmaster and the postman, along with the returning migrants – Eusébio Pinto and Teodósio Rodrigues – subscribe to various different papers published in Panjim in Portuguese (including *O Heraldo*). These subscriptions attest to the men's interest in both regional and global affairs. In fact, events in India are often situated in the world context, in their discussions: 'There's some news out of British India, about those thousands thrown in jail. [...] That's a large number, but it may not stop Ghandi. Of course, there's unrest everywhere. Just look at Germany' (Rangel-Ribeiro, 1998: 11). Through these group gatherings, the novel illustrates the complex local dynamics at play in respect of language, knowledge and political stance. While Portuguese is represented as the language of choice both for gaining access to the news, and discussing it, the men wind up speaking English in order to accommodate Eusébio who has returned from the Persian Gulf: 'the language in which he was most fluent, with forays into Portuguese and Konkani as necessary in search of the mot juste' (Rangel-Ribeiro, 1998: 10). Reading the *Times of India* and listening to the news on the BBC (since radio is a novelty in the village), Eusébio brings new voices and perspectives to the scene of political debate in Tivolem. At the same time, his companions make fun of his 'anglophile' attitude (Rangel-Ribeiro, 1998: 12): 'So now he gets the news, told with a British accent? Most certainly the voice of authority' (Rangel-Ribeiro, 1998: 12) – says the vicar. In highlighting the way these narrative scenes are built around the circulation of news, we seek to demonstrate how the novel's discussion of global power relations and their impact on the political situation in Goa, was a transversal debate across a range of different characters.

While these various elite characters occasionally meet to chat in different social contexts, such as the post office, the group's regular meetings take place at an isolated bridge on the outskirts of the village.

> The bridge spanned the nullah where the lane, leaving the houses and the Goregão road behind, turned toward the church and the outlying government primary school and the Tivolem-Mapusa road. The bridge was hence somewhat isolated; the only vehicular traffic was an occasional bullock cart that came that way, and sometimes a taxi from the city, and every other week or so Dona Esmeralda's gaily painted yellow wooden carriage in which she rode to Mapusa, with Ramu the old retainer perched on the driver's box up front and singing to the bulls that pulled the carriage smartly along. The bridge was thus a quiet place, and though set apart from the village proper, was yet as convenient to the church as to the gentlemen on the hill; and since the knee-high stone parapets on either side were just some ten feet apart, the vicar and his friends could sit comfortably facing each other, discussing

> pressing local issues and matters of international moment without ever coming to blows. (Rangel-Ribeiro, 1998: 9–10)

This image evokes not only movement and circulation but also stasis and immobility. The bridge is a symbol of connection, of linkage between spaces, an anchorage or mooring point, to use the terminology of Hannam, Sheller and Urry (2006), which makes the journey possible, a prior condition enabling mobility to occur. However, the citation above also highlights its geographical isolation, it acts as a bridge only for occasional traffic; it does not in fact serve as a connection point if we look at it from the perspective of transport infrastructure. This bridge is not actually travelled across, but it is visited. And it is through these regular visits by the group made up of the vicar's friends, that the ideas and stories become to circulate and the reader is invited to rethink the political formations which have emerged within the dynamics of opening and closure, isolation and connection, frontier and passage, on this Indian Ocean shore. In its half-closed/half-open state of semi-isolation *Tivolem* seems to metonymize Goa itself, which is characterized to the present day by its socio-cultural and political insularity. Precisely this idea of insularity at a symbolic level was evidenced in various interviews with writers and intellectuals in Goa, conducted under the aegis of the *NILUS* project. According to Frederick Noronha, a Goan writer, publisher and journalist, the ideas of separation and connection, and a questioning of identity, in the context of historical and political processes are clearly discernible in Goan prose, especially as regards the motif of the family (Noronha, 2018). While social relations are indeed foregrounded in *Tivolem*, as is immediately evident with the image of houses and courtyards in the Subodh Kerkar painting forming the cover illustration, the political debates which take place at the uncrossed bridge constitute an important element of background, in both the novel and the painting, for us to understand the role played by the circulation of news and ideas in the formation of national and regional identities.

Tourism and identity

In the introduction to his collection on the relationship between tourism and postcolonial discourse, C. Michael Hall and Hazel Tucker point out that '[t]ourism therefore reinforces and is embedded in postcolonial relationships. Issues of identity, contestation and representation are increasingly recognized as central to the nature of tourism (…)' (Hall & Tucker, 2004: 2). It is clear that this statement refers to tourist trips which constitute movement along/within a centre(semi-)periphery axis, if we deploy Wallerstein's terminology, whereby inhabitants of the Global North visit the frequently less developed territories of the Global South.

Representations of 'post-colonial tourism' also afford the main axis of the two texts that we are going to talk about here, and which are also set along the Indian Ocean, namely, *O outro pé da sereia* [The mermaid's other foot] (2006), a novel by Mia Couto and *Hinyambaan* (2007), a comic novella by João Paulo Borges Coelho. For any reader seeking to compare these two texts, it becomes quite clear that they portray two different kinds of postcolonial tourist trip. What we see in *Hinyambaan* is a 'classic' form of tourism, a South African family holiday to the region of Inhambane, home to one of Mozambique's most famous and most popular tourist beaches. In *O outro pé da sereia,* one of the main storylines concerns the visit that two Afro-American tourists make to Vila Longe, a fictitious city in the Mozambican interior. In this instance, we come across a particular type of 'diaspora tourism', frequently known as 'roots tourism', 'genealogical tourism' or 'personal heritage tourism',[14] the objective of which goes significantly beyond a mere leisure activity. In many respects, it more closely resembles the logic of pilgrimage. In this case, what we have specifically, is 'black-Atlantic diaspora (Gilroy, 1993) tourism' which entails a journey of 'personal discovery that is in many respects unique only to the black race' (Timothy & Teye, 2004: 115). Thus, as Austin remarks:

14 See Timothy (1997).

> For those in the diaspora, an identification with the origin of transatlantic slave trade and with Africa as the ancestral 'home' is increasingly being perceived as the missing link in their quest to find their roots and to understand their collective sociohistorical experience, an act that is considered necessary for their self-realization. (Austin, in Timothy and Teye, 2004: 115)

The difference between these literary representations of tourist travel is important, indeed fundamental, to the two narratives in question. However, before we embark on a more detailed analysis we must, in line with Frederic Jameson's[15] thinking, consider the global geopolitical situation underpinning the phenomena described here, to which these two representations constitute a descriptive response.

In these two texts, this situation finds expression in the centre-periphery relationship which these tourist trips reproduce, whereby the people from the more developed territory travel to the less developed, economically poorer territory that is, the peripheral territory, carrying with them their huge baggage of stereotype, prejudice, hope, fear and pre-formed images. This becomes the lens through which we perceive the multiple mechanisms conditioning the agency of reciprocity in these relationships. At the same time, it is the baggage that has accumulated around these unequal relations of centre and periphery, which enables us to problematize the relationship between seeking and finding (in the tourist sense), coming from the most surprising and unexpected angles and drilling down into the multilayered issues which surround historical and socio-cultural identity.

Using an analogy from Foucault, John Urry talks about the tourist gaze, to designate a particular way in which tourists perceive the places they travel to. According to Urry, the many different manifestations of the tourist gaze are grounded in social experiences and in a non-tourist consciousness and they exist prior to the journey:

> [T]he gaze therefore presupposes a system of social activities and signs which locate the particular tourist practices, not in terms of some intrinsic characteristics, but through the contrasts implied with non-tourist social practices, particularly those based within the home and paid work. (2002: 1–2)

15 See Jameson (1986: fn. 5).

As suggested above, the particularities of the 'tourist gaze' pertaining to the characters in the two books analysed here are shaped not only by their personal social experiences but also by their collective, public, historical and socio-political ones, which are part of the centre-periphery axis. These relations manifest themselves in fundamentally different ways, ranging from the deracination of diaspora, to the complex relationship which Mozambique has with South Africa, its 'substitute hegemon'.[16]

In *O outro pé da sereia*, from the first words which describe the arrival of the Americans, the historian Benjamin Southman and his wife, a Brazilian-born sociologist/psychologist named Rosie, the pilgrim spirit and the quasi-mystical expectations of the trip become evident.

> África, a sua África, ia ganhando desenho, um contorno próximo e real. Por fim ele chegava à terra de onde há séculos os seus antepassados tinham sido arrancados pela violência da escravatura. Era preciso esse regresso para que Benjamin Southman, historiador afro-americano, se reconstituísse, ele que se sentia como um rio a quem houvessem arrancado a outra margem. (Couto, 2006: 161).

> [Africa, his Africa, was gradually acquiring a clearer image, with contours that felt close and real. Finally, he was reaching that land from which his ancestors had been wrenched by the violence of slavery, centuries before. This return journey had to be undertaken so that Benjamin Southman, the Afro-American historian, could remake himself, feeling as he did like a river whose other bank has been ripped away from him.]

In the section cited above, it is immediately clear that for Southman, Africa sits somewhere between Paradise Lost and the Promised Land. The use of the possessive pronoun signals his sense of identification with this territory that he has never previously visited. And this is further reinforced by the word 'return' which indicates an intended continuity, both historical and identificatory, with his hypothetical African ancestors, a return to the cradle, to the origins of his own being. However, although Southman feels a sense of belonging to the continent, as everything indicates, his supposed Africanness clashes with his identity as a person who was born and grew up at the centre of the world system. In this sense, his

16 Thompson (2008), 'hegemon zastępczy'.

behaviour veers between the impulse to kiss the very soil of Africa, and the impulse to make a cautious sign of the cross at the mere thought that the pilot flying him to his ardently desired destination might actually be Mozambican. Southman's inner dissonance, represented in terms of his personal anguish and sorrow, is what drives him to search for a definitive product, an Africa encapsulated in magic and myth, a product that does not ultimately exist. In other words, Benjamin Southman knows perfectly well what he is looking for, a fossilized form of memory that has been crystallized through a (first world) imaginary emanating from 'the centre'.

In this context, Benjamin and his expectations (and the Mozambicans' attempt to fulfil them, a point to which we will return later) represent a typical centre-periphery relationship characterized by a demand for the exotic. It becomes evident that Southman's quest is not focussed on reality or 'truth', if we may call it that. Rather he is seeking some kind of affirmation, the affirmation of what he has imagined and what he has learned (he is after all a historian specializing in Africa). As a result of this, he clings to those fragments of 'Africanness', which confirm his expectations in one form or another, while ignoring or distrusting anything that does not fit with his perspective. The following section is a good illustration of this:

> – Queríamos *que nos dissessem tudo sobre a escravatura, desses tempos de sofrimento…*
> – *Ah, sim, sofremos muito com esses vangunis,* disse Matambira
>
> Os olhos do americano brilharam enquanto procurava uma caneta para anotar no seu caderno de pesquisa.
>
> – *Como lhes chamou, vangumis?*
> – *Vanguni, rectificou o puglista.*
> – *Deixe-me anotar. Portanto, era esse o nome que davam aos traficantes de escravos.*
> – *Exacto.*
> – *E diga-me: há lembrança do nome dos barcos que eles usavam?*
> – *Barcos? Eles não vinham de barco, vinham a pé.*
> – *Como a pé? Como é que transportavam a carga humana lá para aterra deles?*
> – *A terra deles era aqui., eles nunca saíram daqui. Nós somos filhos deles.*

Incrédulo, Benjamin Southman deixou cair o caderno. Casuarino tentou corrigir, mas o americano não permitiu. Aproximou-se de Zeca Matambira e, com tom paternal, quase doce lhe inquiriu:

– *Diga-me, meu amigo, você está a falar dos portugueses?*
– *Portugueses? Naquele tempo nós éramos todos portugueses…*
– *Está a falar dos brancos?*
– *Estou a falar de pretos. Desculpe, de negros.*
– *Mas fale desses negros, desses vangunis …*
– *Esses negros vieram do Sul e nos escravizaram, nos capturaram e nos venderam e mataram. Os portugueses, numa altura, até nos ajudaram a lutar contra eles …*

Com um gesto mecânico, o visitante desligou o gravador. O seu semblante estava deformado pela estupefacção. Duvidaria da sanidade do interlocutor? (Couto, 2006: 174)

[- *We would like you to tell us everything about slavery, about those times of suffering …*

Ah, yes, we suffered a great deal under those vangunis, said Matambira.

The American's eyes lit up while he searched for a pen to write in his notebook.

What did you call them, vangumis?

Vanguni, the boxer corrected him.

Let me write that down. So is that what you called the slave traffickers.

Yes, that is right.

Tell me: does anyone remember the name of the boats they used?

Boats? They didn't come in boats, they walked here.

What do you mean, they walked here? So how did they ship their human cargos back to their own country?

Their own country was here, they never left here. We are their descendants.

Unable to believe what he was hearing, Benjamin Southman let his notebook fall. Casuarino tried to catch it, but the American did not let him. He drew closer to Zeca Matambira, and in a condescending, almost gentle tone, he enquired.

Tell me, my friend, do you mean the Portuguese here?

The Portuguese? We were all Portuguese back then …

Are you referring to white people?

No, I am referring to blacks. Sorry, I mean black people.

> *Tell me about these black people then, these vangunis …*
>
> *Well, these black men came up from the South, and enslaved us, they captured us, sold us and killed us.*
>
> *And there was a point when the Portuguese actually helped us to fight against them …*
>
> With a mechanical gesture, the visitor turned his recording device off. His face was frozen with shock. Could the man who was speaking to him be insane?]

This passage adopts two different angles on what we might call inequality, or asymmetrical relations, something which also, incidentally, applies to Benjamin himself, since the identity of the African American on a diasporic quest for his roots is actually concealing that of a secret service agent. One of these angles is personal and individual, and the other refers to centre-periphery relations on a global scale. So, on the one hand, we have an American academic (coming from the world 'centre') who adopts a predictably condescending tone and refuses to believe the peripheral Other, as soon as he hears something that does not confirm his own preconceived knowledge. On the other hand, it becomes clear that the History of Africa as a whole, holds no value for the Centre. Rather, the aim is to reduce it to a minor historical adjunct of black American History, which is regarded here, to all intents and purposes, as belonging to the History of the West.

The narrative ultimately deconstructs both discourses, thus defending the power of the peripheral. As an Afro-American, Benjamin Southman cannot find the Africa he is looking for in Mozambique, because '[e]ssa África ele tem que procurar na América, ou dentro dele mesmo' [that is an Africa he will have to look for in America, or inside himself] (Couto, 2006: 342). Even in relation to the closure that he has so avidly sought, through the attribution of an African name, Southman cannot follow the country's codes of behaviour. Unable to wait for the ceremony, he bribes his informant and when he finds out what his new name is, he loses interest in 'pertencer a uma identidade maior' [belonging to any bigger identity group] (Couto, 2006: 336). He carries on up the river and gets hopelessly lost. The disappearance of Benjamin Southman, the African American, also signifies the disappearance of Benjamin Southman, the secret service agent, 'que investigava a explosão da nave de espionagem no Norte de

Moçambique' [who was investigating the explosion of the spy ship in the North of Mozambique] (Couto, 2006: 370), the star respectfully buried by Mwadia and Zero Madzero. In this sense, Benjamin is characterized by a degree of falsity, a kind of void which is often central to tourist travel and therefore also to the centre-periphery axis more generally, and this takes the form of a radical disconnect in the codes of communication.

The situation of Benjamin's wife, Rosie, plays out somewhat differently. First and foremost, this trip is not really framed as a pilgrimage to the holy land of her ancestors, and she is not driven by the classic dreams of the diaspora. For her, this excursion falls somewhere between, a wife accompanying her husband and a sociology field trip. Free of the need to achieve any clearly pre-ordained objective, Rosie achieves a degree of openness which, as she listens carefully to the people she is interviewing, allows her to achieve a deeper level of contact with the reality around her and those who experience it on a daily basis. Whether it is the 'woman to woman' conversation she has with her hostess Dona Constança, the almost amorous encounter she has with the former boxer Zeca Matambira, or the confession of Jesustino, Rosie manages to create a degree of reciprocity and authenticity in her communications with people, acting as a kind of intermediary between the world of her husband and that of Vila Longe. Rosie is the person the inhabitants of Vila trust with the true stories of their past. In this sense, she has access not to stories of Africa, but to the stories of Africans. It is also in this experience of a meeting between two worlds that Rosie ultimately rediscovers her own identity and thence herself. 'Regresso ao Brasil. Esta viagem me fez entender quem sou, de onde sou …' [I am going back to Brazil. This trip has made me realize who I am, and where I am from …] (Couto, 2006: 374) she declares as she says goodbye to Constança. And we believe that her origins have a crucial role to play in this process. Thus, because she is Brazilian, she personifies the mediatory function of the semi-periphery, an intermediary point on the trajectory between the centre, represented by her husband Benjamin, and the periphery, symbolized by the Mozambican characters.

While *O outro pé da sereia*, offers a window on intercontinental tourism, the comic novella *Hinyambaan* portrays a tourist trip between two neighbouring countries, which have had a long and intense but profoundly unequal relationship in political and economic terms, affording an example of centre-periphery relations viewed under the microscope. At the beginning of this text, we used Ewa Thompson's phrase, 'surrogate hegemony' which refers in general terms to a situation whereby a territory rejects its relationship with the colonizing country, 'turning' instead towards a territory of superior status in the world system, to seek its Eldorado there. In the case of Mozambique, this country is South Africa. This relationship, characterized by a kind of admiration and fascination was summed up perfectly by Rui Knopfli when he said 'o meu Paris é Johannesburg' [my Paris is Johannesburg] (Knopfli, 1964). This relationship is the central premise for *Hinyambaan*, which is a *road movie* in the form of a book, dealing with the adventures of a South African family, the Odendaals, on a trip from Johannesburg to the Mozambican province of Inhambane.

Although the book is about tourists who come from neighbouring (African) country, which might reasonably suggest some degree of familiarity with the place in question, the novella demonstrates from its opening pages that the opposite is true. From the moment they begin preparing for the journey, as they load their little Corolla with bags and suitcases, the conversations between Hermann and Henrietta reveal a series of assumptions, prejudices and stereotypes. Although none of them knows how to pronounce the place names correctly, ('Hinyambaan', 'Mapiutou'), the Odendaals seem to have some very fixed opinions about their neighbouring country. And these extend to various aspects of history, politics and everyday life in Mozambique, starting with whether or not they can get to buy essential goods such as barbecue charcoal there ('se for carvão de fabrico local é provável que custe muito a acender, ou mesmo que não acenda de todo' [if it is locally produced charcoal it will probably be very hard to light or it might not light at all], Borges Coelho, 2007: 10), through the capabilities of the security guards ('Aposto que o funcionário está a dormir, além de que duvido que saiba ler a matrícula' [I bet the guard is asleep, and I doubt he can read the car number plate anyway], Borges Coelho, 2007: 12), to more significant historical and political issues relating to the colonial past. It is immediately evident that all the 'knowledge' the Odendaals possess about Mozambique 'revela um olhar redutor e homogeneizador do espaço em que a perceção pessoal é indissociável de um imaginário coletivo baseado em preconceito sobre o nível civilizacional do país vizinho' [reveals a reductive, homogenizing perspective in which personal views are indissociable from a collective imaginary grounded in prejudice regarding the neighbouring country's level of civilization] (Falconi 2017: 157). It becomes evident that the typical South African tourist view of Mozambique is profoundly bound up with, and influenced by, the distant and unequal relationship between the two countries, such that the tourists' attitudes towards the place they are visiting and the people who live there, mirror the historical and political positioning of South Africa in relation to Mozambique. As Jessica Falconi (2017: 157) puts it very well, the social and cultural factors which shape the tourist gaze in *Hinyambaan* are represented, not only by 'ideias pré-constituídas cuja origem não é revelada, sugerindo pertencerem a um discurso comum amplamente difuso e consolidado' [preconceived ideas, the origin of which is not disclosed, suggesting that they belong to a commonly held discourse that is widely disseminated and solidified] but also by the information given to the Odendaal's by their friend Joss du Plessis. Du Plessis's connection with Mozambique goes back to the colonial period which he looks back on with nostalgia. 'O Joss du Plessis diz que nessa altura se nadava em cerveja de primeira qualidade e se comiam toneladas de camarões. Isso, claro, antes destes tipos de agora terem destruído a cidade …' [Joss du Plessis said that back in those days you could swim in the very best beer and eat prawns by the ton. That was before the current lot ruined the city, obviously…] (Borges Coelho, 2007: 23). Since independence, du Plessis has maintained his relationship with the neighbouring country, working in road building. We believe that this character, who figures only when mentioned by others, never as a character in his own right, operates as a synecdochic representation of the privileged capitalist centre. He determines not only the dominant perspective on the periphery but all forms of relationship with it, including the way that subjects from the 'centre ' behave in peripheral space, and the relations they establish with the people who live there. This latter aspect of relations is characterized by a mixture of condescension where contact is indirect, and insecurity, where contact is direct. The former is evident, for example, in the way that Henrietta reacts when she spots some children selling flowers at the cemetery gate. 'Coitados (…) Os miúdos. Aposto que se não vendem as flores que têm passarão fome lá em casa. Se calhar, até levam pancada dos pais por isso' [Poor things (…) Those little kids. I'll bet if they can't sell the flowers they have, the household goes hungry. And maybe their parents beat them for it] (Borges Coelho, 2007: 26). There is a similar reaction when she notices a street that has been well maintained.

> Olha lá para fora, interessada no aspeto das primeiras casas da Matola, na altura em que a estrada se transforma como que por encanto numa avenida com um pequeno jardim a meio, separando as duas faixas.
>
> 'Coitados', comenta. 'Vê-se que fazem um esforço para embelezar a estrada, apesar das dificuldades por que passam. (Borges Coelho, 2007: 18)
>
> [Interested in what Matola's first houses are like, she looks out of the window at the point when the road changes, as if by magic, into an avenue with a little garden running down the middle, separating the two carriageways. 'Poor things', she says. 'You can see that, for all their hardship, they are trying really hard to make their roads look nice.]

Apart from being extremely condescending, the first citation above also clearly demonizes the peripheral subject, while the second clearly emphasizes Henrietta's feelings of superiority. To put it another way, Henrietta looks at something which she would have considered quite normal, had she observed it in her home country, and assumes that it would have required a huge effort in this peripheral location. Both reactions possess a common denominator, the absolute impossibility of the South African tourists identifying with the space or the people of the neighbouring country. In this inability to identify, or rather, this imperative non-identification, with the Other, there are clear echoes of the colonial past, combining two traditions identified by Achille Mbembe, the Hegelian one in which the colonized Other, classed as an animal 'could in no way be 'myself'' (Mbembe, 2001: 26), and the Bergsonian one, based on the idea that 'one could, as with an animal, sympathize with the colonized (…)' (Mbembe, 2001: 27).

The situation changes when the Odentaals come across a young Mozambican man Djika-Djika and are taken to 'visit' his village. Or perhaps it would be more accurate to say they are taken to be 'shown' his village. As Jessica Falconi (2017: 159) observes, the young Mozambican 'na perspetiva do olhar turístico representa o informante nativo (...) ponte entre dois mundos e tradutor de culturas' [as seen by the tourists, represents the native informant (...) the bridge between two worlds and the translator of cultures]. Making the most of this role, Djika-Djika engineers an encounter 'no qual emergem as estratégias de cooperação, resistência e manipulação ativadas pelos "nativos" em relação aos turistas, e consequentemente a complexidade das relações de poder em jogo no olhar turístico' [in which strategies of cooperation, resistance and manipulation surface, activated by the 'natives' in relation to the tourists, and as a result we also observe the complexity of the power relations at stake in the tourist gaze] (Falconi, 217: 160). According to Falconi, the tourists' observations before they visit the village may be classified in terms of the 'tourist gaze', 'the visual experience that is produced by travelling in modes of transport such as cars (...) where the passenger remains still and watches moving images unfold before them' (Falconi, 2017: 158). Once they are in the village, this becomes the 'mutual gaze', a situation in which 'os Odendaal "olham para este estranho mundo", mas também estão expostos ao olhar deste' [the Odendaals are 'looking at this alien world' but are also exposed to the alien world looking back at them] (Falconi, 2017: 160). Suddenly outside the comfort zone of their little *Corolla*, their protective box of glass and steel, the South African tourists find themselves in a position of dependency. What this means is that from a position of superiority, and the power they assumed they possessed, they suddenly 'fall' into a position of dependency and insecurity. This inversion of roles is depicted when the Odendaals become involved in the villagers' activities: in Henrietta's case, this is when she offers to help prepare the meal. And with Hermann it occurs with the male camaraderie after dinner. However, for the latter, his 'loss of status' is visible only in his half-hearted resistance to the traditional drink, while the process in respect of his wife takes a more complex form. Thus, '[p]ara as mulheres da casa, a presença da Henrietta no espaço da cozinha representa uma infração dos códigos da hospitalidade que é encarada com curiosidade, humor e uma ponta de paternalismo' [for the women of the household, Henrietta's presence in the kitchen infringes the codes of hospitality, and is greeted with a mixture of curiosity, amusement and a hint of condescension] (Falconi, 2017: 160). The South African woman is relegated to the simplest tasks since the Mozambican women who are cooking 'pretendem que demonstre o merecimento antes que possa intervir em níveis mais elevados, aqueles que verdadeiramente determinam o sucesso de uma refeição' [want her to prove herself before she can become involved in any higher-level tasks that would be vital to the success of the meal] (Borges Coelho, 2007: 63). And when they finally offer Henrietta this opportunity and she thinks that she 'mostrará como se cozinha de verdade uma galinha' [will show them how a chicken is really cooked] (Borges Coelho, 2007: 63), her myrtle sauce reduces them to fits of laughter. This image is further reinforced when the female elder of the village simply cannot believe that her meal has been cooked by the foreigner.

The character of the elder, Grandma Thum, becomes, incidentally, the key figure in this inverted power play when Odendaal's son, little Hendrik, gets lost in the bush chasing a butterfly. It is at this precise moment that all the privilege and the 'centre-based' knowledge of the South Africans becomes useless, and they have no choice but to trust themselves entirely to the methods of the old lady, in order to find their son. This 'knowledge from the periphery' functions here in the manner of the 'first swallow of spring', a warning sign that the dismantling of the entire centre-periphery system will follow.

While the family and neighbours of Djika-Djika are taken by surprise when the foreigners visit them, the inhabitants of Vila Longe make careful preparations to welcome the Americans. If the wishes and expectations of the tourist represent one side of the coin, in this tourist interaction, the response from the local community is the other. In Couto's novel, it is precisely that response, or rather its various ramifications which transmute, as we will go on to demonstrate, into a series of heterogeneities, omissions and aspirations, not only as regards centre-periphery relations but also in respect of the country's relationship with its own history and its contemporary civic positioning.

From the very beginning, the narrator makes it clear that the intentions of Chico Casuarino, the person who orchestrated the Americans' visit are not honest. Casuarino represents the stereotype of the corrupt nouveau riche con artist, determined to 'invent' the product the Americans are looking for and sell at it any price: 'Nós vamos contar uma história aos americanos. Vamos vender-lhes uma grande história' [We are going to tell the Americans a story. We are going to sell them one great big story.] (Couto, 2006: 154). This attitude, alongside the tourists' apparent interaction with the community, when they participate in events such as Rosie's trance brought on by the drumming at the batuque, the planned African naming ceremony for Benjamin, and the visit to the traditional healer's house, are all examples of what Dean MacCannell (2008: 173) called 'staged authenticity'. This is a situation in which the tourist experience supposedly goes beyond mere folkloric observation of an ethnographic nature and the tourist is drawn closer to daily experience and community life within the space that is being visited. In the case of this novel, this staged authenticity conceals plans and intentions about which the narrator delivers an implicit but clear ethical judgement: 'Os preparativos para a chegada dos estrangeiros foram concebidos como se um crime estivesse sendo congeminado. E de um crime, na realidade, se tratava' [they prepared for the foreigners' visit as if they were plotting a crime. And a crime, in truth, is exactly what this was.] (Couto, 2006: 148).

At the same time, the attitude adopted by the impresario figure, Chico Casuarino gives grounds for deconstructing the idea of identity based solely on skin colour, promoted by the Negritude movement, which was criticized, at the time by Frantz Fanon (1961), and more recently by Achille Mbembe (2001, 2012). It effectively enables a symbolic rupture between the 'Africanist' nostalgia of the Black-Atlantic diaspora, and the 'Africanness' that is experienced by the people who live on the continent of Africa, ultimately enabling a negotiation of Mozambican identity in its own right. Within this framework, Casuarino represents the first spark of voracious capitalism in its peripheral form, ready to take advantage of anyone who appears to 'have more', based solely on how the distribution of good appears to be in the world system, without reference to historical processes. 'Eles gostam de pagar [...] Estes gostam porque sentem-se culpados, está perceber? Saíram daqui, deixaram a malta sofrer com o colonialismo e, agora, regressam engravatados, cheios de inglesuras, e a gente ainda passando fome' [They like to pay for things [...] They like to do that because they feel guilty, you see? They left here, they left us guys behind to endure colonialism, and now they come back here with their suits and ties, and their

English mannerisms, and here we still are half starving.] (Couto, 2006: 152). As this deceptive exchange of words reveals, 'Isso que o nosso irmão chamou de "salvei", nós aqui chamamos de patrocínio' [when our brother said 'I saved you' he was referring to what we call patronage] (Couto, 2006: 172).

The character who acts as Chico Casuarino's main adversary is Arcanjo Mistura, the intellectual barber. This former militant for the national cause who was once imprisoned by the PIDE, is living in the post-independence era, as a voluntary recluse in Vila Longe. The barber is faithful to the ideals of independence and he is strongly opposed to the exotic image of Africa which Casuarino wants to construct, and he denounces its (neo)colonialism: 'Você quer-nos apresentar como criaturas exóticas, vivendo de crenças e tradições. Não era essa a imagem que os colonos faziam de nós?' [You want to represent us as exotic creatures, living with our old beliefs and traditions. But isn't that exactly how the colonizers used to see us?] (Couto, 2006: 339). Arcanjo is also the one who deconstructs, in a fairly radical, almost utopian manner, the concept of identity as national and racial. He has a conversation with Benjamin in which he doggedly tries to convince him, that his 'struggle' and his quest should concentrate not on the 'discovery' of Africa, but rather on the de/reconstruction of all thought and affiliation based on racial identification, and on reintegrating himself into American society.

While the Americans' interactions with Casuarino and Arcanjo show two different facets of postcolonial Mozambican identity, namely neocapitalism and the old revolutionary ideals, meetings with other characters bring various other aspects of contemporary Mozambican identity and experience to the fore: the colonial trauma of Zeca Matambira – once a promising boxer, defeated by the fact that he cannot hit a white man; the African thought systems in all their vitality, represented by Mwadia and Constança's links with tradition, and the ethnic complexity of the country portrayed by the figure of Jesustino and his sister Luzmina.

However, the principal element in the staging of Africanness which is undertaken for the tourists, by the people of Vila Longe, relates to the trance sessions during which Mwadia, Dona Constança's youngest daughter pretends to be (or is she really?) possessed by spirits. However, although these sessions are orchestrated, they are not entirely pure invention either.

In fact, this is the point of intersection, not only for the two stories that the novel contains,[17] but also for the two great sources of knowledge: orality and writing. Mwadia 'fabricates' the scripts for her experiences of spirit possession, basing them on real sixteenth-century historical documents. To some people these sessions look like a real 'African experience', and to others they are simply merchandize well sold:

> Os americanos não podiam duvidar de que se tratava de uma encenação. Para eles, o passado estava de visita a Vila Longe por via de nocturnas visitações. Não tardaria que colunas de escravos renascessem e desfilassem pelos transes da filha nova de Dona Constança. (Couto. 2006: 276)

> [The Americans did not suspect that this was purely staged. For them, the past was returning to Vila Longe through these nocturnal visitations. At any moment, columns of slaves would be reborn and would file through the trancelike visions of Dona Constança's youngest daughter]

These sessions function, however, as a fundamental tool for deconstructing any homogeneous vision of the different temporalities which inflect identity construction throughout the narrative. What emerges from them is an unexpected and heterogeneous past, incorporating both the story of Benjamin Southman the African American, and the country called Mozambique.

The Indian Ocean itself operates as the *spiritus movens* both for these complex negotiations of identity and for the deconstruction of the monolithic, western-centred perspective which relegates African space to a complementary function. In the relationship between Southman and Africa, the Indian Ocean is the unknown, it does not figure in his vision of African history, which for him is reduced to the transatlantic black slave trade and a load of folklorized 'traditions'. Thus, in his research, he dispenses with the interview given by the Goan, Jesustino Rodrigues, stating that he only needs 'authentic African statements'. And, once again, the rejection takes a patronizing tone.

17 Events in Vila Longe are interwoven with the supposed diaries of two Jesuit priests, Gonçalo Silveira e Father Antunes, during their journey from Goa to Mozambique.

– Talvez no final. É isso, no final eu o entrevistarei, caro Jesustino. Até porque a sua família esteve ligada ao tráfico de escravos.
– Não sei nada sobre isso.
– Eu vou ajudá-lo a recordar. (…) (Couto, 2006: 322–323)

[Perhaps at the end. Yes, I will interview you at the end, dear Jesustino. Because your family was kind of linked to the slave trade.

I don't know anything about that.

I will help you remember it. (…) (Couto, 2006: 322–323)]

The passage cited above portrays in precise terms the arrogance and myopia of the 'centre' assuming a position from which to define the identity of the periphery, according to the needs and presuppositions of the centre. This monolithic mode of thought is methodically deconstructed by means of this rejected piece of information, since it is precisely the 'Indian Ocean' which shatters the American's painstakingly constructed identification with Afro-Negritude, when Mwadia announces that her ancestor who was forced to cross the Atlantic was not a black African man but an Indian woman. What we see at this point is the distribution of power along the centre-periphery axis being inverted. The centre loses the privileged position from which it defines the *status quo*, and is forced to realign itself by the heterogeneous power of the, now redefined, periphery/peripheries. In this particular case of centre-periphery deconstruction, based on a literary example of Afro-American diaspora tourism in Mozambique, the Indian Ocean resonates with the thinking of João Paulo Borges Coelho. He suggests that we regard the Ocean as 'a ponte heiddegeriana que não liga margens previamente existentes, mas precisa de ser atravessada para que as suas margens surjam como margens' [the Heideggerian bridge which does not connect shores that already exist, but which needs to be crossed for its shores to become shores] (Borges Coelho, 2019: 19). However, in the context of the novel as a whole, the ocean crossing, understood both literally and metaphorically, takes on a causative role, not only for rediscovering different aspects of Mozambicannness but also as regards the construction of narrative aesthetics. These correspond to what Meg Samuelson has termed the 'amphibian aesthetic of coastal form' which 'entwines apparently distinct perspectives, implicating them in the stances they strike while both conjoining and relativizing them through free indirect discourse or unmediated flows of direct speech' (Samuelson, 2017: 20).

In João Paulo Borges Coelho's comic novella, the presence of the Ocean is rather more understated, but we believe the role that it plays in deconstructing the monolithic vision of the centre-periphery relationship is no less relevant here too. Thus, although the coast of the Indian Ocean is defined from the outset as being the Odendaal family's holiday destination, the sea, particularly for the family matriarch, is represented, as Falconi notes as 'uma forma de alteridade – ao mesmo tempo fascinante e aterrorizadora – comparável com o passado mítico de Lost City, ou os animais do Kruger Park' [form of alterity – both fascinating and terrifying – comparable to the mythical past of the Lost City, or the animals in the Kruger Park] (Falconi, 2017: 155). 'Uma coisa é ver o mar e outra entrar dentro dele ' [It is one thing to look at the sea, and quite another to go in it] (Coelho, 2007: 8) thinks Henrietta whose relationship with the sea is 'distante, mas não forçosamente inamistosa' [distant, but not necessarily hostile] (Coelho, 2007: 8) and for whom the prospect of a boat trip suggested by Joss du Plessis, poses a threat to the peace and unity of her family. However, since the du Plessis family unexpectedly cancels the trip at the last minute, the Odendaals are forced to go on their own, and the prospect of an encounter with the ocean fades into the background. Or at least, it does not happen directly in the narrative

The transformative force of the Indian Ocean does, however, manifest itself in a perverse and unexpected way. The Odendaals are very surprised when, at the very moment which might have brought the trail of unexpected, unpredictable events to an end, they find a Jeep parked by the roadside, that 'tem em cima um belo barco de casco branco, com o nome pintado a refulgentes letras vermelhas: *Outrageous Adventurer*' [has a beautiful boat on top of it, with a white hull, and the name painted on in shiny red letters, *Outrageous Adventurer*](Borges Coelho, 2007: 101). It is the du Plessis's boat. Thus, the boat, normally intended to tame the waves of the Indian Ocean, is transformed in this narrative into a very particular form of paradox, a betrayal that is extremely telling. As we see it, the du

Plessis are personal friends of the Odendaals but we are given to understand that they are better off. And if we view the characters in this narrative in terms of a world system in miniature, this places them in a position of centrality, in relation to which the Odendaals are semi-peripheral and the Mozambicans are peripheral. This schematization of the characters is reinforced by the role, described above, which Joss du Plessis plays in constructing the *tourist gaze* of the Odendaals (corresponding, in this instance, to the South African view of Mozambique). In this context, the meeting on the road, which exposes a treacherous lie at the level of personal friendship, also undermines his assumed position of knowledge about the periphery. The fact that the boat (the name of which, incidentally, refers to the unknown and to discovery) provides the metaphorical means by which this betrayal is revealed, clearly points up the specific historical significance of the Indian Ocean as a 'problematizing space' (Hofmeyr, 2019) and occasions an epistemological shift.

Conclusion

As we have demonstrated above, analysing a series of selected literary works in both Portuguese and English, there clearly exists, along with its own, specific readership, a very particular representation of the Indian Ocean that is mediated through the Portuguese language, whether this is viewed from Goa, or projected from Mozambique, in dialogue with neighbouring shores such as those of Zanzibar or South Africa. On the other hand, it is also evident that no region of the Indian Ocean can be thought about in isolation, enclosed within its own national frontiers. This conceptual error is exposed here by the internal (i.e. intra-national) impact of a shared history, laden with interferences and intersections, reflecting the mobility of goods and peoples, and also of news and ideas, which has led to the shaping of local, national and regional causes, in relation to a world system in which these societies have been called, in various ways, both forcefully and spontaneously, to play a role.

The literary works analysed here offer a rich, multifaceted vision of the challenges facing the present day, as they interpret power games and deconstruct fixed ideas and enforced silences, regarding the role of Africa and African societies in the history of the Indian Ocean. And finally, this chapter serves to prove once again and in no uncertain terms, the power that literature possesses to interrogate and reinvigorate us, presenting its readers and critics with a perspective on the Indian Ocean region derived from the works of its writers.

Bibliography

Adejunmobi, M. (2009). 'Claiming the Field: Africa and the Space of Indian Ocean Literature'. *Callaloo*, 32 (4), 1247–1261.

Anderson, B. (1991). *Imagined Communities: Reflections on the Origin and Spread of Nationalism*. London: Verso.

Borges Coelho, J. P. (2002). 'African Troops in the Portuguese Colonial Army, 1961–1974: Angola, Guinea-Bissau and Mozambique'. *Portuguese Studies Review*, 10 (1), 129–150.

Brugioni, E. (2013). 'Literary Cartographies and Humanistic Criticism: The Indian Ocean as a "critical paradigm"'. In A. G. Macedo, C. M. de Sousa and V. Moura (eds), *Humanidades: Novos Paradigmas do Conhecimento e da Investigação*, pp. 87–99. VN Famalicão: Húmus.

Bulletin d' Informations de CONF, unpublished archives, Mário Soares Foundation, available at: <http://casacomum.org/cc/visualizador?pasta=07059.021.008#!10>

——. (2007). *Hinyambaan*. Lisboa: Caminho.

——. (2010). *O Olho de Hertzog*. Alfragide: Leya.

Calafate Ribeiro, M. (2016). 'Literary Voices of Luanda and Maputo: A Struggle for the City'. *Journal of Lusophone Studies*, 1 (1), 88–106.

Can, N. (2020). 'Poetics and Politics of Memory: Notes on João Paulo Borges Coelho's Novels'. In E. Brugioni, O. Grossegesse and P. de Medeiros (eds), *A Companion to João Paulo Borges Coelho*, pp. 69–87. Oxford: Peter Lang.

Couto, Mia. (2006). *O outro pé da sereia*. Lisboa: Caminho.

Devi e Seabra. (1971). *A Literatura indo-portuguesa, Antologia*. Lisboa: Junta de Investigações do Ultramar.

Falconi, J. (2017). 'O imaginário turístico em Hinyambaan'. In Sheila Khan et al. (eds), *Visitas a João Paulo Borges Coelho. Leituras, diálogos e futuros.* Lisboa: Edições Colibri.

Fanon, F. (1961). *Os Condenados da Terra* (trad. Serafim Ferreira). Lisboa: Ulliseia.

Frenz, M. (2014). *Community, Memory, and Migration in a Globalizing World: The Goan Experience, C. 1890–1980*. Oxford: Oxford University Press.

Gilroy, P. (1993). *The Black Atlantic: Modernity and Double Consciousness.* Cambridge: Harvard University Press.

Gupta, G. (2014). *Towards a Decolonial Reading of Tivolem: E ainda há mais mundo, chega lá*. PhD thesis, University of Coimbra.

Gupta, P. (2019). *Portuguese Decolonization in the Indian Ocean World.* London: Bloomsbury Academic.

Hall, C. M. and Tucker, H. (2004). *Tourism and Poscolonialism: Contested Discourses, Identities and Representations*. London: Routledge.

Hall, S. (1990). 'Cultural Identity and Diaspora'. In J. Rutherford (ed.), *Identity: Community, Culture, Difference*, pp. 222–237. London: Lawrence and Wishart.

Hannam, K., Sheller, M. and Urry, J. (2006). 'Editorial: Mobilities, Immobilities and Moorings'. *Mobilities*, 1 (1), 1–22.

Hansen, T. B. (2010). 'The Unwieldy Fetish: Desire and Disavowal of Indianness in South Africa'. In P. Gupta, I. Hofmeyr and M. Pearson (eds), *Eyes across the Water*, pp. 109–121. Pretoria: University of South Africa Press.

Hofmeyr, I. (2007). 'The Black Atlantic Meets the Indian Ocean: Forging New Paradigms of Transnationalism for the Global South – Literary and Cultural Perspectives'. *Social Dynamics: A Journal of African Studies*, 33 (2), 3–32.

——. (2012). 'The Complicating Sea: The Indian Ocean as Method'. *Comparative Studies of South Asia, Africa and the Middle East*, Duke University Press. 32 (3), 584–590.

Jameson, F. (1986). 'Third-World Literature in the Era of Multinational Capitalism'. *Social Text*, 15, 65–88.

Keller, S. and Pearson, M. (2018). *Port Towns of Gujarat*. Delhi: Primus Books.

Knopfli, R. (1969). 'À Paris'. In *Mangas Verdes Com Sal*. (s/l): Edições Europa-América.

Lefebvre, H. (1991 [1974]). *The Production of Space*. Oxford: Blackwell.

Lobo, S. (2013). 'O desassossego goês. Cultura e política em Goa do liberalismo ao acto Colonial' Tese de doutoramento. Universidade Nova de Lisboa. <http://hdl.handle.net/10362/10822>

MacCannell, D. (1973). 'Staged Authenticity: On Arrangements of Social Space in Tourist Settings'. *The American Journal of Sociology*, 79 (3), 589–603.

——. (2008). 'Staged Authenticity Today'. In M. Sorkin (ed.), *Indefensible Space: The Architecture of the National Security State*, pp. 259–276. London: Routledge.

Manifesto do Movimento Anti-colonialista, MAC, Edição do Departamento de Informação, Propaganda e Cultura do CC do PAIGC.

Mbembe, A. (2001). *On the Postcolony*. Berkley: University of California Press.

——. (2012). 'Formas africanas da escrita de si' (trad. Marina Santos), 2012. <http://artafrica.letras.ulisboa.pt/uploads/docs/2016/04/18/5714df750dffd.pdf>. Accessed: 20 May 2020.

Melo e Castro, P. (2012). 'Of Prison Walls and Barroom Brawls: Postcolonial Categories in Mia Couto's "De Como se Vazou a Vida de Ascolino do Perpétuo Socorro" and Laxmanrao Sardessai's "O Barco de África"'. In S. Khan, A. M. Dias Martins, H. Owen and C. Ramos Villar (eds), *The Lusotropical Tempest: Postcolonial Debates in Portuguese*, pp. 127–150. Bristol: HiPLA. University of Bristol.

Mendonça, F. (2014). 'Dos confrontos ideológicos na imprensa em Moçambique'. In C. Braga-Pinto and F. Mendonça (eds), *João Albasini e as luzes de Nwandzengele. Jornalismo e política em Moçambique, 1908–1922*, pp. 17–40. Maputo: Alcance.

Nayak, A. (2010). 'Race, Affect, and Emotion: Young People, Racism, and Graffiti in the Postcolonial English Suburbs'. *Environment and Planning A*, 42 (10), 2370–2392.

Noronha, F. (2018). Entrevista. Arquivo do projeto NILUS; concedida em janeiro 2018 em Saligão, Goa a Ana Mafalda Leite, Ute Fendler e Kamila Krakowska Rodrigues.

Passos, J. (2018). 'Goan Migrant Literature'. *InterDISCIPLINARY Journal of Portuguese Diaspora Studies*, 7, 213–223.

Pearson, M. (2003). *The Indian Ocean*. London: Routledge.

——. (2010). 'Connecting the Littorals: Cultural Brokers in the Early Modern Indian Ocean'. In P. Gupta, I. Hofmeyr and M. Pearson (eds), *Eyes across the Water*, pp. 32–47. Pretoria: University of South Africa Press.

Pécoud, A. and de Gutcheneire P. (2007). *Migration without Borders: Essays on the Free Movement of People*. New York: Berghahn and UNESCO.

Rangel-Ribeiro, Victor. (1998). *Tivolem*. Minneapolis: Milkweed Editions.

Rodrigues, M. E. (2006). 'Sepoys from India or African Soldiers? Dilemmas in the Naturalization of the Portuguese Colonial Army in the 18th Century'. *História: Questões & Debates*, 45, 57–95.

Samuelson, Meg (2017). 'Coastal Form: Amphibian Positions. Wider Worlds and Planetary Horizons on the African Indian Ocean Littoral'. *Comparative Literature*, 69 (1), 16–24.

Sardessai, L. (1971). 'O Barco de África'. In Devi e Seabra, *A Literatura indo-portuguesa, Antologia*. Lisboa: Junta de Investigações do Ultramar.

Sheller, M. and Urry, J. (2006). 'The New Mobilities Paradigm'. *Environment and Planning A*, 38, 207–226.

Smith, D. A. (2017). 'Poetry, Politics and Newspapers, Portuguese: Laxmanrao Sardessai and Goa in the 1960s'. <https://www.mail-archive.com/goanet@lists.goanet.org/msg121001.html>

Souza, A. de (2010). 'The House of Wailing'. In P. Nazareth (ed.), *Modern Goan Literature*, pp. 169–174. Goa: 1556 and Broadway Book Centre.

Stroud, Ch. and Jegels, D. (2014). 'Semiotic Landscapes and Mobile Narrations of Place: Performing the Local'. *International Journal of the Sociology of Language*, 228, 179–199.

Thompson, E. (2008). 'Postkolonialne refleksje'. In *Pórownania*, 5/2008, Poznań.

Timothy, D. J. and Teye, V. B. (2004). 'American Children of the African Diaspora: Journeys to Motherland'. In T. Coles and T. J. Dallen (eds), *Tourism, Diasporas and Space*. PAGINAS. London: Routledge.

Urry, J. (2002). *The tourist gaze*. London: Sage.

Vergès, Françoise and Marimoutou, Carpanin (2012). *Journal of Multidisciplinary International Studies*, 9 (1), January 2012, Lola Sharon Davidson and Stephen Muecke (eds). Sydney, Australia: UTSePress.

Vertovec, S. (1999). 'Conceiving and Researching Transnationalism'. *Ethnic and Racial Studies*, 22 (2), 447–462.

Vicente, F. L. (2017). 'Black Women's Bodies in the Portuguese Colonial Visual Archive (1900–1975)'. *Portuguese Literary & Cultural Studies*, 30/31, 16–67.

Vink, M. (2007). 'Indian Ocean Studies and the "new thalassology"'. *Journal of Global History*, 2, 41–62.

GABRIELA BEDUSCHI ZANFELICE

7 Postcolonial environments: Notes on the relationship between nature, literature and the Indian Ocean in Borges Coelho's *Índicos Indícios*

In an essay entitled 'O Índico como lugar' ['The Indian Ocean as place'], João Paulo Borges Coelho (2019: 13–28) presents an interesting reflection on the notion of place as an autonomous entity, with a life and spirit of its own. Retrieving the concepts of *Spiritu loci* and *Genius loci* that originated in Greek and Roman thought, the author comments on the spirits that accompany and characterize each place, providing them with a unique atmosphere that allows us to be conscious of being in such places. Through material 'traces, clues and evidences' (Brugioni, 2020: 180) such as scents and breezes, the rustle of leaves and the changing of colours throughout the day, the *Genius loci* would announce its presence, warning us that that place was being looked after (Borges Coelho, 2019: 14). Remotely corresponding to the Christian Guardian Angel and the 'genius' theorized by Agamben (2006) – both referring to a kind of spirit that accompanies each human being – the 'spirit of places' for João Paulo Borges Coelho is what remains even with the passing of time, the ways in which places resist and maintain their specificity regardless of human scrutiny and disturbances (Borges Coelho, 2019: 14–15). Places, nature and their materiality, the author reminds us, were always crucial elements for different peoples: since the remote runes and inscriptions made in caverns by the Neanderthals we have had signs that, before looking upwards in search of the divine, human beings looked down to the earth which provided them with food and to the place that surrounded them (13).

The author observes, however, that with the evolution of hegemonic Western culture, human beings began to look at places as dead things, deprived of strength, genius or mystery, transforming them into passive places, reduced to the action and predation of humans (15). In the colonial period this invasive attitude reached its apex, with the subduing of nature by humans reflected in the ways in which colonial societies were also seen and treated: as unworthy of respect or equality due to their status as just another element of nature, mere wild animals to be equally domesticated and dominated. It is only with the recent environmentalist turn that the assumptions of this worldview started to be challenged, with greater attention being directed at topics revolving around one central question: is it possible to maintain the viability of life on planet Earth without radical and structural changes in the ways in which humans relate to the environment?

Since then, attempts to conceptualize nature have been numerous and multifold. Among these, we may cite the 1970s 'deep ecology' movement (Naess, 1989), with its first-worldist and idealistic view of nature as a redemptive, uncontaminated and threatened wilderness (Mukherjee, 2010: 26); the 1980s and 1990s 'social ecology' field (Harvey, 2000; Smith, 1984; Davis, 1998), greatly indebted to Raymond Williams' (1973, 2005 [1980]) dialectical view of the relationship between humans and nature as part of the simultaneously natural and cultural historical trajectory of 'economic, political and biological evolution' (Mukherjee, 2010: 29); and the 'environmentalism of the poor' perspective (Guha & Martinez-Alier, 1997), which called attention to the differences between the ecological movements and environments from the Global South, on the one hand, and from Europe and North America, on the other. The latter, marking a postcolonial turn in social ecology for its alertness to the colonialist and imperialist roots of poverty, pollution and resource exhaustion across the Global South's environments, were further revised and elaborated by contemporary authors such as Pablo Mukherjee (2010), Rob Nixon (2011) and Jennifer Wenzel (2019). These, in turn, emerged as fruitful contributions for ecocriticism, a term coined in the late 1970s[1] to designate a field whose focus

1 According to Cajetan Iheka (2018: 10), William Rueckert was the first to utilize the term, in 1978.

relies on the study of the relationships between culture, literature and the environment. Chengyi Coral Wu (2016: 144–145) makes the case for the existence of an African environmental literature and criticism originating as early as the 1960s, arguing, with DeLoughrey and Handley (2011: 6), how environmental concerns in studies and novels such as Chinua Achebe's *Things fall apart* (1958) predate, in a couple of decades, the institutionalization of ecocriticism within Anglo-American academies. Particularly relevant in these perspectives, in our view, is their attentiveness to the indissociable articulation between environmental, postcolonial and material questions, providing significant conceptual and epistemological tools for reading and uncovering 'the complex and diverse relationships between soil, water, plants, animals, air, humans and their various kinds of labour' (Mukherjee 2010: 18) and their registrations throughout literary works.

As this brief overview attests, the environmental question has occupied the concerns of scholars from several different backgrounds, constituting a powerful multi- and interdisciplinary tool in understanding and interpreting 'the necessarily interlinked devastation of the humans, nonhumans, soil, water, air and crops of [the] postcolonial [environment]' (Mukherjee, 2010: 11). In his *Postcolonial environments: Nature, Culture and the Contemporary Indian Novel in English* (2010), Pablo Mukherjee defines 'postcolonial environment' as 'an integrated network of human and nonhuman agents acting historically' (5), that is, as an integrated network of natural and cultural systems under the historical condition of the ongoing and intensified capitalist exploration (13). Referring to the way in which postcolonial literatures perfectly capture 'the rhythm of the lived experience of contemporary global existence' (9), Mukherjee observes the irreducible literariness of the postcolonial novel. Through a performance of its own story, this literariness would then register the environment simultaneously at thematic and formal levels, destabilizing the cultural and ideological *status quo* by inscribing, in its narratives, the environmental devastation caused to humans and nonhumans (10–11). As the author notes, even though his work focuses on literatures of postcolonial India, its central ideas can be productively applied to other places of the postcolonial world, offering powerful insights into contexts and literatures such as those of Mozambique. Bearing this in mind, this chapter analyses the ways in which the short story collections *Índicos Indícios I: Setentrião* (2005a) and *Índicos Indícios II: Meridião* (2005b) by João Paulo Borges Coelho register the Mozambican postcolonial environment. Seeking to investigate these narratives' dialogue with the environmental impacts on the Indian Ocean and its coastal societies, and with a special focus on the ways in which the capitalist system's dynamics of combined and uneven development aggravate this situation, we argue that Borges Coelho's literature provides singular and innovative imaginaries with which to understand and imagine the relationships between humans and nonhumans. Rendering critical reflections that challenge the dichotomic forms through which we usually think of these domains, *Índicos Indícios*'s imaginaries build complex connections between literary fiction and the Mozambican contemporaneity, pointing towards the inextricable links between history, society and nature.

The registration[2] of animals and elements of nature in *Índicos Indícios* (2005a, 2005b) point specifically towards such ecological preoccupations by its author (Falconi, 2013; Medeiros, 2022). As in other of his texts – such as *Água. Uma novela rural* (2016), *Ponta Gea* (2017) and *Museu da Revolução* (2020) – *Índicos Indícios* presents us with ecosystems where human and nonhuman subjects are deeply integrated, with narratives registering the social and material conditions of Mozambican society without losing sight of its connections to the world system and the world ecology at large. Following the elaborations of the Warwick Research Collective (2015, 2020) and understanding Borges Coelho's works as world literature (i.e. as the literature of the modern capitalist world system), we understand that it inevitably 'registers the transformations in world ecology that have been both cause and consequence of the transition to, and subsequent reorganisations of, the capitalist world-economy' (Niblett, 2012: 16).

The paradigm of the Indian Ocean is, as the books' titles suggests, a crucial one, appearing throughout the stories both as main topic and as

2 My use of the concept of registration is indebted to the Warwick Research Collective's proposition of world literature as the literature of the modern capitalist world system, or the literature that registers, both in thematic and formal levels, the rhythm of its lived experience, whether their authors are conscious of it or not. See WREC (2015, 2020).

more subtle background. The element of water holds a fundamental role, displaying a dialogue with environmental crises at global and regional, macro and micro levels. Opening up interpretative possibilities for reflections on the relationship between Mozambique's national space and the Indian Ocean (Brugioni, 2017) as well as on the interactions between human, nonhuman and more-than-human worlds, the narratives' critical paths question official and hegemonic discourses on Mozambican history, creating imaginaries that go beyond established territorial and political frontiers towards the sea. An attentive perception to Mozambican events such as the spiders, birds, seeds and trees taking over the abandoned Grande Hotel da Beira (Borges Coelho, 2005b: 45–72) emphasizes nature's agency in reclaiming the space of a colonial building whose precarious ruins are inhabited by the poor; the signalling of correlations between the distant coasts of the Indian Ocean and the Mediterranean (Borges Coelho, 2005b: 9) invite us to re-examine how cartographic divisions of seas and ocean basins are usually thought. In becoming the 'inaugural place of narration, index of a bigger history, a physical and conceptual space through which it is possible to (re)define Mozambique's telluric and imagetic geography' (Brugioni, 2015: 103), the Indian Ocean emerges as a 'liquid archive, a repository of memories and traces which are fundamental to give meaning to the plots and stories that edify Mozambique's social and cultural mosaic' (103).

In Borges Coelho's writing, human and nonhuman elements are mutually implicated, suggesting new ways of imagining the relations between humans and nonhumans, built and natural environments. With a writing style filled with metaphors and symbolic references to the stories' aquatic and animal components that reflect simultaneously upon the past, the present and the future, Borges Coelho signals the inevitable enmeshment between postcolonial and environmental topics, challenging the 'continuing imperialist modes of social and environmental dominance' (Huggan & Tiffin, 2010: 2). The articulation of postcolonial and environmental perspectives thus represents a potent tool for the analysis of contexts such as the Mozambican one, for allowing the comprehension that the establishment of hierarchical relations between 'civilized' societies (i.e. former imperial centres) over 'uncivilized' ones (former colonies) was only possible due to its association with another hierarchy that underlies it: that of human communities over nonhuman ones. In this light, the study of how forms of appropriation of nature provided the basis for the emergence and expansion of the modern capitalist world system (Wallerstein, 2011) is indicative of the inexorability of the connections between capitalist development and environmental collapse (Marques, 2018). Fuelled by processes of subjugation of animals, natural elements and certain human societies that are seen as 'inferior' precisely due to their association with nonhuman dimensions, the growing social unevenness around the world ends up being updated through forms of environmental injustices, an aspect further evidenced in coastal societies due to their proximity with seas and oceans (Wenzel, 2019). Environmental events such as the intense and recurrent episodes of droughts and flooding hitting Mozambique every year are a common feature of the lives of peoples all over the world, and must be understood in their totality if we bear in mind that, in the era of the Anthropocene – or Capitalocene, according to Moore (2016) – the effects of human action span the entire earth and marine ecosystems. Understanding literature's role in 'shaping our understanding of the world and the planet', then, may lead us to 'forg[e] new modes of relation among human and with nonhuman nature' (Wenzel, 2019: 7), given that 'things like climate change, fossil-fuel dependence, and resource depletion are not merely technological, economic, or political problems but also narrative problems and problems of the imagination' (Wenzel, 2019: 7).

For Jennifer Wenzel, a deeper understanding of cultural imaginations and narrative logics revolving around environmental questions are important even for areas beyond literary studies, for their potential in promoting 'more robust accounts of the past, present and future of global inequality' (9). Observing that 'Industrialization and consumer capitalism in the Global North have made outsized use of the earth's atmosphere and oceans as 'sinks' for waste products like carbon dioxide (CO_2)' (10), the author highlights that 'imaginative gestures across geographic borders and experiential divides' (25) may offer multiple answers and solutions to the promises of modernization and development when we look at them from different temporal, geographical and experiential points of view. Proposing that we *read for the planet*, Wenzel emphasizes the urgency of investigating 'how to calibrate the world-system of World Literature with the Earth system remade in the Anthropocene – as well as other vectors of environmental justice' (29), pointing to the need for thorough analyses of the production of inequalities and unevennesses that have allowed the expansion of capitalism in a global scale. For her, as well as for us, a central question arises: 'At a moment of authoritarianism ascending, inequality exploding, and oceans rising, what does the future look like?' (31). The elevation of the oceans' levels constitutes, naturally, another point of convergence between topics present in Borges Coelho's works and questions concerning Indian Ocean and environmental humanities scholars, evidencing the relevance of investigating the environmental question, its effects on seas and oceans, its origins in the capitalist system and its narrative parallels throughout literary works.

Throughout the ten narratives that constitute *Índicos Indícios* (2005a, 2005b), transnational imaginaries are built across several dimensions – historical, geographical, symbolic and material ones. These evoke and are situated in several localities within and outside of Mozambique, such as Tanzania, Kenya, Somalia, Djibouti, Yemen, Saudi Arabia and Australia, as well as cities and villages such as Beira, the Island of Xefina, the Province of Zambezia and the Island of Ibo. When located in specifically Mozambican coordinates, the stories are constructed with thorough attention to the different cultural, environmental and socio-economic aspects present in each neighbourhood of such localities. The relevance of rivers and the sea for those around them is a constant motif, moving away from perspectives that might situate such contexts and their characters as members of a homogenous and idealized national collectivity. In narratives such as 'Casas de ferro' [Iron houses] and 'A força do mar de agosto' [The force of the August sea], for instance, questions of social unevennesses and economic dependency are pressing, undermining linear and romanticized ideas of progress and development that so often haunt discourses on Africa's future. Within the field of postcolonial studies, for instance, its very institutionalization in European and North American universities has sometimes led to more or less explicit 'celebration of the new world order of late capitalism and new imperialisms' (Mukherjee, 2010: 41). In order to move away from this, it is imperative to stress how facts such as the social and environmental impacts caused by intensified mineral extraction and the poor work and migration conditions it requires (as the case of Mozambique and South Africa makes evident); oil spills; air emissions; marine life and habitat destruction; introduction of invasive species; disruption of local communities' lives caused by the ongoing traffic of cargo vessels and container ships, etc., are but a few of the numerous examples of ways in which discourses that praise ever-increasing trade rates and capital accumulation fail to see their relation to the overall disruption of the earth's oceans and atmospheres, as well as to 'the production of inequality and unevenness on a global scale' (Wenzel, 2019: 37).

In the short story 'Os sapatos novos de Josefate Ngwetana' ['The New Shoes of Josefate Ngwetana'] (Borges Coelho, 2005b: 49–107), the locations visited by Herculano Ngwetana during his journey along Mozambican and South African coasts are often those that 'as cartas nacionais já não conseguem desvendar' ['national maps can no longer decipher'] (56), situated in peripheral places that share the Indian Ocean as a common horizon. The story, set in a postcolonial period, is imbued with a unique configuration of nature: blending elements typically regarded as belonging to opposing domains, the narrative represents a powerful subject for ecocritical studies. The tale begins in Machangulo, Mozambique, where the young Herculano grew up watching his father Josefate hunt crabs in the mangrove's mud. The place is described as 'uma terra complicada' ['a complicated land'] (51) in which nature spread its dispositions with some confusion:

> Põe-se ali o sol do lado do mar, e não de dentro como seria normal. Podia ser praia e não é porque lhe falta areal, sendo o que existe estreito e cinzento. Onde é já francamente mar crescem ainda umas árvores com a mesma desenvoltura que teriam se estivessem em terra seca (navegando, os barcos afagam por vezes os seus ramos como pássaros que voassem, ou amarram neles os cabos de âncora para descansar). Por outro lado, dentro da terra surgem lagoas altivamente encapeladas – Buti, Mandi, Mangalidje e muitas outras – como se fossem ainda elas próprias o mar. Os pequenos rios que na quase praia desaguam entram pelo mar adentro transportando ainda as suas distintas cores de rio, de um acastanhado pardacento, conseguindo com esse artifício continuar a ser rios já bem dentro da baía, quando há muito deixaram de ter margens sólidas que os sustentem. De tal forma que é possível a um pescador desaparecer com o seu xitatarru no horizonte da baía e voltar depois dizendo que não chegou ao mar, que andou apenas pescando dentro desse rio sem margens que se vejam. E fazê-lo sem estar mentindo.

É inegável, portanto, que a natureza se deixou tomar por alguma confusão quando espalhou as suas disposições neste lugar, misturando os ingredientes de forma pouco habitual. Por isso são do Machangulo os maiores caranguejos que há neste mundo, grandes e gordos como mamíferos. (pp. 51–52)

['There, the sun sets by the side of the sea, not from within as would be normal. It could be a beach but it isn't because it lacks a sandy shore, and what exists is narrow and gray. Where it is already unmistakably sea, some trees still grow with the same ease they would if they were on dry land (sailing, boats sometimes stroke their branches like birds in flight, or tie their anchor cables to them for resting). On the other hand, within the land arise proudly rippling lagoons – Buti, Mandi, Mangalidje, and many others – as if they were still the sea itself. The small rivers that flow into the almost-beach venture further into the sea still carrying their distinct river colors, a murky brownish hue, managing by this artifice to remain rivers well inside the bay, long after they have ceased to have visible banks to support them. So much so that a fisherman can disappear with his *xitatarru*[3] into the bay's horizon and return later saying that he never reached the sea, that he was only fishing within that river with no visible shores. And he can do so without lying.

It is undeniable, therefore, that nature allowed itself to be taken over by some confusion when it spread its dispositions in this place, mixing the ingredients in an unusual way. That is why the largest crabs in the world belong to Machangulo, big and fat like mammals.']

As the passage highlights, nature and animals are described in a vibrant and complex manner, being constantly highlighted throughout the visited locations and shown to possess as much agency as human characters. The main characteristics of this ecosystem are based on the blurring of boundaries between opposing poles and on the interrelation between sea and land, aquatic and terrestrial spaces. Trees grow in the sea in the same way they would on dry land; on dry land, agitated lagoons emerge, resembling the sea; rivers maintain their river-like colours even after flowing into the bay and losing their solid margins; the sea itself looks more like a river, albeit without banks. Even the crabs, despite being taxonomically classified in the phylum of arthropods and the class of crustaceans, are equated with mammals. Nature is depicted as an active and agentive

3 A small boat made out of palm trunk common in Mozambique's capital Maputo.

entity, remarkably mixing its elements in Machangulo. Considering, as Meg Samuelson (2017) does, the coast as a place for the creation of new epistemologies, simultaneously oriented towards land and sea, interior and exterior, Jessica Falconi rightly observes that the description of Machangulo in the story constitutes 'um exemplo marcante desta estética anfíbia própria da literatura litoral' ['a striking example of this amphibious aesthetic characteristic of coastal literatures'] (Falconi, 2019: 83), 'uma área híbrida entre mar e terra, já que na zona costeira persistem elementos típicos do interior e viceversa' ['a hybrid area between the sea and the land, since typical elements of the hinterlands persist in the coastal zone, and vice versa'] (84). The crab itself is significant in this respect, being an arthropod capable of living both on land and in water and which is abundant in the mangrove – an environment characterized by its 'promíscua convivência entre a secura da terra e o molhado do mar, com muito mais do segundo que da primeira' ['promiscuous coexistence between the dryness of the land and the wetness of the sea, with much more of the latter than of the former'] (Borges Coelho, 2005b: 55). Central elements in the story for constituting the main source of sustenance for Josefate Ngwetana, the large crabs are hunted daily by those belonging to the lower class of Machangulo, those who 'receberam mar como quem recebe o seu normal quinhão. Por isso partem todos os dias para o trabalho sentados num pequeno talo de palmeira, remando em cima dele, lavrando a onda com mais ou menos sucesso' ['received the sea as one receives their normal share. That is why they depart for work every day seated on a small palm frond stem, rowing on top of it, plowing through the waves with varying degrees of success'] (Borges Coelho, 2005b: 53). The allusion to the marginalized social status associated with those who depend on the sea and on the mangrove for survival underscores Mozambique's regional unevennesses, since those who own cattle and plows appear as belonging to the upper classes, having received their share of land at a time of *ancient dispositions*, 'em que os deuses distribuíram a terra pelos homens' ['when the gods distributed the earth among men'] (53). Apart from inscribing a non-romanticized view of those who live by and off the sea, the passage further complicates the idea of the more-than-human element of the gods being the one causing such economic disparities, for allowing its critical connection to the Portuguese colonizers who, in 'ancient times', were in fact the ones responsible for dividing the land among the people of Mozambique.

Because he despised his father's profession and hated crabs, Herculano sets off for South Africa in search of work, later returning on foot to Machangulo to bring his father a pair of patent leather shoes as a gift, seen by Herculano as an offering of respect and dignity to his father who had spent his life barefoot, mired in the mangrove. On this journey back home, Herculano completes twelve tasks on twelve Mozambican beachfronts, all of which displaying unique dispositions of nature. In this postcolonial revisiting of the classical myth of Hercules, 'civilized' humans are not faced with a 'barbaric' nonhuman world to be overcome. Instead, both living and non-living matters are shown to flow within and out of each other, completely intertwined and capable of agency. Crickets act as an engine 'para manter acesa a escuridão' ['to keep the darkness alive'] (59), 'fabricando os segredos da noite' ['manufacturing the secrets of the night'] (80); sound flees 'para dentro na noite como uma barata espavorida' ['deep into the night like a frightened cockroach'] (60). The wind carries away and brings back sounds 'ao seu insondável arbítrio' ['at its unfathomable whim'] (72); Lake Piti is 'um mar dentro da terra' ['a sea within the land'] (74); the road at Ponta Dobela 'nascia no mato e se acabava no mar; diferente portanto das outras estradas, que têm partida e destino' ['began in the bush and ended in the sea; different, therefore, from the other roads, which have a point of departure and a destination'] (75). The men of Majumbo struggled 'como insectos presos numa teia, peixes numa rede' ['like insects trapped in a web, fish in a net'], ravaging the place and 'o povo, esse indefeso cardume, com fúria desmedida, como faziam no mar os malditos tubarões' ['the people, this defenseless school of fish, with immeasurable fury, just as the damned sharks did in the sea'] (94).

Herculano passes through the twelve beaches after crossing the border from South Africa to Mozambique: 'Atravessou-a só desconfiando que a atravessava, sem disso ter a certeza, pois há fronteiras que apenas são importantes quando se fala nelas, não quando as atravessamos: havia praia antes e continuou a havê-la depois, sem transição' ['He crossed it only doubting that he was crossing it, without being sure of it, because there are borders that are only important when we talk about them, not when we cross them: there was a beach before, and there continued to be one afterwards, without transition'] (57). Throughout this journey, he faces monsters in between the state of man and beast and men that, having forgotten their human condition, learned to act like sharks by eating people; he is guided by a gazelle that teaches him about the importance of restraint in hunting and the respect for places; he wanders into a starving herd that makes him reflect on the 'dureza dos tempos actuais, capazes de esfarrapar até a tradição' ['hardness of the present times, capable of tearing tradition to shreds'] (84); he meets an enormous white elephant (one of the last representatives of this species due to the cruel hunting at the Maputo Natural Reserve) and gives up on attacking him. At the penultimate beachfront, Ponta Gomeni, he encounters desolation, with a description that evokes the environmental devastation caused to the area due to social conflicts and the unrestricted use of nature:

> a mesma extensão infinita de terra queimada pelo fogo e pela violência. Estreitaram-se os rios, as suas águas caram cada vez mais escuras, cada vez mais espessas, até deixarem de todo de correr. Veio depois a fétida lama, acabando também ela por secar. Transformaram-se em amplas e poeirentas estradas sem utilidade alguma. As árvores morreram de pé, primeiro o maldito vento norte tendo-lhes varrido as folhas, depois a terra avara secando-lhes nas veias os sucos vitais. Os capins perderam viço e seiva, zeram-se em pó. Até as rochas, antes secretamente alimentadas por subterrâneas humidades, tendo estas secado, ostentam agora superfícies baças e mortiças como peles doentes. (p. 99)

> [the same endless stretch of land burnt by fire and violence. Rivers narrowed, their waters becoming darker and thicker until they ceased to flow at all. Then came the foul-smelling mud, which also dried up. They turned into wide, dusty roads with no utility. Trees died standing, first the cursed north wind sweeping away their leaves, then the stingy earth drying their vital juices. The grass lost its vitality and sap, turning into dust. Even the rocks, once nourished by underground moisture, now displayed surfaces as dull and lifeless as sickly skin]

Having faced the three-headed dog of death and hell and narrowly escaping with life, Herculano is struck by a feverish dream that offers further insights into the racial conflicts initiated by the white colonizers: 'Engana-se quem diz que é negra a cor da morte. É branca, cor da espuma e do sal, da areia e da seca. Dos ossos. Do sol' ['Those who say that the color of death is black are mistaken. It is white, the color of foam and salt, of sand and drought. Of bones. Of the sun'] (100). He finally reaches the last beachfront, Ponta Abril, where he is able to recover and eat in an orchard filled with bananas, oranges, mangoes, guavas, papayas and *ocanhos*[4] (103), before crossing the gateway to his village to deliver his gift to his father.

Mixing the realms of the real and the imaginary, the story rewrites the Greek hero's story to investigate the relationships between humans and nature around them, the interdependence between human and nonhuman elements, and the necessity of preserving both animal lives (as in the episodes of the gazelle and the white elephant) and plant life (exemplified by the reference to the Maputo Natural Reserve and the issues brought about by severe drought). Its focus on nonhuman agency further highlights how human actions or elements are not the only protagonists of the environments of the African coast and the Indian Ocean, thus dismissing the hierarchy between humans and nonhumans that are typical to grand narratives of the Western civilization in order to rethink such totalizing discourses.

Borges Coelho (2019: 15–16) observes that facts such as the decline in the amount of places to be explored and that the demographic explosion at a global level led human beings to realize that the Earth is, after all, finite. This perception, characterizing a way of rescuing the former relation of respect, enchantment and attention to places, propelled current revisions about the ways in which the concept of place is understood, ultimately leading to the emergence of resistance movements against the growing and rampant neoliberal exploration that characterizes our modern capitalist world system (WReC, 2015, 2020). In this re-elaboration of concepts such as environment and nature, for instance, the relationship between humans and the environment – be it a natural or a built one – is reevaluated, taking into account the need for a radical shift regarding ethic, aesthetic and conceptual paradigms with which to interact with and understand the world. Cajetan Iheka's question for those in the field of African literary criticism thus becomes central: '[I]f colonial modernity elevated the human as the

4 *Sclerocarya birrea*, popularly known as marula, a fruit-bearing African tree.

avatar, the centre of the universe, often to the detriment of the nonhuman worlds, both seen and unseen, how has African literature reinstated the nonhuman in relation to the human?' (2018: 22). According to the author, although liberation movements and discourses in Africa have primarily emphasized the human dimension, a redefinition of postcolonial studies and African literary criticism that intends to seriously challenge neoliberalism and late capitalism must necessarily 'include a recognition of nonhumans as companions in a precarious world' (5). In this regard, the ecocritical perspective emerges as an important element in emphasizing the role of literature in imagining other ways of being in and inhabiting the world: not only due to the eminently political or combative potential of such imaginations but also because of its ability in unmaking the very historical and epistemological hierarchies that provided the basis for past and present ideologies of domination and exploitation. As demonstrated by Graham Huggan and Helen Tiffin in *Postcolonial ecocriticism: Literature, Animals, Environment* (2010: 22), human liberation will never be fully completed without a challenge to the ways in which human societies built themselves in hierarchical relations in face of other human and nonhuman communities, and without imagining new ways through which these groups, inevitably connected by ecological factors, might be creatively transformed.

The short story 'A força do mar de Agosto' ['The Force of the August Sea'] (Borges Coelho, 2005b: 123–139) explicitly addresses some of the social and environmental issues that accompany development initiatives in contexts located on the 'peripheries of the world-system' (WReC, 2020: 34). The narrative reveals traces of historical events in the Indian Ocean when the apparent solidification of the sea in Maputo Bay unveils a series of objects previously hidden in its depths, such as airplanes, skeletons, Austrian soldiers, fabrics, crockery, sailors, boats and ships, as well as the bodies of poor fishermen, among others. Acting as 'a reminder of colonial (and postcolonial) encounters in and through the sea' (Miranda, 2020: 112), the 'desertification' of the sea represents, on one hand, a violent threat to the lives of fishermen and inhabitants of Inhaca Island and the Mozambican coast, and on the other hand, an opportunity for 'development' and capital expansion from the perspective of an openly neoliberal government. Mozambican police, technicians and government officials; fishermen and their wives; South Africans and North Americans all come together to try to unravel the mysterious disappearance of the sea, relying on the use of technological resources such as cameras, satellites, GPS, compasses, mapping devices and nautical charts: tools primarily mobilized to serve the interests of those situated 'above' the Mozambican people. As Rui Miranda (2020) has brilliantly observed:

> The impossibility of thinking beyond the paradigm of the appropriation of virgin land, 'vasto espaço branco por preencher' [vast blank space to be filled in], lays bare the motivations of the 'Governo de todo nós' [Government of all of us] that closes off any possibility for other scientific efforts and endeavours by going all in on 'soluções mais tangíveis, as ditas nossas soluções' [more tangible solutions, the so-called our solutions]. The appeal to rely on 'as nossas próprias forças' [our own strengths] translates into a 'Plano Director' [master plan], drawn 'numa grande folha branca' [on a big white page], with a view to urbanizing as much of the area as possible. Through the Government's configuration of the sea as a blank space for neocolonial filling in, the crisis in the sea has patently become – as the neoliberal mantra holds – an opportunity, predictably, for the expansion of capital. The Government will repeat in the sea, now as farce, a tragic history of colonization of the land.

Within the context of a reflection about the environmental impacts on the Indian Ocean (understood as a literary trope but also as a very real and material space), João Paulo Borges Coelho's essay 'O Índico como lugar' (2019) provides fruitful questions with which to think about the role of literature in its relation to place, or, recalling a subtitle from the author's essay, to think about 'the places of literature' (16). Due to its being authored by a renowned historian and professor of Mozambique and Southern Africa, Borges Coelho's literary project is frequently read through perspectives that seek to frame it within a historical perspective; that is, that seek to understand it as a relatively faithful representation of the historical facts that occurred within Mozambique's political frontiers, given this is the country in which the author resides and writes. Although the importance of the dialogue between his literary work and Mozambique's history and contemporaneity is undeniable, it is equally relevant to observe the attention given by his literature to the geography of places, as the author himself has affirmed in an interview with Carmen Tindó Secco when observing that, besides history, he feels, 'com o meson peso a geografia, a respiração dos lugares' ['with the same intensity, the geography and the breathing of places'] (Secco, 2010).[5] Another topic addressed by the author in some of his non-fictional works (Borges Coelho, 2009, 2019; Zanfelice & Brugioni, 2020) relates to the inventive and imaginative character of literature as the distinctive features which provides a writing with its literariness, thus distinguishing it from other written forms such as documental or journalistic texts, for example. Hence, the singularity of the relationship between literature and places remains precisely in the invention of places operated by literature through language. This invention can be achieved through countless aesthetic strategies and narrative procedures, but, according to the author, the central objective always remains the same: namely, that of 'chegar, por via da imaginação, ao contacto com o *Genius*. Só desta maneira a literatura cria os seus lugares' ['reaching, through imagination, the *Genius*. Only then can literature create its places'] (Borges Coelho, 2019: 17).

And the Indian Ocean – this 'great sea' (17), as the author calls it – is one of the places frequently created by João Paulo Borges Coelho's literature. However, as several studies focusing on bodies of water have pointed out, characterizing such a wide space like an ocean as a *place* is not an easy task, bringing about several challenges and problems regarding the question of unity (Bentley, 1999; Pearson, 1998; Alpers, 2014). Places are usually imagined in smaller scales, distinctively delimited and minor areas such as a lake, a forest or a city, while broad and liquid spaces such as the Indian Ocean – occupying nearly 70 million kilometres with its waters whose coasts reveal a myriad of diverse cultures, languages and histories – complexify the task of reaching a unifying element. Investigating what would then be the *Genius* of the Indian Ocean, Borges Coelho addresses important historical aspects of this ocean such as the long-distance travels and the contacts between Arab, Persian and Hindu civilizations initiated more than five thousand years ago (Pearson, 2003: 3). In *Índicos Indícios*, connections between Mozambique and the Indian Ocean World are constant, with objects and characters whose familiar, religious, personal or professional (hi)stories allude to India, China, Australia, South Africa

5 All translations are my own unless otherwise noted.

and other countries on the Eastern African Coast. 'The enchanted cloth' (Borges Coelho, 2020: 11–35), for instance, the opening story of *Setentrião*, departs from Macaripe's neighbourhood in the Island of Mozambique and makes way for an extensive transnational travel towards Mecca, registering the historical memory of Islam in Northern Mozambique through the imaginary religious pilgrimage made by the tailor Jamal. The story registers aspects related to the (semi-)peripheral condition in the modern world system, articulating important issues stemming from combined unevenness in several instances of ethnic, religious and class conflicts. Addressing the tensions involved in the plurality of social, religious and political perspectives of Mozambique's Muslim community, Jamal's pilgrimage in 'The enchanted cloth' creates an interesting play between the ocean and the desert, where 'the Indian Ocean replaces the desert as a space of epiphany' (Can, 2020: 56).

But human actions or elements are not the only ones granting unity to the Indian Ocean. The monsoon seas, popularized by the literature of Australian sailor Alan Villiers (1940, 1952), were greatly responsible for providing adequate sailing conditions in and across its waters, 'criando um mecanismo altamente favorável às ligações entre o subcontinente indiano, a península arábica e a costa africana' ['creating a highly favourable mechanism to the connections between the Indian subcontinent, the Arab Peninsula and the African coast'] (Borges Coelho, 2019: 20) against which humans could hardly fight in terms of maritime traveling. In the works of João Paulo Borges Coelho analysed here, the attention to natural agents like winds, storms, etc. is quite common, underscoring the crucial role of atmospheric agents in the Indian Ocean region, the ocean of monsoons. Highlighting what he calls the Indian Ocean's deep structure, Michael Pearson similarly observes that it is not only people that travel and make connections (Pearson, 2003: 8). As the author reports, the southern bluefin tuna – fish that can weigh approximately 25 kilograms and live up to forty years – go through an immense journey among the waters south from Java Island, where they reproduce, swim around the Australian continent and South Africa, to then go back to Java to spawn.

This kind of complexity and magnitude of the Earth's ecosystem throughout all of Borges Coelho's literary project. The story 'Casas de ferro' ['Iron houses'] narrates the eviction of the poor inhabitants of the Grande Hotel da Beira and their attempt at transforming the remnants of stranded ships into new housings, giving rise to an innovative perspective on the different power relations established both among humans and among humans and nature. On the one hand, the Força Municipal [Municipal Force], representing the State's and the Businessmen's power and authority, displays a clear hierarchy when facing the inhabitants of the Hotel's ruins, who can do little or nothing against the ostensive police apparatus of helmets, face guards, shields and batons, resignedly obeying the threat of force. On the other hand, nature also reveals its agency,[6] creating obstacles to delay the Municipal Force's meeting with the inhabitants through its abundant sandpits, the overwhelmingly hot sun and the shallow waters that demanded great locomotion efforts on the part of the policemen, who, equipped with heavy boots and tight uniforms, sought to inflict a second eviction upon the people. After receiving the first eviction order to move out of the Hotel – significantly qualified as '[uma ordem] *seca*, espremida num parágrafo só, um ponto único' ['(a) *dry* (order), squeezed into a single paragraph, a single point'] (Borges Coelho, 2005a: 70) – the population is forced to move to the abandoned rusty boats sitting on the beach, subject to furious storms and tempests in their attempt to survive and reside. By the end of the story, when the storm has become less violent, the narrator affirms: 'Há já dias que o vento deambula resmungando, e que as águas, agora cinzentas, se recusam a abrir um caminho seco por onde a Força possa ali chegar' ['For several days now the wind wanders around muttering, and the waters, now turned gray, refuse to open up a dry path through which the Force might get there'] (72). Although the future of those living in the fragile boats is uncertain, the disposition of nature (Wenzel, 2019) as an ally to the People showcases a sensible alliance between those that, united

6 Understood in its conception elaborated by Cajetan Iheka (2018), who proposes the idea of 'distributed agency', that is, 'the idea that humans possess and share agency with the landscape and animals, among others' (4). In this view, agency is not something based exclusively on the intentionality of humans, but also in the actions and effects generated by nonhumans (3).

by their vulnerability to the State's and Businessmen's brutality, challenge the power of those who seek their destruction.

In the posthumous narrative of 'O hotel das duas portas' [The two-door hotel], the erosion caused by nature in the hotel of the Island of Santa Carolina – a grandiose one-hundred-room building established by colonial administration in the 1950s and left to fall into ruins still in the 1970s – sheds light on the racial and social changes and conflicts faced by Mozambican society since then. The story is narrated by Marta, a vain and erudite upper-middle class white prostitute that finds herself dead on the island's shore. Her kidneys, waist, face and nose are compressed by the beach's rocks in such a way that allows her to see even the rock's 'pequenas nervuras, o rendilhado que a maré foi burilando nela, os minúsculos grãos de areia transparentes que polvilham a sua superfície' ['small nervures, the tracery engraved by the tide on it, the minuscule transparent sand grains that powder its surface'] (Borges Coelho, 2005a: 73–116). In this tale of memory, Marta describes the events preceding her death, and which derives from a conflict between her client Teodoro, a secret agent of the Direção-Geral de Segurança da Polícia Internacional e de Defesa do Estado de Moçambique (PIDE), and Duda, a communist, anticolonial and pro-independence young man also killed by Teodoro. The narrative inscribes a powerful critical reflection about the colonial government and the authoritarian practices of repression and torture utilized in attempts to contain the imminent revolution, raising questions regarding the destruction and collapse of an old world as well as the construction of a new one.

Nature seems to accompany the transformations happening in Mozambique and inside Marta, being simultaneously immutable and transformative: 'O mesmo mar imutável no seu repetido movimento; a mesma areia pesada, imune ao tempo; o mesmo sol lento e monótono. E, todavia, tudo se transformando velozmente em redor da rapariga e dentro dela' ['The same immutable sea in its repetitive movement; the same heavy sand, immune to time; the same slow and monotonous sun. And yet everything rapidly changing around the girl and inside her'] (Borges Coelho, 2005a: 94). Besides (post)colonial conflicts, racial and class conflicts are also central, stemming from the contact between Marta and Carlos, the island of Santa Carolina-born black barman that had just come back from a period of experience in a bar in South Africa to work in the Island's hotel. The relationship between them is loaded with a level of mutual mistrust. When Carlos starts telling Marta 'os pormenores da sua vida e do lugar' ['the details of his life and of the place'] (99), Marta concludes that he was not telling her everything he could, noticing the presence of a violent gaze in his eyes. Trying to interpret this gaze, Marta asks herself: 'O que pensa um negro de uma branca?' ['What does a black man think of a white woman?'] (99). Through references to the behaviours of animals such as crows, crabs, ants and dogs, the narrative characterizes 'a fantastic fable' (108) rooted in reality, resorting to the use of metaphors between human and nonhuman worlds and ultimately demonstrating how both worlds are, in fact, inextricably linked. The actions and arrogance of the island's rich tourists have a devastating impact on the beach, the natural landscape and the local inhabitants, all of which having been historically obliged to open up space for the hotel as it grew. In order to allow for the luxury tourism on the island, seen as a natural paradise, many local inhabitants were displaced and transferred to other regions of the country's mainland such as Vilanculos, Inhassoro and Govuro, making human and nonhuman lives extremely difficult due to the deterioration of situations such as the shortage of fishery and the exponential growth of hunger and diseases. With the approximation of the armed conflicts that brought the country's independence in the 1970s, the hotel ceases all of its activities and gradually becomes a ruin, a moment in which the narrative registers an inversion in its actors' agency. If the island was once 'dócil e submissa à invasão do hotel e dos seus turistas' ['docile and submissive to the invasion of the hotel and its tourists'] (91), a kind of 'premonição topográfica' ['topographic premonition'] (93) signals the transformative fury of nature, whose elements – the sea, the sand, the rocks and the Sun – devour the narrator's corpse and the hotel's walls.

The wrath and fury expressed in the climatic manifestations of the Indian Ocean are characteristic aspects of this ocean, with naturally frequent phenomena such as cyclones, typhoons and earthquakes being exponentially aggravated by anthropogenic actions and climate change. On the one hand, encompassing over 100 hydrographic basins, Mozambique's territory constitutes a crucial one for the distribution of hydraulic resources for several populations (within and outside the country's borders) that live around the Maputo, Incomáti, Limpopo and Zambezi basins (Filipe, 2003: 13). On the other hand, being bathed by the Mozambique Channel through an extension of around 3,000 km, the Mozambican coast also represents a gateway to Indian Ocean cyclones which frequently attack it. Jossias Filipe (2003: 21) cites, for instance, the tropical cyclone Mayotte, which killed and destroyed many in the province of Nampula in February 1934; the 1950s January and February cyclones and torrential rains that wreaked havoc in the provinces of Inhambane, Gaza and Maputo, as well as the districts of Chókwe, Xai-Xai, Manhica and Magude; the 1984 Domoína Cyclone striking Maputo; the 1988 Filão Cyclone in Beira and Quelimane; and the Nadia Cyclone ravaging Nampula and Cabo Delgado; all of these had disastrous consequences for coastal populations and environments.

As Borges Coelho reminds us, oceans such as the Atlantic and the Pacific are filled with devices and researches that allow for a deepened study of their conditions and characteristics. In the Pacific Ocean, for instance, since 1994 there has been a complex system of sensors that allow for the collection of data in order to prevent damage caused by maritime storms and hurricanes before these reach its margins (Borges Coelho, 2019: 23). The case is, unfortunately, not the same for the Indian Ocean, whose hurricanes that annually strike the Eastern African coast are discovered only too late. The scarcity of information regarding the Indian Ocean is in fact a frequently observed gap across several disciplines of knowledge – such as in International Relations (Gupta, 2010), History (Pearson, 2003) and Literature (Brugioni, 2019) –, a gap which is nonetheless being gradually fulfilled with the recent acknowledgement of the Indian Ocean as a 'strategic arena for the 21st century' (Kaplan, 2009; Hofmeyr, Kaarsholm & Frederiksen, 2011). Trade between Dubai and Africa, the ocean's strategic location within the petrol route, the 'String of Pearls' strategy financed by China, as well as the growing influence of this country in the region are some of the events pointing towards movements of expansion of capital and power disputes currently occurring in the Indian Ocean. If, on the one hand, such events might attract attention and resources for a necessary investment in the ocean's knowledge in domains such as Marine Biology and the Humanities, for instance, on the other hand they also deepen dangerous patterns of economic unevennesses, environmental destruction and ocean pollution, exacerbating climatic and social issues that threaten the lives of the ocean's human and nonhuman agents.

For Borges Coelho, the singularity or the *Genius* of the Indian Ocean can be found precisely in its physical and material characteristics. These, in turn, call for our imagination, both in its 'natural' or 'unaltered' aspects and in its human-fostered configurations: 'na periferia feita de praias amarelas e objectos aguçados (os coqueiros, os minaretes das mesquitas, os velhos padrões portugueses e as pequenas velas triangulares), nos seus ventos constantes e tão característicos, nos seus amuos destruidores' ['in the periphery made of yellow beaches and sharp objects (the palm trees, the mosque's minarets, the old Portuguese patterns and the small triangular candles), in its constant and characteristic winds, in its destructive sulks'] (Borges Coelho, 2019: 24). The two literary works by João Paulo Borges Coelho discussed here similarly invite us to imagine which are the places of the Indian Ocean in the context of Mozambican literature. His writing inscribes, with sophistication, the importance of the sea for the characters and the events situated in its margins, registering aspects encompassed by several maritime cultures (such as eating habits, dress codes or family itineraries), as well as the specificity of water itself as an element of nature. As a parcel of the author's critical reception has already been observed regarding the body of his work, *Índicos Indícios* challenges simple binaries among categories such as human and nature, nature and technology, past and present, history and literature, and tradition and modernity, undoing the oppositions through which we usually understand such terms (Medeiros, 2022: 262). Engaging with current climatic questions through the inscription of the Indian Ocean in several narrative levels, *Índicos Indícios* suggests new ways of comprehending the relation between human beings and the environment.

Bibliography

Achebe, C. (1958). *Things Fall Apart*. London: Heinemann Educational Books.

Agamben, G. (2006). *Profanações*. Lisbon: Cotovia.

Alpers, E. A. (2014). *The Indian Ocean in World History*. Oxford: Oxford University Press.

Bates, C. and Carter, M. (2021). 'Kala Pani Revisited: Indian Labour Migrants and the Sea Crossing'. *Journal of Indentureship and Its Legacies*, 1 (1), 36–62.

Bentley, J. H. (1999). 'Sea and Ocean Basins as Frameworks of Historical Analysis'. *Geographical Review*, 89 (2), 215–224.

Borges Coelho, J. P. (2005a). *Índicos Indícios I: Setentrião*. Lisbon: Editorial Caminho.

——. (2005b). *Índicos Indícios II: Meridião*. Lisbon: Editorial Caminho.

——. (2009). 'E depois de Caliban? A história e os caminhos da literatura no Moçambique contemporâneo'. In C. Galves, H. Garmes and F. R. Ribeiro (eds), *África-Brasil: caminhos da língua portuguesa*, pp. 57–68. Campinas: Editora da Unicamp.

——. (2016). *Água. Uma novela rural*. Lisbon: Editorial Caminho.

——. (2017). *Ponta Gea* [*ebook*]. Lisbon: Editorial Caminho.

——. (2019). 'O Índico como lugar'. In A. M. Leite, E. Brugioni and J. Falconi (eds), *Estudos sobre o oceano Índico: Antologia de textos teóricos*, pp. 13–28. Lisbon: Edições Colibri.

Bowen, N. (2002). 'Os chupa-sangue na província da Zambézia'. *Estudos Moçambicanos*, 19, 89–112.

Brugioni, E. (2015). 'Por detrás de tantos nomes, o mar. Reflexões em torno da relação entre Moçambique e o Oceano Índico: narrativas, imaginários e representações'. *Via Atlântica*, n. 27, 93–110.

——. (2017). 'Territórios sobrepostos, histórias entrelaçadas. Itinerários comparativos e literaturas africanas: o Oceano Índico como paradigma crítico e estético transnacional'. In Torres Pereira, Marcos Paulo, et al. (eds), *Pós-colonialismo e Literatura: Questões Identitárias nos Países Africanos de Língua Oficial Portuguesa*, pp. 17–35. Macapá: UNIFAP.

——. (2019). *Literaturas africanas comparadas: paradigmas críticos e representações em contraponto*. Campinas: Editora da Unicamp.

——. (2020). 'Of Clues, Traces and Minor Histories: Theorizing the Genre of Historical Novel in Rainhas da Noite'. In Elena Brugioni, Orlando Grossegesse and Paulo Medeiros (eds), *A companion to João Paulo Borges Coelho*, pp. 179–202. Oxford, Bern, Berlin, Bruxelles, New York, Wien: Peter Lang.

Can, N. A. (2020). 'The Translation of the East in Mozambican Literature'. *Anuari de Filologia. Literatures Contemporànies*, 10, 53–68.
Carson, R. (2002). *Silent Spring* (Fortieth Anniversary Edition). New York: Mariner Books.
Davis, M. (1998). *Ecology of Fear*. New York: Metropolitan.
Deloughrey, E. M. and Handley, G. B. (2011). *Postcolonial Ecologies: Literatures of the Environment*. Oxford: Oxford University Press.
Falconi, J. (2013). ' "Para fazer um mar". Literatura moçambicana e Oceano Índico'. *Diacrítica*, 3 (27), 77–92.
Fendler, U. (2018). 'O "Black Ocean" azul'. *Remate de Males*, 38 (1), 42–53.
Filipe, J. (2003). *Moçambique 2000: As águas da morte*. Maputo: Moçambique Editora.
Guha, R. and Martinez-Alier, J. (1997). *Varieties of Environmentalism: Essays North and South*. London: Earthscan.
Gupta, M. (2010). *Indian Ocean Region: Maritime Regimes for Regional Cooperation*. Nova York: Springer.
Harvey, D. (2000). *Spaces of Hope*. Berkeley: University of California Press.
Hofmeyr, I., Kaarsholm, P. and Frederiksen, B. F. (2011). 'Introduction: Print Cultures, Nationalisms and Publics of the Indian Ocean'. *Africa*, 81 (1), 1–22.
Huggan, G. and Tiffin, H. (2015). *Postcolonial Ecocriticism: Literature, Animals, Environment*. London: Routledge.
Iheka, C. N. (2017). *Naturalizing Africa: Ecological Violence, Agency and Postcolonial Resistance in African Literature*. New York: Cambridge University Press.
Kaplan, R. D. (2009). 'Centre Stage for the Twenty-first Century: Power Plays in the Indian Ocean'. *Foreign Affairs*, 88 (2), 16–32.
Marques, L. (2018). *Capitalismo e colapso ambiental*. Campinas: Editora Unicamp.
Medeiros, P. (2022). 'O afogamento do tempo: Catástrofe ecológica, dialética e realismo alegórico em Ponta Gea'. Trans. Gabriela Beduschi Zanfelice. In E. Brugioni, F. Gallo and G. B. Zanfelice (eds), *A obra literária de João Paulo Borges Coelho. Panorama Crítico*, pp. 261–287. Campinas: Editora da Unicamp.
Miranda, R. (2020). 'History, Literature and the Indices of the Ocean: Force of Signification in Borges Coelho's "A força do mar de Agosto" '. In E. Brugioni, O. Grossegesse and P. Medeiros (eds), *A Companion to João Paulo Borges Coelho*, pp. 107–130. Bern: Peter Lang UK.
Moore, J. W. (ed.). (2016). *Anthropocene or Capitalocene? Nature, History, and the Crisis of Capitalism*. Oakland: PM Press.
Mukherjee, U. P. (2010). *Postcolonial Environments: Nature, Culture and the Contemporary Indian Novel in English*. London: Palgrave Macmillan.
Naess, A. (1989). *Ecology, Community and Lifestyle*. Trans. David Rothenberg. Cambridge: Cambridge University Press.

Nixon, R. (2011). *Slow Violence and the Environmentalism of the Poor*. Cambridge, MA: Harvard University Press.

Pearson, M. N. (1998). *Port Cities and Intruders: The Swahili Coast, India, and Portugal in the Early Modern Era*. Baltimore: Johns Hopkins University Press.

——. (2003). *The Indian Ocean*. London: Routledge.

Secco, C. L. T. (2010). 'Entrevista a João Paulo Borges Coelho'. *Buala*, <www.buala.org/pt/cara-a-cara/entrevista-a-joao-paulo-borges-coelho>. Accessed: 15 September 2022.

Smith, N. (1984). *Uneven Development: Nature, Capital and the Production of Space*. New York: Basil Blackwell.

Villiers, A. (1940). *Sons of Sinbad*. New York: Charles Scribner's Sons.

——. (1952). *Monsoon Seas: The Story of the Indian Ocean*. New York: McGraw-Hill.

Wallerstein, I. (2011). *The Modern World-System*. California: University of California Press.

Wenzel, J. (2019). *The Disposition of Nature: Environmental Crisis and World Literature*. New York: Fordham University Press.

Williams, R. (1973). *The Country and the City*. New York: Oxford University Press.

——. (2005 [1980]). *Culture and Materialism*. London: Verso.

WREC. (2015). *Combined and Uneven Development: Towards a New Theory of World-Literature*. Liverpool: Liverpool University Press.

——. (2020). *Desenvolvimento combinado e desigual: Por uma nova teoria da literatura-mundial*. Trans. Gabriela Beduschi Zanfelice. Campinas: Editora Unicamp.

Wu, C. C. (2016). 'Towards an Ecocriticism in Africa: Literary Aesthetics in African Environmental Literature'. In Fiona F. Moolla (ed.), *Natures of Africa: Ecocriticism and Animal Studies in Contemporary Cultural Forms*, pp. 141–165. Johannesburg: Wits University Press.

Zanfelice, G. B. and Brugioni, E. (2020). 'O Índico, uma promessa de horizonte – Entrevista com João Paulo Borges Coelho'. *Via Atlântica*, 38, 437–469.

JESSICA FALCONI AND GIULIA SPINUZZA

8 Fishing for stories: Audiovisual narratives of the Indian Ocean: Goa, Mozambique and East Timor

In this chapter, we shall deal with a set of audiovisual narratives centred on territories of the Indian Ocean, namely Goa, Mozambique and East Timor, with the aim of exploring some of the elements that unite and divide this area. In this itinerary, we will be guided by Michael Pearson's reflection on littoral societies, allowing us to undertake a reading in counterpoint of these narratives.

The narratives selected for the analysis have in common the topic of fishing, an activity that develops its own characteristics and has important social and cultural meanings in each of the territories under scrutiny. This topic allows us to explore historical, social, environmental dynamics, etc., and themes related to preservation and safeguarding, while their audiovisual representation provides the opportunity to address the coasts, islands and oceans – in this case the Indian Ocean – as spaces for imaginaries and aesthetic creation (Samuelson, 2017; Verne & Verne, 2017). In fact, there is a long tradition of documentaries about fishing that goes back to the classics of documentary cinema, from films like *Drifters* (1929) by John Grierson and *Man of Aran* (1934) by Flaherty up to the creation of the Festival des Pêcheurs du Monde, which has been held in Lorient[1] since 2008 and which screened Sonia Filinto's documentary, which we will discuss below.

In the first part of the chapter, Jessica Falconi traces the theoretical framework deriving from the concept of littoral society, and tackles the documentaries *Shifting Sand* by Sonia Filinto and *Why are they here? Chinese stories in Africa*, by Yara Costa, while in the second part Giulia

1 The Festival reached its 11th edition in 2019. For more information, see <https://www.pecheursdumonde.org/>

Spinuzza focuses on the documentary *Wawata Topu – Sereias de Timor-Leste* by David Palazón and Enrique Alonso and presents final considerations about the topics discussed.

Constructing the Indian Ocean: The littoral societies

The town of Calangute in Goa, India; the Island of Mozambique, on the east coast of Africa, Mozambique; and the village of Adara, on Atauro Island, East Timor, are the locations under consideration, respectively, in the audiovisual narratives considered in this chapter. These places belong to territories formerly colonized by Portugal and part of the circuit of exchanges and circulations of the Indian Ocean, the seats of ancient littoral societies that underwent significant changes down the ages. It is precisely the concept of littoral societies, formulated by Michael Pearson, that allows us to establish a dialogue between the visual representations of fishing in these places, so as to reflect characteristics, imaginaries and narratives of these shores of the Indian Ocean so far removed from one another.

Because of the strength of their connection, this concept has become one of the structuring elements of the Indian Ocean as an analytical unit, on which Pearson worked for three decades and which today represents an unsurpassable legacy for studies in this area. As Edward Alpers (2018) illustrates, Pearson addressed this concept in various publications and conference papers, among which we are selecting for this chapter three seminal articles, namely, 'Littoral Society: The Case for the Coast' (Pearson, 1985) in which the first formulation of the concept appears; 'Littoral Societies: The Concept and the Problem', published in 2006, where the main characteristics of littoral societies are systematized; and the more recent 'Places in the Indian Ocean' (2017), where the authors ask about the present and future of littoral societies.[2]

2 The 2017 article was translated into Portuguese and included in Leite, Brugioni and Falconi (2019).

In the first article on the subject, Pearson proposed the concept of the littoral society as a tool for going beyond the limitations of three main fields of study, all interrelated: (1) Maritime historiography, which in Pearson's view, appears too centred on the nautical aspect of human activities in the oceans, dealing mainly with types of ships, shipwrecks or naval battles; (2) Studies of the port cities which, in his opinion, were more concerned with the dynamics proper to the interior than those related to life in the sea; (3) Historiography of the colonial empires, which focuses mostly on the impact of the Europeans on the oceans, as a historical rupture. In all of these fields, according to Pearson, there lacked a true 'whiff of ozone', referring, with this expression, to a more in-depth and somehow tangible treatment of human experiences of life at sea. Thus, the concept of littoral society emerged as a tool to surpass the limitations pointed out, and then, introduce this 'whiff' into the historiography of the Indian Ocean, strengthening it as a historiographical paradigm and specific field of study (Alpers, 2018: 15). For Pearson, dealing with the littoral means focusing on 'the continuum between land and sea activities, with the strip of littoral in the middle, acting as a hinge or mediator'(Pearson, 1985: 4). Seeking to narrow down the definition of this object of study, Pearson points out what would be the limit of the littoral society, that is, a frontier that cannot be fixed like a point on the map, but rather an ideal demarcation line beyond which human activities remain outside the sphere of influence of those activities related to the sea:

> It is always a matter of interaction between the affairs of land and sea. This interaction will never cease at sea, for no one lives totally on the water and completely uninfluenced by the land. The land frontier is porous, elastic: all we can say is that when a land activity is in no way influenced by the sea, then we are not interested. This then is the land frontier, the end of the littoral. (Pearson, 1985: 6)

The littoral is thus characterized by being a permeable porous, flexible and transitional area, constituting a continuum between land and sea, whose relationship is founded on the complementarity between coast and interior. In the final part of the article, Pearson points to some characteristics that define the littoral societies, that he would take further in successive reflections: the pluralistic nature of these societies, due to the presence of traders and foreigners and other social and ethnic groups, etc.; the greater consumption of fish in relation to the interior; the peculiar impact of the monsoon on the littoral zones; other cultural and religious differences (Pearson, 1985: 7). These aspects are taken up again and reworked in the 2006 article, in which the central hypothesis is that despite the different land influences, the societies of the littoral zones have more in common with other littoral societies than with their own neighbours in the interior. The connective and comparative potential of this concept emerges more clearly in this hypothesis, able to build bridges between different and distant communities, on the basis of observation of their interactions with the sea.

As for the characteristics common to these societies, Pearson explores three main categories: location, occupation and culture, recognizing that if the first is generally evident, the last is more problematic. Regarding the location, Pearson defines what he understands by littoral: 'the coastal sea zone, the beach, and some indeterminate frontier on land', clarifying, however, that 'The place I am thinking of is not only the beach, for this is a very narrow zone and has no permanent people' (2006: 354). Pearson also singles out the islands as privileged places for observing the presence of littoral societies; some islands are 'purely littoral' (358). It is littoral peoples who inhabit the coastal zone, whose lives are intimately linked to the sea and influenced by it, but they also maintain links with the land. The littoral communities live on the coast between land and sea. The fisherfolk's families are perfect examples of the interconnection between land and sea: in the majority of cases, both the demands of the interior and of distant places overseas condition their lives and activities. For Pearson, recurring models and elements in common emerge in the littoral communities that relate to the material cultural, dietary habits, religion, gender relations, and a more general condition of risk and precariousness linked to the nature of the activities at sea in comparison with the ones on land. More recently, different littoral societies have been facing the impact of international tourism, technological innovations, changes in the practices of global consumption and environmental pollution. In the historian's opinion, the littoral societies continue to exist but their characteristics are affected by radical transformations of the symbiotic relationship between land and sea typical of the coastal bands.

In 'Studying the Indian Ocean World. Problems and Opportunities' (2007), Pearson flags up an aspect that will lead him to a different view of the actual condition of the littoral societies; this is the so-called 'territorialization' of seas and oceans that have been transformed into graves for the extraction of resources, requiring as a consequence the regulation and delineation of frontiers by States and international platforms. For Pearson this is a phenomenon that began two centuries ago, through which the land ecotone had imposed itself on the maritime, having a profound impact on the current configuration of the littoral societies. In fact, in the article 'Places in the Indian Ocean', through the concept of the ecotone, Pearson reconsiders the existence of littoral societies, stating that we are witnessing the end of the ecotone of the coast. Before this end, the littoral societies lived in an essentially amphibious dimension; the fisherfolk simultaneously practised fishing and agricultural activities, and the sailors lived for long periods at sea maintaining strong connections to the land perpetuating this amphibious culture through family transmission. Pearson does not go so far as to openly declare the disappearance of littoral societies but he does invite us to reflect on this process of territorialization of the littoral in order to illustrate the spatial and temporal dynamics that impact on the Indian Ocean also as a unit of analysis, in a problematizing attitude that has never ceased to test the scope and limits of this same category.

Thus, Pearson's concept of littoral society provides a solid framework for us to analyse audiovisual narratives about fishing in Goa, Timor and Mozambique, with the aim of evaluating elements in common and differences and reflecting on the existence and the current configurations of the littorals, the people who inhabit them and, above all, of the representations and narratives that depict them.

First shore: Calangute, Goa

Shifting Sands (29', 2013) is Sonia Filinto's first short film,[3] and, as the opening credit indicates, it focuses on the little town of Calangute, located on the tourist coast of Goa,[4] and the fishing community that lives there. The topic of fishing is broached exclusively in the oral testimony of the members of this community who represent the only source of information accessible to the public. If, on the one hand, this option reveals the intention to explore the personal stories and perceptions of the subjects involved, on the other hand, they are the ones who inform about the fishing situation based on their own lived experience, while the camera captures 'movements' and 'directions' of a world undergoing a profound transformation.

Tourism, emigration, the industrialization of fishing and intergenerational family relationships stand out in the testimony as the main factors that impact on the actual configuration of this community. In particular, the modernization of small-scale artisanal fishing, as well as the introduction of the great trawlers are phenomena that crossed the Indian Ocean from shore to shore, affecting the littorals and their communities. As Robert Newman emphasizes, after the integration of Goa into the Indian Union, both the government and the international aid programmes incentivised the mechanization of fishing and the importation of huge trawlers, which linked this sector more and more to the interests of the global market (Newman, 1984).

3 The documentary was shown in Dubai; in the Festival International de Films Pêcheurs du Monde 2015 in Lorient; in various places in Goa, and more recently in the Fundação Oriente in Lisbon.

4 Bathed by the Arabian sea, with a population of approximately 14,000 inhabitants, Calangute is called 'Queen of the Beaches', constituting an established domestic and international tourist destination. Since the end of the 1960s, the tourism market in Goa has been facing great changes and diversification, which has caused impacts and reactions in the local populations in general, and the coastal communities in particular (Sreekumar et al., 1995: 4).

The testimony of the Calangute fisherfolk focuses on these aspects, referring to the conflict with the agents of the great trawlers, as well as their impact on the crisis in artisanal fishing, and the consequent choice to dedicate themselves to other activities, linked to trade and tourism. The people interviewed also mention that the ambition of the young people and the current opportunities for education and access to other jobs have led to the abandonment of fishing and/or emigration to other places, which has caused increasingly greater importation of labour from other Indian states. These circumstances have significantly altered the structure of the community which has in fact been opening up more and more to the interests and dynamics of the global economy (Rubinoff, 1999).

As for environmental questions, the fisherfolk also mention the phenomena of coastal erosion and loss of beach, which deprives them of the possibility of drying their nets, affecting the activity carried out on a small scale. To this must be added the progressive emptying of the coastal strip caused by the trawlers and the frequent destruction of the nets of the local fishermen, who see their activities severely impoverished (Pearson, 2006).

These topics are addressed in the documentary according to a structuring in thematic micro-blocks of varying duration, indicated by inter-titles in the frame, which do not coincide with closed narrative units, privileging a linear segmentation of the testimonies, some of which take up more than one block. The inter-titles of each micro-block allude to the topic to be addressed but does so in a connotative way, sometimes poetic, sometimes ironic, making use of the lexical and syntagmatic polysemy and suggesting other meanings. Besides the more denotative inter-titles like The *NRI Gulf's Story*, which introduces the topic of the emigration of Goans to the Persian Gulf, in other words, a contemporaneous transit in the Indian Ocean, one example of poetic usage is that of the first micro-block, *Labour of love*, which introduces a set of takes of a boat preparing to go out to sea. An example of ironic use is the inter-title *The Big Fish*, which introduces the fisherman talking about the conflicts with the great trawlers. This use of inter-titles helps define the voice and tone of the documentary, marked by expressiveness and affectivity, therefore suggesting the use of the poetic mode of documentary representation (Nichols, 2001).

The categories used by Pearson to define the littoral societies in the 2006 article – location, activity, culture, are shown to be useful tools for the approach of the documentary since they make it possible to set up a dialogue between the topic explored and the salient aspects of the staging, as well as their effects on the representation. Thus, with regard to location and activity, it should be highlighted in the first instance that despite tourism being a central aspect of Calangute's identity, mentioned right away in the opening title, it is only in the last narrative block (*A retired fisherman*) that tourism is focused on in a set of shots that capture the beach full of sunshades. The position of the semi-subjective camera, the use of closeups and the raccord-gaze suggest that the point of view on the beach is that of Marceline Fernandes (Figures 8.1, 8.2 and 8.3),[5] a key figure in the narrative of the short film, to which we will return below.

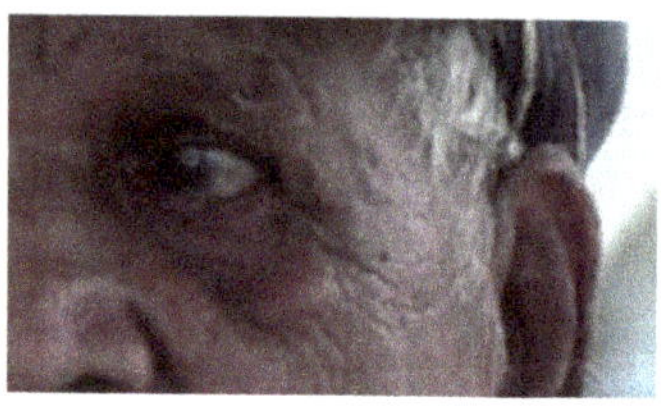

Figures 8.1, 8.2 and 8.3 Marceline Fernandes looking at the beach

5 All images from *Shifting Sand* are taken from <https://ruralindiaonline.org/articles/shifting-sands-and-changing-tides/>. ©Sónia Filinto.

Therefore, the coast and the beach we see throughout the film are staged into a filmic space dominated mainly by wide shots in which fishermen and boats appear, suggesting a full identification between the community and the place. Taking up Pearson's observation, according to which there would not be a permanent population, the documentary seems to argue the opposite, excluding from the field of vision the floating population of tourists and somehow 'giving back' the beach to the fishermen.

If, on the one hand, the testimonies seek to individualize the members of the community, focusing on memories, trajectories and personal opinions about fishing, on the other hand, the wide shots on the beach depict the fishing as a collective and choral activity that transforms the landscape into a natural/cultural montage, that is, a composition of nature, objects, men and fish. An example of this choral dimension of fishing and the resulting visual montages are given by the shots that portray the placing and removal of the enormous nets used in traditional fishing methods. This is a montage of human and nonhuman elements between which relations of affectivity and correspondence are established throughout the documentary. Besides the testimonies, the individualization and dimension of the affectivity are created by aesthetic choices that evoke other sensorialities, in addition to vision, mainly the tactile corporeal sense. Laura Mark defines this sensorial quality of the images as 'haptic visuality', flagging up some strategies that produce this effect, including the focus on the surface of the objects, the emphasis on texture more than on form (Marks, 2000). It is in this perspective that we can relate the closeups of Marceline Fernandes (Figure 8.4) and his wife, to the extra big closeup of an old boat or trunk (Figure 8.5). In these shots, both the faces and the material and the objects transform the screen into a surface on which are inscribed touches, memories and affects.

Figure 8.4. Marceline Fernandes

Figure 8.5. a detail of a boat

In the long shots and wide shots, the frames are made from different camera angles and heights, and the composition lines of the picture privilege the shore as the point of contact between sea and land, between water and earth, articulated also with the editing technique of shots and reverse shots which in fact set up a constant dialogue between these two elements and then, between exterior and interior. In fact, the spatial homogeneity of the documentary, dominated by the beach and the sea, is fragmented in the editing through the insertion of shots that focus on landscapes of the interior, such as fields, streets, markets, the theatre and the church.

In the wide shots, the montage of human/nonhuman results in 'amphibious' images that mix the natural, the biological, the material, configuring the coast as a zone of hybrid construction (Figures 8.6 and 8.7). This and other shots, as well as the editing choices and the search for multiple camera angles and positions, seem to profile what Meg Samuelson defines as the amphibious aesthetic: 'It is characterized by the perspectival ambivalence and optical bifocality that arises from littoral proprioceptivity in which sensory reception is oriented simultaneously towards land and sea, interior and exterior, here and there' (Samuelson, 2012).

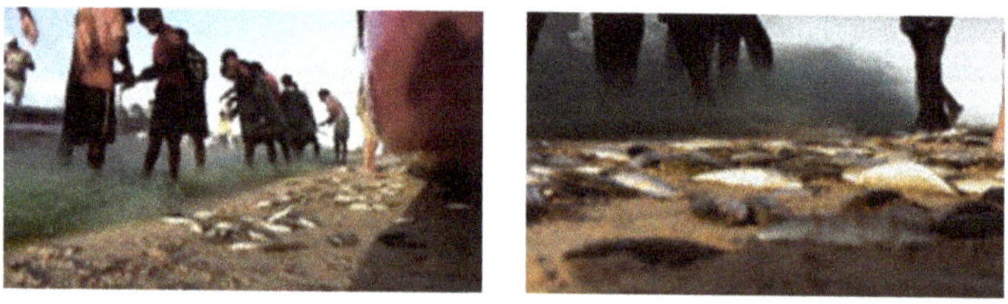

Figures 8.6 and 8.7 Fishing activities on the beach

As material objects, the boats and nets play a central role in the audiovisual representation of the Calangute fishing community; in various scenes and shots the narrative centres on the entry and departure of the boats in the sea, as well as the placing and removal of the nets according to a temporal organization that is neither linear not univocal, suggesting an almost circular temporality, that is, continuous repetition of the fishing activity. However, this representation does not seem to convey a romanticized or mythical narrative of the fishing but points, on the contrary, to the profound changes that affect the rhythm of this activity and the world related to it. In fact, as told by Josephina Cardoz, member of a family of fisherfolk, if before they fished just when there was no monsoon, soon after the Feast of the Virgin Mary, either to avoid atmospheric risks, or so that the fish could grow adequately, in the present day they fish at all times, as long as the weather permits, to meet the demands of the market and the needs of subsistence.

Josephina Cardoz, Luizinha Silveira and Ursula Luizita Fernandes bear witness to the role of women in the economic, social and cultural system protagonized by the fishing community. We look once more to Pearson and his characterization of the fisher families in the littoral societies of the Indian Ocean. According to the historian, in the majority of these societies, the women cure, dry and salt the fish, functioning also as the link between the littoral and the 'frontier' in the interior, in so far as it is they, generally speaking, who move around in order to sell the fish in local markets (2006), as is the case of Luizinha Silveira, briefly depicted in her movements between the church and the market, in the narrative block entitled *Faith's follower*. As we shall see below, the case of East Timor presents significant differences concerning gender roles in the realm of fishing.

The documentary also tackles the identity and socio-cultural aspects of this community, alluding to its identification with the Kharvi community. This aspect is introduced in the third narrative block, entitled *A matter of identity* whose initial shots depict an artistic-musical performance about fishing. In particular, in a performance scene, the dialogue between a man and a fishwife summons up the social prejudice associated with the Kharvi community. This is a community traditionally placed in an inferior position in the caste system and marginalized in Goan society because it is associated with poverty, manual labour and a lack of political influence (Rubinoff, 1991). In the dialogue, the fishwife reacts to a scornful comment by the man: 'Don't say kharvi, say fishing community!' The raccord of the editing connects this statement to the laughter of Josephina Cardoz, who reflects on this aspect from her kitchen, alleging that the Kharvi identity is not a question of names, but of origins. Domício Silveira's testimony points to the prejudice and social discrimination suffered by the community, indicating, however, that the situation has been changing as fishermen have been conforming to the dominant lifestyles because they have more economic resources.

In addition to the testimonies that refer to the problems faced by the community, the focus of the camera on the material and technological differences of the boats points to the changes and the differentiation underway in the sector, while the movement of boats and bodies before the camera refers to a world in transformation whose directions and destinies are multiple and uncertain. On the other hand, it should be emphasized that in the first narrative block in which Marceline Fernandes appears, introduced by the inter-title *It was a different time*, the wide shots filmed on the beach function as representation of the memory of the elderly fisherman who relates – in voice on and off screen – his beginnings in the fishing business still under Portuguese colonial domination. The use of images that document the present to make visible the narrative of the past points as much to the strategies of fictionalization used in the documentary as to the intention to capture continuities and ruptures, between past and present, in the world of fishing and in the heart of the community.

Marceline Fernandes constitutes a link of continuity and unification, albeit fragile due to his age, between fragments of time and space in the construction of the audiovisual narrative. The account of his memories bestows meaning on the image of the present and vice versa, and his gaze on the beach at the end of the documentary points to a time to which he no longer belongs. With no possibility of obtaining a pension because he no longer had the fisherman's licence obtained during colonial times, Marceline Fernandes died in 2012, according to the information in the final caption. In conclusion, the themes addressed by the testimonies, as well as the narrative structure indicated by the inter-titles, point to a general context of change, marked by opportunities but also by the precariousness and risk that fishing entails, the sea being considered the source of life and death. The documentary gives us back a fragment of the Indian Ocean and a coastal community whose future is uncertain, impacted by the routes and maps of the global economy. On the other hand, the aesthetic choices and narratives provide a representation of the Calangute beach as an affective and poetic space, where men, boats, nets, fish, waves and sand participate together in the ritualization of daily life between land and sea.

Second shore: Island of Mozambique

Why are they here? Chinese Stories in Africa by the Mozambican film-maker Yara Costa was launched in 2011 and deals with the journeys of three Chinese emigrants, respectively, in the Island of Mozambique, in Semonkong, Lesotho and Accra. In this paragraph we will briefly address the episode pertaining to the Island of Mozambique, centred on problems related to fishing, so as to reflect on another declination of this activity in an insular community in the Indian Ocean.

A UNESCO heritage site, the little Island of Mozambique is a place where controversial transnational cultural memories converge, a symbol of the Portuguese presence in Africa, but also a crossroads of the commercial, religious and cultural circuits of the Indian Ocean. For this and other reasons, the representation of the Island has given rise to an extensive corpus of literary texts, photographs, news articles and documentaries, a good part of which are centred on the division between colonial city and African city; on the architectural heritage left there by the Portuguese; and on the cultural and religious traditions of the population. Among the audiovisual narratives centred on the Island, Licínio Azevedo's documentary *A Ilha dos Espíritos* (2010) stands out in particular. The main theme is popular religion of an animist bent where the sea is depicted as a material and symbolic archive of stories, beliefs, transits and cultural contacts (Leite & Falconi, 2018). It should also be pointed out that Yara Costa's second documentary, entitled *Entre eu e Deus* (2018) is set on the Island of Mozambique, focusing on the conversion of a young Mozambican woman to Islam. Thus, both films by this film-maker work on the less popular aspects and stories of the Island, contributing to deconstruct mystified images of this pivotal place in the African Indian Ocean.

In *Why are they here? Chinese Stories in Africa*, the episode dedicated to the Island of Mozambique develops the general argument of the documentary, whose aim is to investigate the presence of Chinese people on the African continent following the migratory waves that occurred after the first conference of the Forum on China-Africa Cooperation in 2000. The episode, segmented into four sequences, tackles the figure of Min Nan Wang, a young Chinese man aged 23 who arrived in the continent in 2006 to work in the exportation of sea cucumbers to China – the country where the sea cucumber is considered a gourmet foodstuff. As already mentioned, the interaction with far-away markets is one of the central characteristics of the littoral societies described by Pearson, currently constituting a pressure factor that has had a significant impact on different communities of the Indian Ocean, especially as regards the consumption and exportation of foods and their consequences for local sustainability (Pearson, 2003; 2006). Although the captions of the documentary frame the China-Africa relations in the temporal limits of the last two decades, in the case of Mozambique and, in particular the exportation of sea cucumbers to the Chinese and Asiatic market in general, it should be emphasized this activity was recorded in previous eras (Medeiros, 2007; Alpers, 2019), which connects the trajectory of the young person viewed in the documentary to a broader history of relations, transits and traffic on the stage of the Indian Ocean.

The first sequence of the episode introduces the young Chinaman, locating him in the context of the Island of Mozambique, thus providing a representation, albeit partial, of the location of another community, this time insular. The juxtapositioning of the shots of this initial sequence points to the correlation between space and social relations, working the ideas of exterior/interior; open/closed; inclusion/exclusion. In fact, the young man is framed in an exterior space delimited by metal bars that outline the sea, while he tries to talk on his mobile phone, without success, to his distant family. In the following shots, while the voice-off screen refers to feelings of homesickness, a series of shots in a closed space show moments when the inhabitants of the island and, in particular the women, come together and socialize (Figures 8.8, 8.9 and 8.10).[6] The internal composition of the shots and the interrelationship reveals a complex articulation of the binomial exterior/interior, open/closed, light/shadow. The spaces where the African women socialize are marked simultaneously by vivid colours and

6 All images from *Why are they here? Chinese Stories in Africa* ©Yara Costa Pereira.

shadow, while the space of the young Chinese man's loneliness is open and full of light, but cut by the bars.

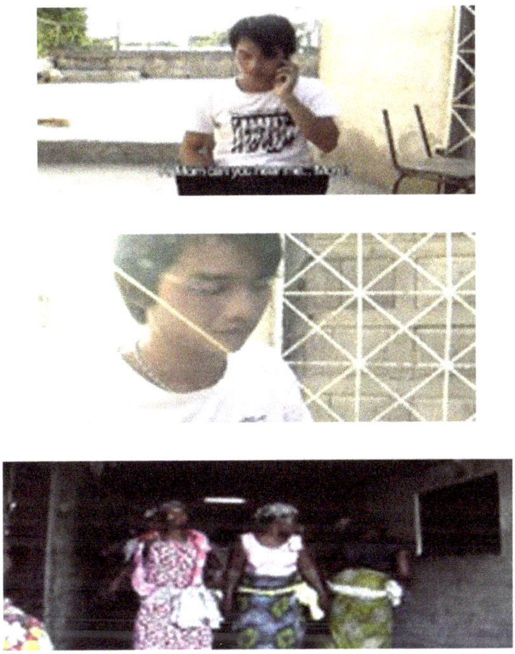

Figures 8.8, 8.9 and 8.10 Min Nan Wang's loneliness in contrast with a group of African women

Attention should also be paid to the presence of a closed wooden door in both spaces, which suggests symbolically a possible opening that does not, however, materialize in the narrative. Also, the shots that depict the young man in the streets of the Island register his isolation and his lack of integration into the social fabric of the Island (Figure 8.11). This is confirmed by his testimony, which alludes to the unfriendly relations with the inhabitants of the island, who, in his words, are afraid of the Chinese and step out their way whenever they see them in the street. Thus, for Min Nan Wang, looking lingeringly at the sea is the way for him to feel closer to his country, which transforms the ocean into a connecting link, more than separation, between countries and continents (Figure 8.12).

Figures 8.11 and 8.12 Min Nan Wang looking lingeringly at the sea

If this imaginary connection acquires positive connotations in the young man's discourse, the real inter-economy of which he is a mediator has impacts and problematic consequences in the insular community, as is explored in the second sequence of the episode, throughout which there is a closer focus on the impact of the sea cucumber business on local sustainability. Different shots focus on the search for the sea cucumbers in the sea, showing that this is a kind of artisanal fishing, carried out through diving by individuals or small groups, with the help of little boats. It should be pointed out that along with other activities, artisanal fishing continues to be one of the main economic activities and sources of income and subsistence for the insular community of the Island of Mozambique. Through the captions and interventions of the fishermen, the documentary depicts a fragile environmental, economic and social balance, since the sea cucumbers were abundant along the Mozambican coast, but the growth in demand in Asia, the lack of regulation and the illegal trade has radically altered the availability of this species beside the Island of Mozambique. According to the testimony of biologist Célia Macamo, many families from coastal locations depend on the sea cucumber business, because of which the current shortage puts their food security and subsistence at risk. On the other hand, the Mozambican fishermen complain about the conditions in which the business takes place, since the value of the cucumbers is fixed by the Chinese entrepreneurs, leaving the fishermen without a great margin for negotiation. Thus, the episode of the documentary shows how the sea cucumbers, from being organisms in the Mozambican coastal ecosystem, are transformed into transnational merchandize through the dynamics of the economy, power and desire, incorporating stories of work, displacements and unequal relations, as well as the failures of the postcolonial State.

From the aesthetic point of view, we note that in this documentary what was referred to above as haptic visuality is also privileged, where the capacity of the images to evoke tactile and, more generally, bodily sensations is explored. In fact, various shots capture the inhabitants of the island involved in activities in which the movement of the body is central, as, for example, the dance; the religious rituals in the mosque; the preparation and placing of the m'siro,[7] and the actual diving to catch sea cucumbers. In these shots, too, the emphasis on the texture of the bodies and material recreates a universe experienced through the senses, beside the visible and the audible, which suggests to us a possible aesthetic feature common to the audiovisual representations of the Indian Ocean. In addition to this aspect, the approach adopted in these two documentaries has made it possible to identify as a common topic the existence of environmental, socio-cultural and economic conflicts around the fishing and communities involved in these activities. In both narratives, the relation exterior/interior is declined through the topic of migration. In *Shifting Sand*, the focus is on the emigration of the fishermen of the Kharvi community to other places in the Indian Ocean, namely the Persian Gulf, while immigrants from other Indian states have entered the fishing business, thus producing an alteration in local community structures. On the Island of Mozambique, the presence of Chinese people involved in the sea cucumber business also has an impact on the life and sustainability of the families of local fishermen, generating problematic intercultural relations as well.

Third shore: Atauro Island, East Timor

The documentary *Wawata Topu*,[8] which portrays the fisherwomen of the little coastal village of Adara, Atauro Island, East Timor, was made

7 Beauty mask made with natural dust.

8 The documentary *Wawata Topu* had great resonance on the international scene. For example, the first Forum of Fisherwomen of East Timor took place in Dili in October 2018, organized by World Fish and centred on the sector of small-scale women's fishing <https://fish.cgiar.org/impact/stories-of-change/gaining-voice-first-women-fishers%E2%80%99-forum-held-timor-leste>. Considering that it is

in 2013 by David Palazón and co-directed by the anthropologist Enrique Alonso. The documentary focuses on the fisherwomen of the village of Adara, whose particularity is that they practise underwater diving in the coral reefs of the island (maximum depth about 3 metres), hunting the fish with a kind of harpoon. The Adara women are the only ones in the whole country to practise this kind of fishing and in a certain sense they remind us of the *ama* in Japan, who also traditionally dedicate themselves to underwater fishing, as the acclaimed film Ama-San by Cláudia Varejão shows.

In two scientific articles (Palazón, 2016; Alonso, 2016) the directors analyse the preparatory phase of the documentary and its immediate impact on the local population; these texts are extremely interesting because they refer to details of the filming and present the theoretical framework of the ethnographical research that underpins the production of the documentary.

The fisherwomen represent various generations, from 15 to 70 years of age. They are clearly the centre of attention of the invisible interlocutor behind the camera, these women who have struggled against hunger and poverty and who, for this reason, decided to dedicate themselves to this kind of small-scale fishing. However, this short documentary (just thirty-three minutes) does not limit itself just to fishing activity, because it manages to encompass a great variety of social and economic themes, without neglecting the aesthetic level, with the stunningly beautiful images that show the underwater fishing, as we shall see below.

In a perfect balance between the images filmed in the village and the underwater scenes, we observe, then, that in *Wawata Topu* there exists a perfect harmony between the informative material (which predominates in the interviews) and the aesthetically more interesting sequences, as is the case of the underwater images.

This documentary shows the lived experience of the population of Adara, who between forest and sea, find the indispensable means of subsistence, because in this little village everyone is dedicated to fishing: the men, women and children. But an uncertain future seems to dramatically

---

normally the men who take governmental and local decisions in the field of fishing, the Forum propose to develop the empowerment of fisherwomen.

threaten this apparent timelessness. It is for this reason that, along with the women, another great protagonist of this story is the capacity for resilience of the inhabitants of this island, from the children (who walk along a long path beside the sea in order to get to school), to the women (many of them do not have nets, or boats, and for these reasons they began to practise underwater fishing) to the older inhabitants, who live in a context where there are no other means of subsistence. We might then ask how far the insular communities will be able to survive in these conditions, bearing in mind that the main natural resource (the fish) may be threatened by large-scale fishing practices, by climate change, among other things.

The documentary leaves many questions unanswered, above all in respect of the uncertainties of the future, the younger generations, emigration, the protection of the coral reefs, the tourism potential of the village, among other aspects, but it provides us with precious testimony about this little community. *Wawata Topu* must therefore be considered an extremely important audiovisual repository because through the interviews and the filming of these women and children's everyday life, it shows us the daily life of this little community. In addition to this, despite not having an explicit narrator (voice-over), in this documentary there emerge both the economic and social reasons that led these women to dedicate themselves to this kind of fishing, and other problems that interest the fisherwomen (namely the dynamics of family and marriage,[9] gender issues, economic difficulties, health, tourism and education.)

Many of these aspects have been broadly analysed and articulated with the filmic practice in the directors' texts, mentioned above; we see that the documentary was based on wide theoretical research and field work (in fact, one of the directors, who is an anthropologist, led the interviews, in which the stories of some of the women emerge through the evocation of their memories) but, on the other hand, this documentary must not be considered just as an anthropology tool. For that reason, we do not intend to analyse the relationship between anthropology and the audiovisual product or go more deeply into the questions raised by the directors. Rather, we

9 For example, the question of the *barlaque*, a kind of dowry offered by bridegroom to the family of the bride.

intend to privilege an aesthetic analysis that will allow us to highlight especially the relationship between the women and the aquatic milieu.

Accordingly, we have to mention the poetic aspect of *Wawata Topu*: in fact, the images of the fishing take on a dramatic character and create two levels of reading of the documentary, one that is informative and more linked to ethnographic research – with the interviews of the women and the other dwellers in Adara; the other is poetic – with the images of the fishing at the bottom of the sea and the music or local songs accompanying the fluid movements of the fisherwomen and their lances.

In *Wawata Topu*, the fisherwomen, who are the central but not exclusive subject of the documentary, look and speak directly to the camera (most times in closeup). We then observe that the director frequently uses the participatory model defined by Nichols (participatory mode, 2001: 115), which is mainly comprised of interviews – in which we neither see nor hear the voice of the interviewer (in this case the anthropologist Alonso). The participatory model is often used in the social sciences in field research (as in ethnography) and is a tool which, in this documentary, makes it possible to avoid using voice-over which would intensify the gaze of the women as object of study.

Besides the participatory model, the director also opts for the 'observational mode' (Nichols, 2001: 34), above all in representing daily tasks (going to school, diving, cooking, selling the fish, etc.). The observational model predominates in the underwater images, and it is in these moments that the documentary takes on an unmistakably poetic tone. Generally speaking, we can say that the director combines the two techniques, achieving greater aesthetic and poetic efficacy when representing everyday life in Adara in a non-intrusive manner, without having recourse to the narrating voice.

With regard to the underwater images, in the first part of the documentary, as well as the opening scene in which we hear the song of the 'mermaids', there is a kind of approach to the aquatic world in a shot in which the women are filmed outside – we only see parts of submerged heads – and successively we dive, like them, into the Adara sea. In this first introduction to the universe of the women divers, the horizon of the water's surface is established as the point of intercommunication between the two worlds: that of the land and that of the water, synthesized in the mermaid metaphor.

From the documentary we would like to flag up three longer underwater sequences: in our opinion these are the most poetic images of the documentary, accompanied by music or local songs.

In the first sequence a man and his wife (who is waiting in the boat) are going fishing. In this the rhythmic and hypnotic of the *rama*, an instrument in the form of a long metal bow (a kind of *berimbau)* creates an underworld parallel to the real world. In the second sequence, the children dive and fish in the shallow water of the coral reefs, accompanied by a song they themselves sing. In the third underwater sequence, the women are the protagonists; these stunningly beautiful images are accompanied by a song sung by the women, while at the end instrumental music seems to intensify the dramatic movement of the underwater weapon. With a camera mounted on the actual harpoon, we see the fisherwomen's perspective of their prey, thus in a game of shots and reverse shots the outside view is added to the fisherwomen's point of view.

Figure 8.13 A timorese fisherwoman

Figure 8.14 Marine flora and fauna

©D. Palazón and A.Enrique

These images capture the interaction between the fisherwomen and their prey, showing both violence of the hunt and the underwater dance of the women. In this regard it is important to point out an aesthetic that shows the dramatic side of this interaction, resorting in particular to the sound track. When we consider the most memorable sequences we understand that the traditional instrumental music or song has unquestionable relevance, and is one of the elements that heightens the poeticity of the scenes being represented, because the mermaids' song is, along with the mermaids, as important as they are. The soundtracks of these underwater sequences also relate to the approach chosen by the directors, who do not impose external commentary by using voice-over, but reproduce local sounds and voices. In this way, then, bearing in mind the three sequences mentioned above, we note that the directors privilege the more sensorial side, because the viewers can hear the natural and human sounds of the island (as in the scenes in which we hear local songs sung by the children and women). In the case of underwater fishing, this process becomes even clearer in those final shots of the women's fishing in which we hear the sharp sound of the harpoon hurled against the prey.

In addition to this, in the absence of a narrative voice that interlinks the sequences of this documentary, on an aesthetic level, the thread that unifies these fragments is the horizon of the sea and the water's surface. The horizon of the sea appears behind the women who are interviewed, with their back to the beach, and the water's surface delimits the space outside and inside the water in the underwater images that depict the children and women who are fishing. In this regard, it should be noted that there are various shots in which the camera is placed on the same horizontal axis of the water's surface. This kind of framing allows us to see the portion of the aquatic space in which the women are moving; these images also emphasize the texture of the water and the interaction of this element with the women's bodies.

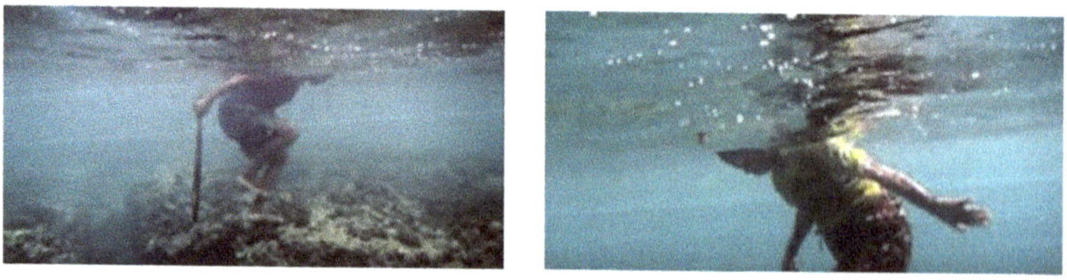

Figures 8.15 and 8.16 Fisherwomen

The underwater world is a nonhuman world and initially it was the documentaries that, capturing the depths of the sea, opened the doors of this unknown universe to the popular imagination (Mikkola, 2018: 5). However in *Watata Topu*, the underwater scenes contrast clearly with the typology of the documentary about natural and animal life at the bottom of the sea:[10] instead of contemplating the marvellous coral reefs and their coloured fishes, we witness the women's hunt; think in particular about the brief shots filmed with the camera placed on the harpoon, making it possible to see and hear the sudden sound of the lance hurled against the fish. On the other hand, it is impossible not to notice the immense coral reefs that cover the bottom of the ocean. In fact, Atauro Island is located in a zone of coral reefs that present one of the greatest marine biodiversity in the world. However, the documentary does not explore eventual ecological or environmental questions in depth but points to the fishing as being practically the only means of subsistence of this population.

Thus, the underwater images are not intended to show this very rich natural world, but rather its relationship with human beings. The sea is part of the human habitat, and, vice versa, the human being is part of this aquatic world in a process of bringing land and water closer together. In this way, not only the vertical axis inside/outside the water which is unified by the line of the horizon and the surface of the sea, but also the horizontal axis land/water that is overridden by the incursions of these 'mermaids' through those of the two worlds.

In addition to this, the directors have managed to avoid an aesthetic that would privilege the potentially 'touristic' side of the island,[11] with golden beaches, palm trees and transparent waters, and this aspect is even more obvious through the representation of the sea as a means of subsistence and not a paradise on earth.

10 In this way the documentary reverses an approach more usual in cinema that focuses on the ocean or the bottom of the sea, that is, disconnecting the existence of these underwater worlds from historical, social, gender or racial dynamics (Starosielski, 2012: 150), which on the contrary has marked the history of the oceans.

11 Despite this, as the authors of the article make evident, the documentary has given a certain visibility to the island, which has had positive repercussions for the tourist dynamic (Alonso, 2016: 268).

To sum up, this documentary reveals that the women, children and other inhabitants of this island are no longer beings who live just on the land but who, in order to maintain their own survival, become amphibious beings, dividing their very existence between land and sea. These women (just like the children) are beings who belong both to the land and the water, and this idea is reinforced by the many images in which the camera is placed at the height of the fisherwomen with their bodies submerged and their feet stepping delicately on the coral reefs. These images become even more unusual because of the vivid colour of the traditional cloths tied round their waists, brightly coloured tee-shirts and beach sandals.

Figures 8.17 Fisherwoman on the coral reefs

It is important to recognize that an anthropocentric perspective predominates in this documentary, centred on human activity and its relationship to the maritime and coastal territory. It is clear that the inhabitants of this coastal community on Atauro Island depend on the resources of the ocean. This documentary focuses on the local stories of this community of women divers, however, they are not depicted as belonging to a tradition that is distant from the global dynamics and problems of the contemporary age and, more specifically, from those that interest the coastal communities. As is indicated in a review of *Wawata Topu*:

> The film's strength is that without relying on voiceover or commentary, through the use of only interviews and extremely well-shot visuals, it manages to convey the very real sense of a community caught in the cusp of change. It may be set in a remote equatorial island village but the film speaks of the universal dilemmas that accompany the condition of modernity, and raises questions about the survival and future of small-scale fishing communities everywhere. (Biswas, 2013)

It is in this sense that the problems revealed in this documentary affect both the women divers of Adara and the fishermen of Goa, it is not a binary opposition tradition/modernity or local/global, it is the struggle of these coastal populations for their own survival and interdependence with the ocean.

In conclusion, this documentary must not be considered just an anthropological object or an ethnographical tool, because there is a clear poetic and aesthetic dimension, above all in the underwater images; and these women cannot be reduced to repositories of tradition, because a private dimension emerges from the characters, their stories and personal memories, their individuality. The lived experience of these women is profoundly intertwined with the sea, at heart, the history of the relationship between the sea as these women coincides with their own life stories, because without the sea, for the fishermen of Goa, this 'amphibian' life could not exist. For these women the space of the sea is also the space of resilience in the face of an increasingly uncertain future.

The Indian Ocean and the challenge of audiovisual narratives

The erosion of the coast and coral barriers, as well as the progressive emptying of the coastal strips bathed by the Indian Ocean, are phenomena strictly related to the current fragility of the littoral societies and the little coastal communities faced with a future in constant and rapid change. In fact, if coastal societies have undergone constant change, in this century we will witness an increasingly rapid increase both in climate change that profoundly marks the activity of these communities and the economic changes related to tourism, fishing, etc.

As we have seen, the documentaries analysed above show littoral societies in a specific temporal space, representing an extremely important cultural repository in a chronological perspective, bearing in mind that certain practices and dynamics that characterize those little communities may come to disappear in a short space of time. The forms of artisanal fishing practised in the three places examined in the documentaries are expressions of local knowledge on the different shores of the Indian Ocean, strongly impacted, for different reasons, by exogenous and endogenous factors of an ecological, economic, social and cultural nature. These factors intervene in various ways in the transformation of the littoral and, sometimes, at their end, in fact they configure paradigm changes. Keeping this aspect in mind, we understand the reason why Pearson attributes great importance to the chronology of the littoral space and we observe that through the practice of the audiovisual narrative the documentaries help to 'crystallize' a process of irreversible change, thus seeking to leave a memory record of these universes.

Taking up the concepts formulated by Pearson and Samuelson that were addressed in the introduction, we observe that the littoral examined in the documentaries may in fact be understood as an ecotone, a space of transition, in this case between sea and land, but also a space of meeting (Pearson, 2017) that bears witness to the persistence of 'amphibian lives', where the communities represented live in symbiosis with the sea. What we intend to corroborate in this essay is the emergence of these cultural forms also from an aesthetic and narrative point of view. Having considered the audiovisual practice of these three documentaries (framing, shots, scenes, topics, sound, etc.), we note that this not only proves that thesis of a continuum between land and sea, but extends this interdependence to the relationship interior/exterior, inside/outside, surface/depth (of water), seeking to record ruptures and continuities between past and present. We can therefore state that these documentaries, through aesthetic and narrative choices, point to an amphibian aesthetic, as well as establishing an affective link – not only linked to a question of survival – between the inhabitants of the coast and the sea. These common elements bring closer the distant geographies of the shore of the Indian Ocean since, as the Mozambican writer João Paulo Borges Coelho (2019: 21) asserts, 'não é só dos grandes cruzamentos oceânicos que se vive, quando se vive junto à costa. Há características próprias que se instalam nesta faixa humedecida por tudo isto. Cumplicidades entre quem partilha essa mesma condição, contrastes com os do interior' [it is not only from the great oceanic crossings that one lives, when one lives beside the coast. There are characteristics that establish themselves in this strip moistened by all of this. Complicities between those who share the same condition, contrasts with those of the interior]. We believe we have detected such complicities by the identification of common affective and aesthetic poetics that invest in recreating the human, material and liquid texture of this ocean, represented as a space of sensorial experiences.

In conclusion, we underline the importance of 'relational' poetics articulated with Indian Ocean Studies, developed by Ute Fendler (2013) on the basis of Glissant's theoretical concept about the relational potential of the world as 'tout-monde'. It is in this context that the cultural productions of the Indian Ocean may be articulated with the notion of insularity, not only in the sense of an in-between-space between land and sea but also between multiple memories and cultures (22). Such an approach makes it possible to identify imaginaries, perceptive sensibilities and cultural practices common to the coastal communities of the Indian Ocean, but above all, to the artistic representations of them that are produced. We can therefore demonstrate, as Torabully proposes, that the sharing of cultural, poetic and affective imaginaries bears witness to the existence of a coralline identity that connects and brings together the worlds of the Indian Ocean.

Documentaries

Costa, Y. (2011). *Why Are They Here? Chinese Stories in Africa*, documentary, 35'.
Filinto, S. (2013). *Shifting Sands*, documentary, 29', [Konkani, subt. English]
Palazón, D. and Enrique, A. (2013). *Wawata Topu: Mermaids of Timor-Leste*, documentary, 32', HDV [Rasua, Subt. English].

Bibliography

Alonso-Población, E., Fidalgo-Castro, A. and Palazón-Monforte, D. (2016). 'Ethnographic Filmmaking as Narrative Capital Enhancement among Atauro Diverwomen: A Theoretical Exploration'. *Development in Practice*, 26 (3), 262–271.

Alpers, A. E. (2018). 'From Littoral to Ozone: On Mike Pearson's Contributions to Indian Ocean History'. *Journal of Indian Ocean World Studies*, 2 (1), 12–24.

——. (2019) 'Moçambique Marítimo' (séculos XIV–XXI)'. *Revista de História*, (178), 1–32.

Beja, H. (2013) 'Há muitos filmes para fazer em Timor'. *Ponto final Macau*. <https://pontofinalmacau.wordpress.com/2013/09/05/ha-muitos-filmes-para-fazer-em-timor/>

Biswas, N. (2013). *Wawata Topu: Mermaids of Timor-Leste* A film by David Palazón & Enrique Alonso, ICSF, *Yemaya Magazine*, n. 44, December.

Borges Coelho, J. P. (2019). 'O Índico como lugar'. In A. M. Leite, E. Brugioni and J. Falconi (eds), *Estudos sobre o Oceano Índico. Antologia de textos teóricos*, pp.13–28. Lisbon: Colibri.

Fendler, U. (2013). 'Narrating the Indian Ocean: Challenging the Circuits of Migrating Notions'. *Diacrítica*, 27 (3), 17–28.

Leite, A. M. and Falconi, J. (2018). 'Island of Mozambique. Narratives from a Contact Zone'. In D. Wieser and F. Prata (eds), *Cities of the Lusophone World*, pp. 69–92. Bern: Peter Lang.

Marks, L. (2000). *The Skin of the Film*. Durham: Duke University Press.

Medeiros, E. (2007). 'Os Sino-moçambicanos da Beira. Mestiçagens Várias'. *Cadernos de Estudos Africanos*, 13/14, 157–187.

Mikkola, H. (2018). 'Movements beyond Human: Ecological Aesthetics and Knowledges in Underwater Wildlife Documentaries'. *Trace Journal for Human-Animal Studies*, 4, 4–26.

Newman, R. (1984). 'Goa: The Transformation of an Indian Region'. *Pacific Affairs*, 57 (3), 429–449.

Nichols, B. (2001). *Introduction to Documentary*. Bloomington: Indiana University Press.

Palazón, D., Enrique Alonso, E. and Fidalgo Castro, A. (2016). 'Linking Gender, Diving and Filmmaking: Conceptualising Film Outcomes as Narrative Capital Gains in the Making of *Wawata Topu* (Women Divers)'. *West Atauro, Timor-Leste'* Asian Fisheries Society *Gender in Aquaculture and Fisheries: The Long Journey to Equality* Asian Fisheries Science Special Issue, 29S, 73–92.

Pearson, M. N. (1985). 'Littoral Society: The Case for the Coast'. *The Great Circle*, 7 (1), 1–8.
——. (2003). *The Indian Ocean*. London: Routledge.
——. (2006). 'Littoral Societies: The Concept and the Problem'. *Journal of World History*, 17 (4), 353–373.
——. (2007). 'Studying the Indian Ocean World: Problems and Opportunities'. In H.P. Ray and E. Alpers (eds), *Cross Currents and Community Networks: The History of the Indian Ocean World*, pp. 15–33. Oxford: Oxford University Press.
——. (2017). 'Places in the Indian Ocean World'. *The Journal of Indian Ocean World Studies*, 1 (1), 4–23.
Rubinoff, J. A. (1999). 'Fishing for Status: Impact of Development on Goa's Fisherwomen'. *Women's Studies International Forum*, 22 (6), 631–644.
Samuelson, M. (2012). 'Abdulrazak Gurnah's Fictions of the Swahili Coast: Littoral Locations and Amphibian Aesthetics'. *Social Dynamics*, 38 (3), 499–515.
——. (2017). 'Coastal Form Coastal Form: Amphibian Positions, Wider Worlds, and Planetary Horizons on the African Indian Ocean Littoral'. *Comparative Literature*, 69 (1), 16–24.
Sreekumar, A. et al. (1995). *The Tourism Critique and Tourism Movements in Goa*. Bangalore: Equations.
Starosielski, N. (2012). 'Beyond Fluidity: A Cultural History of Cinema under Water'. In S. Rust, S. Monani and S. Cubitt (eds), *Ecocinema Theory and Practice*, pp. 149–168. London: Taylor and Francis.
Verne, J. and Verne, M. (2017). 'Introduction: The Indian Ocean as Aesthetic Space'. *Comparative Studies of South Asia, Africa and the Middle East*, 37 (2), 314–320.

UTE FENDLER

9 The Indian Ocean: A relational imaginary

Over the last two decades the Indian Ocean has turned into a major field of research, mainly in history (Pearson, 2003) and anthropology (Gupta et al., 2010), but more recently, in literary studies as well (Hofmeyr, 2015; Arnold et al., 2020). Very often, the imaginary of oceans is dealt with in terms of tropes and metaphors in literature (Lavery, 2021) and art history. Shanty Moorthy and Ashraf Jamal suggested first steps towards an 'Indian Ocean poetics' (2010), as the area is 'constantly in circulation and recirculation, in flux, suppliant and yet enormously resilient' (2010: 3), offering 'the possibility for new imaginative conjunctures' (2010: 13). The idea of imaginative conjunctures is the starting point for reflections on a relational imaginary in the art works of three artists from Mozambique and from Goa (see Vicente, 2001). I draw on the writings of the Martinican philosopher and author Édouard Glissant, as relationality of local and global spaces is central in all his writings as well as the convergences with questions of identity, relating to others and of being in the world. In his essay *Traité du tout-monde* ('Essay on the All-World', 1997), he defines the 'relation' as follows:

> Et j'appelle *Poétique de la Relation* ce possible de l'imaginaire qui nous porte à concevoir la globalité insaisissable d'un tel Chaos-monde, en même temps qu'il nous permet d'en relever quelque détail, et en particulier de chanter notre lieu, insondable et irréversible. L'imaginaire n'est pas le songe, ni l'évidé de l'illusion. (1997: 22)

> [And I call Poetics of Relation this possibility of the imagination that leads us to conceive of the elusive 'worldness' of such a Chaos-World, at the same time as it allows us to pick up some detail from it, and in particular to sing the praises of our place, unfathomable and irreversible. Imagination is not a dream, or the emptiness of an illusion (2020: 12)]

In one of his five poetics, *Philosophie de la Relation. Poétique IV* (2009), he gives a tentative definition of imaginary:

> L'imaginaire pressent, devine, trouve, il ne prévoit rien en termes de rapport, il n'accompagne ni l'avoir ni le savoir. Il ne conclut à rien. Il suppose en archipel. (2009: 109)

> [The imaginary senses, guesses, finds, it does not foresee anything in terms of rapport, it accompanies neither the having nor the knowing. It concludes to nothing. It supposes in archipelagoes.]

Glissant highlights the meandering way of imagining, bringing things together, connecting them, so that something different or new might emerge or shine through to hint at potential directions of the imaginary. It is a process of linking suggestions, assumptions, ideas and images, bringing up new combinations. He uses the metaphor of archipelagoes to explain his conceptualization of imaginaries: islands of different size, form and nature are scattered in the ocean and can form clusters of varying size and connections depending on the perspective, the criteria of selection for the combination of various aspects that might be based on colours, on size, on age, on distance, on fauna or flora, etc. The linkages can change and with them the archipelagic setting.

In other disciplines, we find two major lines of research on imaginary: one is placed in a sociological context, the other in an iconological one. John B. Thompson, for example, defined the social imaginary as the 'the creative and symbolic dimension of the social world, the dimension through which human beings create their ways of living together and their ways of representing their collective life'. For Cornelius Castoriadis, 'the central imaginary significations of a society [...] are the laces which tie a society together and the forms which define what, for a given society, is "real"'.[1] In these two definitions, we see the importance of the intertwinement of social settings with the imagination of what this community is and might be as well as how they should live together. It is an entanglement of

1 John B. Thompson, *Theory of Ideology* (1984: 6). Thompson referring to Castoriadis (1984: 16–41).

given facts and settings on the one hand, with the imagination of the potential and the modalities of this very specific realization of an imagined community on the other hand, that comes into being by the ongoing negotiation of a collective imaginary and the challenges of the present built on the past, projected to the future.

The 'creative and symbolic dimension' can be linked to Moorthy and Jamal who were most probably thinking of imaginative conjunctures in the sense of the real and the imagined, possible encounters in this vast contact zone (Pratt, 2007) that is the area of the Indian Ocean. I will link this with the ideas developed by Stephen Muecke in his article 'Fabulation: Flying Carpets and Artful Politics in the Indian Ocean' published as a contribution to Moorthy's and Jamal's book (2010: 32–44):

> Fabulations are the kind of storytelling that takes off on a flight of becoming rather than repeating the available discourses and genres. It is 'fabulations' that enable a subject to begin to feel engaged or dis-engaged, to either belong or depart, but in any case to continue the process of becoming. (2010: 41)

His definition of fabulation highlights the process of becoming – and this is very close to Glissant's concepts of 'relation', 'chaos-monde' –, an ongoing exchange between various agents and objects, 'actors in the network' as he explains further on giving some examples for complex exchanges that could take place between and with concepts, monsoon winds, songs, ceremonies, slave beads' (2010: 41). Further on, he speaks of the oscillation 'between description and fictionalization, between speech and writing' (2010: 41), which links the actors not only with the media of fabulation and the modus (Fendler, 2014), but it also underlines that the distinction between documentary and fiction is obsolete, as the process is fluid linking different epochs, regions, genres and modes crossing all borders. The reading of Muecke's article brings us back to Glissant's poetics of the Relation.

I would like to combine these various approaches as multiple points from which to view the process of coming into being of a relational, archipelagic imaginary. The fabulations contribute to the creation of a collective imaginary that is based on the 'actors of the network', as Muecke put it, that draw on shared cultural, socio-historical experiences as well as on imagined elements, thõse being objects, narratives, images, etc., as is the case for the artists from Mozambique and Goa relating to the ocean as a relational space that is built upon shared imaginaries.

How to approach the poetics of a relational imaginary? Focus on the 'space in between' or 'putting in tension of images'

For this article, I would like to focus on the iconological elements in photos, sculptures and installations, taking some inspiration from the procedures of the German art historian Aby Warburg (1866–1929) who suggested building an atlas of tropes and motives in images. He developed a procedure that would allow the tracing of recurrent motives in different art histories trying to follow the entanglements of images, stories and myths that accompany them. All these elements together constitute an imaginary based on the collective memory of entangled images and stories and their respective meanings and narratives.

Furthermore, Warburg talked about the iconology of the 'space in between' (*Ikonologie des Zwischenraumes*), the space in between various pictures that 'tell the same stories', but that can be linked to each other. This approach recalls the explanations by the Russian film-maker Sergej Eisenstein explaining the importance of the *montage*. He drew on the Japanese sign system to explain how each single take has its own meaning, while a sequence might have an additional meaning that arises from the combination of two signs. The example he used was the sign for tears that combines two signs, namely the sign for water and the one for the eye. The combination of the two signs means 'tear'. He used this technique of montage to create additional symbolic meanings in his films. Warburg's approach is similar as he is looking for the space in between two images, for the surplus of meaning that arises from the linkage of two or more images that use the same motive. He creates a kind of map of dialogue between various pictures, moments or spaces of remembrances, what he calls 'Mnemosyne'. Philippe-Alain Michaud (2006) describes Warburg's mnemosyne as a map of a region of art history, but also as a sequence of pictures that represent at the same time a chain of thoughts, which could contain the spaces in-between as moments of ruptures and/or relations, so that 'archipelagoes' of pictures emerge from a larger corpus:

das, was auf jenen unregelmäßigen schwarzen Feldern, die die Bilder auf den Tafeln vereinzeln und eine rätselhafte, prädiskursive Funktion erfüllen, die Motive einander annähert oder voneinander trennt, als Introspektion und zugleich als Montage zu begreifen: in *Mnemosyne* stellt jede Tafel die kartographische Aufnahme einer Region der Kunstgeschichte dar, aufgefasst als objektive Sequenz und zugleich als Gedankenkette, in die das Netz der Zwischenräume die Bruchlinien einzeichnet, die die Abbildungen in Archipele oder auch, wie Werner Hofmann es nennt, in 'Konstellationen' aufteilen oder gliedern. (2006: 4)

[to conceive of that which, on those irregular black fields that separate the images on the panels and fulfill an enigmatic, prediscursive function, brings the motifs closer together or separates them from one another, as introspection and at the same time as montage. In *Mnemosyne*, each panel represents the cartographic recording of a region of art history, conceived as an objective sequence and, at the same time, as a chain of thought into which the network of interstices draws the fault lines that divide or divide the images into archipelagos or also, as Werner Hofmann calls it, into 'constellations'.]

This rapprochement of images, so that the space in-between, of the relation, becomes 'visible' and can invite a reflection on the archipelago of the imaginary, can be a means of drawing closer to the ever ongoing process of relating in the world of chaos Glissant spoke of.

Michaud mentions the 'tension of images' by putting two or more images next to each other, or in relation to each other:

La mise en tension des images, et de la mise en mouvement des surfaces que Warburg produit dans Mnémosyne: Jean-Luc Godard, dans ses *Histoire(s) du cinéma* cherchant « à rapprocher les choses qui ne sont pas disposées à l'être », travaille le matériau filmique comme Warburg celui de l'histoire de l'art, faisant surgir le sens de l'actualisation des images par révélation réciproque que seule permet la technique du montage. (2006: 9)

[The putting in tension of images, and the putting in movement of surfaces that Warburg produces in Mnemosyne: Jean-Luc Godard, in his *Historie(s) of cinema* seeking 'to relate things that are not inclined to be', works the filmic material as Warburg works that of the history of the art, making appear the sense of the actualization of the images by reciprocal revelation that only the technique of the editing allows.]

The approach to a relational imaginary is the analysis of the emergence of a tension that might result from the combination of various images. The process of relating images to each other could create a surplus of meaning contributing to the imaginary of the Indian Ocean.

Example 1: The photographer Sergio Santimano, Mozambique

Sergio Santimano is a Mozambican photographer of Goan descent. He left Mozambique in 1989 and has lived in Sweden since then going back and forth between Portugal, Mozambique and more recently Goa. The three countries are linked through colonial and postcolonial history, so his personal life routes connect complex histories of family stories entangled with waves of migration between the three continents. Having worked as a photographer in the then young, independent Mozambique of the late 1970s and the 1980s, his work was influenced by the striving for local representation and the construction of a national identity in a multiethnic and multicultural country. Documenting the cultural diversity and the richness of the country under the influence of tendencies of the 'cinéma de vérité', a democratization of the construction of images and the integration of the voices and perspectives of the people, Santimano captured moments of the daily life of peasants, fishermen and children. His photos – in black and white or in colour – succeeded in getting very close to the person depicted so that the images seem to freeze intimate moments that turn into memorable ones due to the qualitative values of intimacy and iconicity.

One of his most popular photographs taken in Mozambique in the 1980s is the photo of a young boy holding a camera made of mud, pretending to take a snapshot while facing Santimano's camera. The camera is made of the same material as the walls of the house in the back so that they create continuity between the incomplete rectangle of the house with the rectangular box of the camera. The link between the house, the perspective and the camera angle effectively creates a direct connection between the local space, the inhabitant and the photographer who is taking the photo. The boy's broad smile attests to his enjoying his game as well as his gaze – from behind his 'fake' camera – into Santimano's camera. This photo is emblematic for the process of photographic documentation of the Mozambican people at the time. They look at themselves and at each other creating a shared imaginary. The strong connection between the space, the two cameras and the gaze of the photographer who is becoming part of this scene as well as the gazes going back and forth bring forth a cohesive moment of recognition of the other: the two persons could easily switch their perspectives. In this manner, it illustrates the new gaze turned towards oneself in a loving and respectful looking-out for the other. The photo captures the 'tension' between two pictures in the process of becoming that are directly related to each other. One could describe it as a relational approach that is inherent to many of his photos.

Over the last seven years, he has undertaken a personal exploration of his family history, proposing an exhibition entitled 'Legado' (Maputo, 2020), the legacy and spending lengthy periods of time in Goa taking pictures of interiors, houses and landscapes. In 2021, an exhibition entitled 'Porta para Goa' followed in Portugal with a slightly different corpus. It is a mixture of personal memories, remembrances of the family that are present in the house, intimate photos of family members, but also of collective memory, rendered by taking photos of old houses in Goa. Finally, there are photos that are linked to crops and spices and to the ocean with fishermen and boats, before photos of persons and spaces follow in the narrative arrangement of the catalogues. The imaginary proposed by these two exhibitions is mainly linked to the history of Santimano's family. However, it goes beyond this focus by connecting the history of the family to the ocean as the bond between Goa and Mozambique.

Example 2: The Mozambican sculptor Pekiwa

Pekiwa (Nelson Augusto Carlos Ferreira) is a sculptor from Maputo, Mozambique. A large part of his oeuvre are sculptures using old boats and pieces of wood from the Island of Mozambique. Getting the material involves not only a trip to the far North from Maputo but also a journey to the past. The Island of Mozambique was the most important connection on the route from Portugal to Goa. The sixteenth-century fortress is one of the largest in colonial history of Africa and together with the enormous governor's palace, the monuments testify to the connections built during four centuries of colonial expansion and settlement. At the same time, fishing and boats belonged to the local tradition of connections along the long East African Coast for centuries before the Europeans used the winds and currents for their exploration and trade routes. The cultural connections along the coast and across the ocean are built on centuries of contact between cultures. Using old boats or pieces of boats connects with the material history of the coastal regions that carry along the practices of sailing, of encounters that have taken place from precolonial times to the present day.

Pekiwa uses the boats as a sort of blackboard, a medium that he uses to 'write down', to visualize oral history. He explains his objective of transmitting stories of the past:

> 'Para mim esculpir objetos antigos é ir de encontro ao passado, uma forma de preservar-o, uma tela para contar histórias é quase uma obrigação de contar ao mundo atual estas vivências.
>
> [For me, working on an old object is an encounter with the past, a form to preserve it, a canvas on which I can tell stories. It is almost an obligation to tell these lived experiences to the contemporary world.] (my translation)

The boats are a canvas for painting as well as paper for writing. The figures that appear seem to stem from stories about the sea: there are silhouettes of fishes, mystical figures that are part of the human and the mystical universe of hybrid creatures, human faces.

Figure 9.1

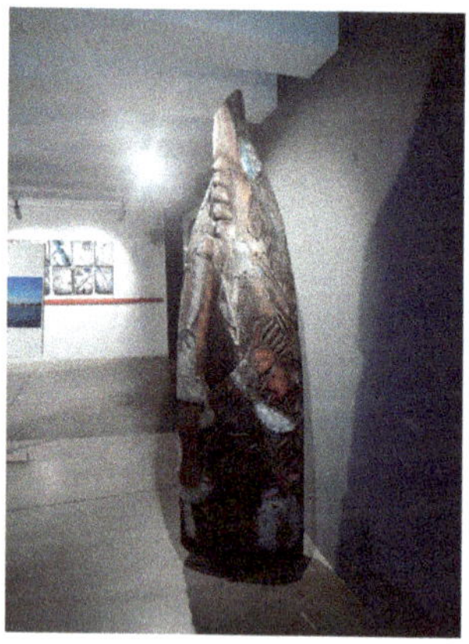

Figure 9.2

Sculptures finished in November 2022 for the multimedia exhibition curated by Ute Fendler at Galeria, Maputo, 22 November 2022

Figure 9.3

Figure 9.4

The details of each side of the keel (taken at the atelier of Pekiwa, nov. 22, Matola)

The left side of the boat carries two heads that almost seem to form one: the animal-like head is next to the back of the head of a human figure, as if one would chase the other or both moving together as if they were one body the upheavals beneath them seem to be waves carrying the figures on their race through the water. From the other side, the 'waves' are the curly hair of a human head resting on branches and leaves, a garden scene. In this way, the two faces of the boat complement each other, intimately combining the land and the sea, mystical and human figures that are all bound together across time and space. A frontal perspective turns the waves/curls into the dorsal fins of a bigger fish that could host all of the figures and plants in its belly, bringing to mind stories of whales and arks that carry species and worlds with them.

The metal plaques are reminiscent of glittering fish scales, but at the same time, they are – inorganic – fragments that connect wooden pieces that break open or apart. The shiny stitches seem to display the bonds between different spaces and epochs, but also make reference to painful stories of slave ships with iron chains, silver bullets. The pattern of the holes that pierce the planks of the boat create a paravent or screen that protects the other side from being seen, allowing just parts of it to be viewed through the holes. The structure is that of old wooden windows in Arabian style.

The partial visual access gives rise to a fragmented perception of the events taking place on the other side. Pekiwa's creation of a fragmented vision of the stories reminds us of the construction of history that draws on historical visual and material pieces to put a larger narrative together. The sculpture does not only carry the coastal history and the daily practices of fishing with it, it also tells the stories of the fishermen and the ghosts of the sea they have to struggle with or by whom they are supported. The position of the spectator/listener changes the storyline as different parts become visible while others disappear. Dimensions seem to change with the perspective as well as the connections between various pieces.

In this way, the sculpture also raises awareness of the positionality of the person who tells or reads the story. Pekiwa succeeded in inscribing the history-making process into the boat along with the flux of stories that might change directions, protagonists and constellations.

The boat is from the Ilha de Mozambique but it could also be from Zanzibar or from Mauritius, carrying the multilayered storylines of a complex interrelated zone of contacts that have taken place over centuries. Carving the wooden boats creates a certain permeability that allows time and space to enter and transcend the present moment via this medium. They are three-dimensional and invite us to go beyond the mere object by creating intersections and merging stories and images.

Example 3 The Goan oceanic artist Subodh Kerkar

Subodh Kerkar is a trained medical doctor who left his profession behind to focus on his artistic work that he had practised from a young age onwards under the supervision of his father who took him for long walks on the beach. In 2015, he founded the Museum of Goa, an art gallery with space for outreach workshops and collaborative work as well as a gallery.[2] He describes himself as an 'ocean artist', someone who is inspired by the ocean and who works next to and with the ocean.

2 <https://museumofgoa.com/about-us/>

Figure 9.5

The anchor

Figure 9.6

The porcelain

Both photos taken at MOG, February 2018

Covered in shells, the anchor is emblematic of the work of the ocean. The artist exhibits the ocean's artwork turning a utility object into one that has been completely absorbed by the ocean. The form of the anchor alone brings to mind the original object that has metamorphosed into a colony of shells. In a very similar way, Kerkar puts a porcelain plate recovered from a sunken ship into a historical shrine, so that it reflects the time the ocean had to work with the remnants of the sunken ship. The old wooden shrine that was used for sacred objects sets up a tension with a mercantile object but one that was nevertheless part of a 'sacralized process', and not questioned as the trade routes brought wealth to the trading countries.

The combination of work done by the ocean with that of the artist transforms objects into quasi-sacred relics of colonial trading history. The last example of oceanic work is the torrom, a piece of wood that is used to push the boat into the sea. This piece of art, 'Torrom II' (2016), is used as the body of a rowing boat. Old Chinese spoons are placed in a row on each side turning the old wood gliding support for boats turns itself into a boat that is driven by the mercantile force of trading between India and China. The artwork echoes the waves of historical migration and of the ocean.

Figure 9.7
The torrom, MOG, 2018

Besides the recreation of historical episodes by using objects that could testify of past encounters via the oceans, Kerkar works with performative installations. The series 'Fishermen' is a performative work for which fishermen enact daily practices. They would gather on the beach in such a way that their bodies form the silhouette of a boat, performing the act of rowing, fishing, etc. The performance of the fishermen can be considered as a 'reenactment of daily practices and of oral history'. In this way, the human body turns into the medium of a communicative memory, of a shared and lived history.

Figure 9.8

Figure 9.9
Fishermen and the ocean, MOG, 2018
A relational imaginary: spaces in between the oeuvres

Bringing some of the pieces of art together as a mnemosyne will make visible parallels in motives and aesthetics that could be part of or contribute to a shared imaginary that does not use language but visual formats for the transmission of myths and narratives building a relational imaginary. The spaces in-between two or several symbolic representations create the surplus of meaning in the archipelagic setting of the mnemosyne.

While Santimano focuses on a very personal, individual perspective that is imbued with nostalgia, Kerkar uses the same elements to visualize major historical topics with which he sets up a broader diachronic perspective of the cultural exchange of the last five centuries creating the 'Indian Ocean' as a shared space of experience. He suggests the concept of 'inseparability' to capture the essence of this mnemo-space: the past is intrinsically entangled with the present.

Both artists work with pepper and chili. In Santimano's photo, the work of cleaning and drying the fruits is at the centre of his interest. The frame he chooses makes the aesthetic arrangement visible in everyday practices which turns the visual aspect of it into an iconic image that can easily be remembered and transmitted. Similar to Santimano's approach, Kerkar turns to the iconic value of the chili that symbolizes the trade of spices on the East-Western trade routes. However, he adds the oceanic dimension in the arrangement of the image: the practice of spreading the chili on the ground in a perfect circle – which echoes the daily practice captured in Santimano's photo – is taken to a more abstract, meta-discursive level by adding an artificial chili, an object created to draw attention to the important significance of this tiny fruit. He turns it into a monument based on the daily practice of the cycle that he exposes on the beach. In doing so, he links the importance of the production of chili to the transoceanic trade that also includes indentured labour and slave trade as constituent parts of the exchange on a global scale.

The relational construction of a visual imaginary brings together the various layers of production and distribution with the colonial and postcolonial implications it carries as a historical burden. The photo of chili that might also appear in a travel guide seems to be reduced to the aesthetic beauty of the colourful spatial arrangement, but the individualized memory of daily practices in combination with the monumentalized chili statue against the backdrop of the ocean create this oceanic memory based on archipelagic fragments that gain their signification via the activation of the relations between the two of them.

Figure 9.10

Santimano: chilli

Figure 9.11
Kerkar: chilli

When it comes to fishermen, the gatekeepers of knowledge about the sea, movements and connections, there is a similar disposition in the relation between the two artistic works.

Santimano places the human being in its living and working condition in the centre of his photo fig. 9.12. The men are dressed in working clothes. Three of the four men cover their heads with hats to protect themselves against the fiery sun they are exposed to. The horizontal line is slightly inclined in comparison to the lined-up human bodies placed on a boat that therefore has to be in movement as it is not parallel to the sea line and the horizon. The movement is an integral part of the photo indicating the difficulties of keeping balance on a boat in continuous movement. Three of the men turn their gazes to the left, while the man in the middle turns towards the right, bending slightly and shielding his eyes from the light with his raised arm. His body forms a natural elegant bow in the centre of the photo bringing movement to the frozen take of the moment. The movements back and forth of the water are inscribed in the picture by the arrangement of the bodies and the lines of the rim of the planks, the sea and the horizontal line forming one unit. The photo captures the moment of the balanced movement of the bodies, the boat and the sea as the essence of the daily practice of fishing.

Figure 9.12
Santimano

Figure 9.9
Kerkar

Kerkar also dedicates some of his work to fishermen, paying tribute to their intimate knowledge of the ocean and its connections. He arranges the daily practice of rowing as a performative act on the beach so that it is aesthetized and musealized by the mise-en-scène of the practice: the beauty of the rhythm in the movement and the visual arrangement dominates the installation but it still also conveys the hardship of the work fig. 9.9 that needs joined efforts to confront the sea.

Santimano captures the essence of the object in a momentous 'natural' or spontaneous performance while Kerkar rearranges and resets the objects in such a way that the essence of the relation between the fisherman, the boat and the sea becomes part of the artwork. Setting both works next to each other creates a continuous exchange between the multilayered aesthetic pieces so that the individual experience becomes intimately intertwined with the historized abstraction of the importance of fishermen in the long history of the Indian Ocean.

The last example will focus on the representations of boats as the icon of the contact zones and trade routes, as well as an icon of various historical layers of different linguistic, cultural and epochal storylines of the Indian Ocean.

Figure 9.13 Kerkar

Boat, MOG, 2018

Figure 9.14 Pekiwa

Boat, Reitoria UEM, 2018

Placing the works of the three artists next to each other in a Mnemosyne, the iconic quality of the boats differs in each one according to the artistic approach of the artist. Santimano's photo of the fishermen pushing the boat into the sea trying to get over the strong waves breaking on the shore, focuses on the daily confrontation between human beings and the sea – and eventually their working together. The dynamic comes from the boat lifted up by the waves. The boat turns into an arrow-like object cutting across the parallel lines of the spindrift, the deeper waters and the horizon.

Kerkar's sculpture of the boat functions as a synecdoche, a *pars pro toto*, as this part of the boat does not only stand for the whole object but also for the historical and social context in which it was used. Inside the boat, Kerkar places forms made out of old wires that bring to mind nets with fish, but also human bodies, or unidentifiable figures that might appear in the sea or at the horizon, mirages of vague and fluid appearances, spectres of mystical creatures that are part of the narratives linked to the sea and to history. As the bodies are made out of old tyres, it also refers to the contemporary period, being on the road and stories of migration. In this way,

Kerkar links the past and the present, the sea and the land, and the mystic and crude daily life experiences of migration and hard labour.

Figure 9.15 Figure 9.13

Kerkar, 2018, MOG

With Pekiwa, the boat could also be considered a synecdoche, but it is highly individualized by the carvings that adapt themselves to the wooden foundation. The slightly curved form of the boat defines the form of the mystic and human figures that he 'carves out' of the wood as if they had always been there and just needed to be made visible. A view of the whole boat shows its great dimensions that gives space to various figures and stories on the wooden 'canvas'.

Figure 9.16

Figure 9.14

2013 Reitoria (photo 2018) Pekiwa

Figure 9.17 Boat, detail, Reitoria, Pekiwa

Figure 9.18 Boat, detail, Reitoria, Pekiwa

Figure 9.19 boat, detail, Reitoria, UEM, Pekiwa

Taking a closer look at the boat, he uses a mixed approach combining carving, perforations and metal bindings. There are parts that perforated the wood, giving the boat a more fragile aspect. These parts also connect the two sides of the wood, as if the spectator might see the various narrative layers at once. Furthermore, the boat seems to be reassembled from bits and pieces as iron hinges reconnect the fragments to build a whole body again. The transparency of some parts of the boat underlines the storytelling and memorizing processes. The arm of a figure becomes the shoulder of another figure, as well as the outline of the rim of the boat. The changing quality hints at the variability of the stories depending on the storyteller and the context. The face of a human being whose mouth seems to be sealed by the metal hinge reminds of the epoch of slavery. The carving of lines creating the pattern of a net above and underneath the face evokes the context of imprisonment, nets to bind bodies, just like animals, and it also recalls bodies gone or thrown overboard and turned into dead meat.

Putting the three of them next to each other creates a dialogue between them that seems to complement the iconographic value of the boat: the boat as the medium to confront the sea, the boat as the medium to carry nets and objects as well as stories so that it fills the void.

Fabulation stirred by the tensions in between the various boats. All together, individual, historical and collective memory aspects complement each other. The human figure right above the suffering body, is a body gracefully floating in the water, it recalls rather swimming and bathing, showing intimacy with the water, a mermaid figure of fairy tales. At the top of the boat is a face that is more like a fish than a human being. On the left hand side, there is another face, mouth wide open, that resembles more of a fish. The two faces seem to be the fish at the bottom of the boat and the humanoid face at the upper side of the boat, as if the creature can change its appearance according to its place of being, underneath or above the water.

In a conversation with Pekiwa in February 2020 and again in November 2022, he speaks about his stays on the island where he spends time with the fishermen who share with him their experiences and traditional tales as well as their own encounters with the sea and its creatures. He cannot give details about stories and the connections with his work, but he says that the stories he heard all flow into the process of sculpting, creating a storytelling piece that carries the essence of the stories of the fishermen and the island. His work is a process of abstracting from a multitude of stories and experiences to create a piece that represents the life stories of the fishermen and their relationship with the ocean. His boat-sculpture-story literally incorporates the experiences and stories in the boat as a medium between the human being and the ocean.

Figure 9.20

Figure 9.13

Figure 9.17

The three works set next to each other exponentiate the meaning of each of them. The photo of the daily practice taken by Santimano (fig.9.20) contextualizes the boat sculpture of Pekiwa (fig. 9.17), while the boat sculpture tells the stories of the encounter between the ocean, the humans and the boat. The boat of Kerkar opens the perspective towards aspects of migration, of suffering bodies trying to escape (fig. 9.13). The three works together create a complex universe of the interaction of the human being with the ocean using the boat as a medium of transport and communication. The human, the historical and the contemporary dimensions are raised in the spaces in between the works creating a relational imaginary, an archipelagic memory and vision of the world.

Conclusion

The procedure of the three artists – as different as they are in their work – goes from real objects towards individualized narratives with a meta-discursive dimension. The themes, tropes and objects turned into mnemosynic objects form archipelagos of images that are the basis for a relational imaginary. They also visualize the embeddedness of storytelling in the socio-historical context of the Indian Ocean.

I would like to link these observations with a concept suggested by the Mauritian writer Khal Torabully who speaks of a 'coral identity' (Brugioni & Fendler, 2021). In an interview, he explained that the coral is a hybrid and symbiotic being that evolves in and with the ocean in the continuous flow and movement of the water, fragile and strong at the same time.

> De plus, le corail est tronc, branche, racine, il est mou, dur, multiforme … C'est une créature hybride qui se propage sur un monde multidirectionnel et se développe non plus sur le mode de la connectivité errante seulement mais aussi agglutinante – c'est l'idée de concrétion transitoire d'un processus que je développais en amont. De plus, le corail est une créature symbiotique, et ne fonctionne pas sur un mode fusionnel mais dans le cadre d'une association entre un zooplancton et un phytoplancton dont la négociation est sur un mode égalitaire absolu et permanent. Je le trouvais plus dynamique que l'image du rhizome, d'autant que le corail, tout en étant ancré,

est ouvert aux courants marins. Il est le socle de la biodiversité et peut servir de support à l'idée de mouvement, de migration et d'ancrage (le corail génère la migration des planctons, la plus grande migration d'une espèce vivante sur Terre), tout en étant un processus sans fin. Ce socle corallien de biodiversité n'est pas sans lien avec l'idée d'une diversité culturelle. Il est fragile, comme cette beauté du monde que nous prônons tous deux, mais avec des majorations de certaines composantes et des métaphores différentes.

[Moreover, the coral is trunk, branch, root, it is soft, hard, multiform ... It is a hybrid creature that spreads over a multidirectional world and develops not only in the mode of wandering connectivity but also agglutinative - this is the idea of transient concretion of a process that I was developing before. Moreover, the coral is a symbiotic creature, and does not function in a fusional mode but within the framework of an association between a zooplankton and a phytoplankton whose negotiation is in an absolute and permanent egalitarian mode. I found it more dynamic than the image of the rhizome, especially since the coral, while being anchored, is open to marine currents. It is the base of biodiversity and can serve as a support for the idea of movement, migration and anchoring (the coral generates the migration of plankton, the largest migration of a living species on Earth), while being an endless process. This coral base of biodiversity is not unrelated to the idea of cultural diversity. It is fragile, like the beauty of the world that we both advocate, but with increases in certain components and different metaphors][3]

It is a metaphor for being lost and anchored at the same time, for being in between and in continuous multidirectional movement. Kerkar suggests the concept of 'inseparability' which can be found in the objects that are co-created by the sea, like the anchor or the porcelain covered with shells. This process of coming into being is very close to the idea of a coral being: the transformation of an object or a creature under the influence and the continuous and changing contact of the ocean. All the aforementioned artists approach the question of being in the world by suggesting reflections on fluid and relational identities. Each of them focuses on a different aspect, on nostalgic memory, entangled histories, or communicative construction of memory and history via art to build a relational imaginary of the Indian Ocean.

3 <https://blogs.mediapart.fr/pierre-carpentier/blog/040216/khal-torabully-le-poete-mauricien-lorigine-de-la-coolitude-evoque-edouard-glissant>

Bibliography

Arnold, M. et al. (2020). *Borders and Ecotones in the Indian Ocean: Cultural and Literary Perspectives*. Montpellier: PULM.

Brugioni, E. and Fendler, U. (2021). 'Islands, Theory and the Postcolonial Environment: Reading the Work of Khal Torabully'. *Portuguese Studies*, 37 (2), 165–177.

Carpentier, P. (2016). 'Khal Torabully, le poète mauricien à l'origine de la Coolitude évoque Édouard Glissant', Morne Brabant, mai 2012. Billet de blog 4 février 2016. Le Club de Mediapart: <https://blogs.mediapart.fr/pierre-carpentier/blog/040216/khal-torabully-le-poete-mauricien-lorigine-de-la-coolitude-evoque-edouard-glissant>

Fendler, U. (2014). 'Fabulating the Indian Ocean – An Emerging Network of Imaginaries'. In M. Mann and I. Phaf-Rheinberger (eds), *Beyond the Line: Cultural Narrations of the Southern Oceans*, pp. 179–198. Berlin: Neofelis.

Glissant, É. (1997). *Traité du Tout-Monde. Poétique IV*. Paris: Gallimard.

——. (2009). *Philosophie de la Relation. Poésie en étendue*. Paris: Gallimard.

——. (2020). *Treatise on the Whole-World*. Trans. Celia Britton. Liverpool: Liverpool University Press.

Gupta, P., Hofmeyr, I. and Pearson, M. (2010). *Eyes across the Water: Navigating the Indian Ocean*. Pretoria: UNISA Press.

Hofmeyr, I. (2015). 'Styling Multilateralism: Indian Ocean Cultural Futures'. *Journal of the Indian Ocean Region*, 11 (1), 1–12.

Lavery, Ch. (2021). *Writing Ocean Worlds: Indian Ocean Fiction in English*. Cham: Springer.

Michaud, P. A. (2008). 'Zwischenreich'. *Trivium* [En ligne], 1 2008, mis en ligne le 10 avril 2008, consulté le 02 décembre 2022. <http://journals.openedition.org/trivium/373>; DOI: <https://doi.org/10.4000/trivium.373> [Zwischenreich. Mnemosyne, ou l'expressivité sans sujet, *Les Cahiers du Musée national d'art moderne*, Nr. 70 (Winter 1999–2000), 42–61].

Moorthy, S. and Jamal, A. (eds). (2010). *Indian Ocean Studies: Cultural, Social, and Political Perspectives*. London: Routledge.

Muecke, S. (2010). 'Fabulation: Flying Carpets and Artful Politics in the Indian Ocean'. In S. Shanti Moorthy and A. Jamal (eds), *Indian Ocean Studies: Cultural, Social, and Political Perspectives*, pp. 32–44. London: Routledge.

Pearson, M. (2003). *The Indian Ocean*. London: Routledge.

Pratt, M. (2007). *Imperial Eyes: Travel Writing and Transculturation*. London: Routledge.

Santimano, S. (2020). *Legado*. Maputo: Kulungwana. <https://legacy.myportfolio.com/legacy-legado>

——. (2021). *Porta para Goa*. Castelo de Vide, 2021. <https://plataforma9.com/noticias/exposicao-fotografica-de-sergio-santimano-porta-para-goa.htm;jsessionid=888C825DAFB9A9895F1BBD6D493AD815>

——. (2021). *Goa-Mozambique*. Cineplex, Iwalewahaus, University of Bayreuth, (curated by Ute Fendler).

Thompson, J. B. (1984). 'Ideology and the Social Imaginary: An Appraisal of Castoriadis and Lefort'. In J. B. Thompson, *Studies in the Theory of Ideology*, pp. 16–41. Los Angeles: University of California Press.

Vicente, F. L. (2014). 'Viagens entre a Índia e o arquivo: Goa em fotografias e exposições'. In F. L. Vicente (ed.), *O império da visão: fotografia no contexto colonial português (1860–1960)*, pp. 319–342. Lisbon: Edições 70.

MEG SAMUELSON AND PAMILA GUPTA

10 Moving frames and circulating subjects: Reflections on Capital Art Studio, Zanzibar

Moving frames and circulating subjects: Reflections on Capital Art Studio, Zanzibar

Located on Kenyatta Road, which is now the main tourist artery through Zanzibar's Stone Town, Capital Art Studio was founded in 1930 by Ranchhod Oza and has remained in constant operation right up to the present, moving just a few metres down the road in 1972 after the premises in which it was lodged collapsed. Today, it is run by Rohit Oza, Ranchhod's youngest son, who continues to reproduce his father's images while adding new ones of his own. This longevity and continuity distinguishes Capital Art Studio: it is unique among the many that were in operation when Ranchhod Oza began to practice for having survived the revolution and its austere and oppressive aftermath, as well as the subsequent turn to the tyrannies of the free market and the digital deluge that has transformed the nature and function of photography in Zanzibar as it has elsewhere. The studio stands at the centre of our collaborative enquiry into photographic culture in Zanzibar in a project that seeks to trace a 150-year arc from the first studio, which was established by the Goan emigrant AC Gomes in approximately 1868, to a new one that was about to be opened by Goan Zanzibari Robin Batista when we last visited in 2018. In this chapter, we reflect on Capital Art Studio's distinctive practices of 'moving frames' to convey the 'circulating subjects' of this African Indian

Ocean island while relating the story of the studio's unusual history and accounting for how we were ourselves moved to undertake this project.

We were both attending a conference on 'Indian Ocean Africa' in Zanzibar in 2011 that had been organized by Preben Kaarsholm (Roskilde University, Denmark) and hosted by Abdul Sheriff, Director of the Zanzibar Indian Ocean Research Institute. Before the conference commenced, Meg was wandering through the labyrinthine streets of Stone Town when she found herself being drawn into a photographic studio by a window display featuring 'Zanzibar Views' from the 1950s. The magnetic pull that these images exerted resided perhaps in their resonance with the paper she would present in the coming days on Abdulrazak Gurnah's novel *By the Sea* (2001). *By the Sea* references Zanzibar's long entanglement in Indian Ocean trade routes and depicts Stone Town in the years leading up to the Revolution of 1964 and through the subsequent decades after the archipelago amalgamated with the mainland state of Tanganyika to form the United Republic of Tanzania. Gurnah had fled Zanzibar shortly after the revolution and *By the Sea*, like the rest of his oeuvre, displays a tempered nostalgia – or what Svetlana Boym describes as 'reflective nostalgia' (Boym, 2001) – for the town of his youth. The images that drew Meg into Capital Art Studio seemed to reanimate that bygone era, while the tonal complexity that she encountered inside its ornately carved door resonated with the mood she had discerned in Gurnah's fiction. Leafing through a box of prints, she selected two that seemed to manifest some of the ambivalence she had noted in Gurnah's depiction of the circulations that had composed the town and which – in her reading – present them as simultaneously cosmopolitan and consuming, simultaneously gracious and rapacious (see Samuelson, 2012, 2013, 2016). These prints included one of Ranchhod's earliest surviving images, which was taken in 1948 and features a dhow approaching Zanzibar under a portentously moody sky.

Figure 10.1: Dhow approaching Zanzibar, 1948 © CAS

Redolent of the Indian Ocean traffic that animates Gurnah's novelistic portraits of Zanzibar, the image was evocative also of the paper that Pamila would present during the conference that we attended over the following days. Entitled 'Monsoon Fever', it addressed the lyrical monsoons in their discursive and aesthetic registers. The paper used Alexander Frater's elegiac travelogue Chasing the Monsoon (1990) as a conceptual centrepiece to capture the monsoon as a fevered energy, affect, sensorium and weather pattern that transforms the lives and livelihoods of a large swathe of Indian Ocean peoples on an annual basis (Gupta, 2012). This reading of monsoonal elements (of wind, water and sky) would shape how Pamila would later reflect on the 'darkness' of monsoonal images from the Capital Art Studio collection (Gupta, 2020a).

Having broached the possibility of developing a research project on the studio with Rohit on her initial visit, Meg met with Pamila for dinner at a popular tourist restaurant fittingly called Monsoon to discuss collaborating on it. The resonance that we discerned between the prints she had purchased and the narratives with which we were concerned is what initially caught our joint attention, but what has held it since is the story of the studio itself, including its particularly distinctive photographic practice. During that first encounter, Rohit Oza had explained to Meg that he had inherited the studio from his father and that his vocational preoccupation was keeping his father's work alive. This he achieved by maintaining the studio's operations and reprinting his father's images for sale, as well as through the intriguing practice of re-taking scenes that his father had framed up to a half century before. An example of this practice in which the photographic frame moves, as it were, across time is presented in Figures 10.2 and 10.3 – these photographs of the Stone Town market were, respectively, shot by Ranchhod Oza in 1956 and Rohit Oza in 2004.

Figure 10.2: Stone Town Market, Ranchhod Oza, 1956 © CAS

Figure 10.3: Stone Town Market, Rohit Oza, 2004 © CAS

Hooked by the poignant practice of the retake, which we reflect on in the final section of this chapter, we returned to Zanzibar in the following year for what became the first of three visits (2012, 2015, 2018) during which we gathered images and interviewed Rohit and several other photographers. In the process, we have pieced together a story of the photographer Ranchhod Oza from the fragmentary images tendered by his taciturn son and from those that he himself framed and developed. It is a story in which the photographer, the camera and the photograph's compositional elements are all in states of circulation peculiar to the Indian Ocean world.

Ranchhod Oza's circulating subjects

A young Ranchhod Oza, accompanied by his bride, boarded a dhow in Gujarat and sailed for twenty-eight days across the Indian Ocean before disembarking in Stone Town in 1925. The route was a well travelled one: lateen-sailed dhows had been harnessing the reliable monsoon winds for over a millennium, knitting the Indian Ocean into an arena through which merchants, scholars, sailors, artisans and slaves circulated (Sheriff, 2010). Ranchhod had no family in East Africa, and his son is not sure why he chose this destination. 'Many Gujaratis were coming then', he simply says, before appealing to the evidentiary function of photography and its ability to record circulating subjects: 'You can see in the old photos – many Gujaratis.' The lively photographic scene that Oza found in Stone Town was itself an expression of Zanzibar's pivotal place in the Indian Ocean and imperial networks. Erin Haney (2010) notes that a 'fervor' for photography swept through the urban enclaves of the Indian Ocean in the late nineteenth century – and Zanzibar in particular became a magnet for Goan photographers, starting with AC Gomes, who arrived in Zanzibar in approximately 1868 after first having set up a studio in Mombasa, and who was followed by others such as E. C. Dias, the Coutinho Brothers and A. R. P. de Lord in the 1890s (Haney, 2010: 49–50). These early Goan entrepreneurs in East Africa were important in developing Zanzibar as picturesque, an ideal type (of place) for the production of a range of images, including studio portraiture, native life, and flora and fauna. They catered to Europeans who arrived in the bustling port via steamship tourism and local elite populations (Arab, Indian, African) (Haney, 2010: 49–50). On arriving in Zanzibar as an aspirant photographer, Ranchhod Oza apprenticed himself to AC Gomes & Sons and would have developed many of his compositional techniques in their venerable practice before striking out on his own to found Capital Art Studio.

Just as Ranchhod had apprenticed with Gomes, he and his son Rohit can also be inserted into a historical genealogy that includes not only a group of 'global Goans' (Frenz, 2014), but a larger community of diasporic South Asians (including Goans, Parsis and Gujaratis) that arrived in Zanzibar less as British colonial subjects but rather as 'imperial citizens' (Metcalf, 2007), transformed in the act of crossing the Indian Ocean, and choosing this destination to fit their own sense of cosmopolitanism. They created a niche for themselves, operating a range of businesses (tailoring, general stores, bakeries, photo studios) that are evidenced in the many photographs taken by Ranchhod Oza of these shop placards dotting Stone Town's narrow streetscapes, and that show a thriving South Asian community in Zanzibar during the first half of the twentieth century (Gupta, 2018b). In documenting these businesses, Ranchhod also included himself as a member of this minority Asian group, and into a longer genealogy of (largely Goan) Zanzibaris involved in adornment work, including two generations of tailor fathers and photographer sons (Gupta, 2018b).

Figure 10.4: A shop placard, Ranchhod Oza, 1950s © CAS

Ranchhod was drawn to the heart of Zanzibar's image production when he became the favoured photographer of Sultan Khalifa II bin Harub Al-Said – the incumbent of what had been the seat of the expansive Sultanate of Oman and Zanzibar but which had in Ranchhod's time been stripped of all but ceremonial power under the now-dominant British Empire. Ranchhod was commissioned to magnify the reduced Sultanate on large format prints that Rohit can still recall from his childhood

(Oza, 2012), and he produced and circulated an iconography of symbolic power, tinting the town with the Sultan's ceremonial red in a series of royal family and palace portraits. Many of Ranchhod's images were themselves entered into circulation in the form of postcards, and some of them only exist today by virtue of this dissemination (see Samuelson, 2018).

Ranchhod's popularity also extended well beyond the Sultan's palace. Rohit, who was born in 1950, remembers how, when he was 6 or 7 years old, his father would cycle to the studio in the morning and return again only at midnight. 'You won't get space [...] to sit here', he tells us in 2015, waving his arm across an empty room as he remembers the queues that would await his father at the door, and how – on special occasions such as Eid – people would continue to stream in until 11 o'clock at night to have their portraits taken against one of the sepia backdrops that Ranchhod had ordered from London in the 1940s.

But Ranchhod's most alluring backdrop was provided by Stone Town itself. Walking incessantly through its streets, his roving camera received the Indian Ocean world as it circulated through the town on the seasonal rhythms set by the monsoon. It conveys a sense of this fluid motion by attending to technologies of transport: bicycles, cars and carts are ubiquitous in these frames, as are the boats that attend Stone Town's harbour (Samuelson, 2016). Propelled by rotation, these technologies suggest ways of thinking about circularity in the Indian Ocean, which Sugata Bose has helpfully defined as an 'interregional arena of human interaction' composed of 'circular migration' (Bose, 2006: 3, 73).

Bicycles in particular stand out as leitmotifs. Ranchhod Oza's focus on bicycles presents them less as objects of desire for Western modernity, even as they were always that too in Zanzibar, and might even be said to 'domesticate' them in much the same way as the clocks that were set to Zanzibari time (Prestholdt, 2008). Bicycles here are as much a type of human transport as they are a form of attire to match an equally fashionable tailored outfit. They also function as gathering points of sociality and workhorses to carry commodities newly arrived by dhow to the interior of the island, circulating through Zanzibar in much the same way that Ranchhod did on his much loved bicycle taking photographs. Lastly, they act as stage props under Ranchhod's discerning eye, their compositional circularity offsetting the squareness of Stone Town's historic Arab architecture and ordered colonial urban planning that the British aspired to (Sheriff, 2008). In other photographs, bicycles are curiously situated at the edge of the frame, standing in for absent owners and ideas of personhood. All throughout the archive, they are 'moving still', with notions of filmic and the photograph compelling and propelling them down another narrow passageway to a chosen destination (Gupta, 2020b).

Figure 10.5: Bicycles, Ranchhod Oza, 1950s © CAS

People, too, are shown to be in constant motion in this photographic archive. This distinguishes Ranchhod Oza's work from that of his predecessors. Surveying what we have been able to piece together of his archive, we see little evidence that he had any interest in reproducing the kinds of taxonomies that AC Gomes & Sons, and their contemporaries, had produced for imperial consumption. While he does appear to have staged some studio portraits that typecast population groups and would occasionally fix his mobile subjects with a caption, these are the exception rather than the rule in his practice. Certainly, the accumulative effect of the surviving images that he took during the 1950s is that of motility – of subjects circulating through a moving frame. If they intimate some of the fault lines of Zanzibari society, photographs such Figure 10.6, which features what Rohit described as 'local people' (Oza, 2018), are expressive also of its intricate entanglements.

Figure 10.6: Ranchhod Oza, 1950s © CAS

Rather than inviting viewers simply to register who is in the frame by reading off a semiotics of race and ethnicity, these compositions call attention to the way in which these subjects move through it. As such, they convey the fluidity that animated this hub of the Indian Ocean world while refraining from fixing its circulating subjects in the damaging manner of the colonial census, which helped to ignite the revolution that broke out in January 1964 (see Sheriff, 2001; Gurnah, 2001).

Many Zanzibaris who were identified as Arab and South Asian were killed during the revolution and many more were expelled or fled in the aftermath, including 17-year-old Abdurazak Gurnah and his brother (Gurnah in Steiner, 2013). A number of Gurnah's novels represent this tumultuous period in Zanzibari history, with one noting that '[a]mong the many deprivations inflicted on those towns by the sea was the prohibition of the Muslim trade' (Gurnah, 2001: 16). One of the first actions of the revolutionary state was to close its ports to the circulating subjects who had moved through Ranchhod Oza's frame (Gilbert, 2004). It also decreed the destruction of all images of the Sultanate. Rohit tells us in 2015 that the 'Revolutionary leaders of Zanzibar' visited the studio after seizing power and pointed to the pictures of the Sultan adorning its walls: 'all these photos you should remove', they advised, and he has a distinct memory of his father burning a heap of prints and negatives. A series of postcards that were being printed in Germany arrived after the conflagration. Ranchhod managed to secure their release because he knew the clerks at the post office, which was (and still is) located just across the road from Capital Art Studio. One postcard from this consignment that remains in Rohit's collection today depicts the Sultan's Palace – which was subsequently renamed 'The People's Palace' – with the words 'The Sultan' struck out on the verso caption. It stands in as trace of the piles of images that were erased during this time (see Samuelson, 2018).

Nonetheless, Ranchhod Oza seemingly negotiated the changing political landscape with remarkable canniness. He apparently did so by circulating between different subjects with his moving frame. Even as he continued to commemorate the last years of the Sultanate, he had begun taking photographs at Afro-Shirazi Party rallies from as early as 1955. When Abeid Karume acceded to the presidency of the recently declared People's Republic of Zanzibar and Pemba, Ranchhod Oza was thus able to replace the portrait of the deposed Sultan that had hung above his front desk with that of the archipelago's new head of state. As had been the case under the Sultanate, he was again called upon to photograph state functions – his subject matter now comprising Karume's military parades rather than the Sultan's ceremonial bands – and also to record the arrival of visiting dignitaries. Indicative of the post-revolutionary regulation of travel, these choreographed photo-ops are typically framed by the airport rather than by Stone Town's bazaars and its seafront (see Samuelson, 2018). The subjects, too, have changed: politicians from mainland Africa, the Eastern Bloc and China replace the drifting merchants and sailors of the Indian Ocean world.

Yet while Ranchhod Oza's camera was trained on a state apparatus that restricted transoceanic travel in order to orient Zanzibar towards its new geopolitical allies, he also continued to photograph quotidian circulations through Stone Town's streets. It is during this period also that his photographic frame begins to move across time, as it were, searching for scenes that convey the conviviality of Zanzibar's bazaars and the fluidity of movement through its streets on the other side of the temporal rupture that the revolution effected. These movements are represented here by Figures 10.7 and 10.8, which were, respectively, taken in the 1950s and the 1980s by Ranchhod Oza on Gizenga Street. In part, the aftershot bears witness to the neglect that Stone Town suffered under the revolutionary government, as well as to the population shift that it affected. But the insistent motility of the subjects of this composition disrupts any such tidy narrative – as does the practice that Rohit would later come to develop from such models.

Figure 10.7: Ranchhod Oza 1950s © CAS

Figure 10.8: Ranchhod Oza 1980s © CAS

The post-revolutionary years were also the period during which Ranchhod Oza was training his son in the techniques of image-making. Rohit remembers that he was given his first camera when he was 12 and tells us that his father taught him everything he knows about photography. Through the 1980s, he began to assume responsibility for the studio, taking over entirely from his father in 1989 when Ranchhod formalized his retirement by travelling to India (Oza, 2018). During this visit, Ranchhod fell gravely ill and Rohit went to bring him home. He tells us that his father died in Stone Town on 3 May 1993 (Oza, 2018), having completed his final circuit around the Indian Ocean before coming to rest on the island whose circulating subjects had passed before his lens.

Rohit Oza's photographic 'retake'

Rohit tells us that his father never staged his compositions but instead received them 'naturally' (Oza, 2015), waiting through time for a composition to appear. Now Rohit does the same. Revisiting an altered place, he waits patiently for the walker, the cyclist, the vendor or shopper who will step into the frame and magically fuse the 'then' of his father's image with his 'now'. These images he describes as 'retakes'. What moves Rohit to retake certain images? When we asked him about his photographic practice, he simply said: 'When I see something is changing, I go and take photographs' (Oza, 2015). This is borne out when he extracts from beneath the desk in his front room a substantial private collection of photographs through which he maintains a record of buildings that are in various states of collapse or which have been earmarked for demolition or renovation. The phrase 'this has changed, they have renovated' recurs also on each of our visits as we talk through the separate box of prints that he presents for retail.

What do the re-taken photographs that he stores and sells say about Rohit and what moves him to invest in certain images? Perhaps a form of 'punctum' (Barthes, 1980) is at work here that opens up and homes in on these compositions (both his own, and those of a father) that are favoured by a son, with certain beloved photographs standing in for much more about history, life and affect? Perhaps, as well, it suggests a more sensuous understanding of the textures, colours and patinas of old photographs in new ways (Gupta, 2018a). These preserved images also showcase Rohit as a heritage practitioner of sorts, delving into his father's archive and visually preserving what materially remains from Zanzibar's baroque past in the face of an uncertain touristic future where high rise luxury hotels are allowed to be built by corrupt politicians, thus only allowing the privileged few a perfect Indian Ocean view (see Gupta, 2019).

Suggestively, Rohit says at one point that he retakes images in order to 'see' rather than simply record changes in the town. There are intriguing intimations here of photography's disclosive rather than evidentiary function – of what the photograph brings to light, as much as of the luminous trace of its referent. But what is it that the photograph as retake enables us to see, and what 'ways of seeing' might it foster (Berger, 1972)? In reflecting on these questions, we return to a series taken across a half century by father and son, and which initially moved us to undertake this enquiry and has since repeatedly drawn us back to it. It is a series that Ranchhod started in the 1950s and carried through to the 1980s (see Figures 10.7 and 10.8 above) and which Rohit has since extended into the twenty-first century.

Figure 10.9: Ranchhod Oza, 1950s

Figure 10.10: Rohit Oza, 2004

Parsed sequentially, the first and final takes of this series by father and son seem to exemplify Roland Barthes's reflections on photography's future anterior tense – on the way in which the medium thickens time on the surface of the image to produce the future anterior of the this-will-have-been (Barthes, 1980: 96). Looking first at one image and then the other, we are struck by the mortality of the subjects and of the urbane way of life that Ranchhod recorded, and then by the precarity that is in turn imposed on those who walk this street today. This gloomy prospect is enhanced during our last visit when Rohit tells us a bit about the subjects of the first image: that the Boharan cyclist and his twin used to run a silversmith in the front of their house, but that they have since died; and, that the man in white shorts behind him is one of the many Gujaratis who used to populate the town before they were dispersed by the revolution (Oza, 2018).

Placing the two images side by side, however, we see the intimacy between past and present coming to light and as well as their transport into the future. While perhaps inviting their viewers to apprehend and anticipate death, the images seem simultaneously to affirm what Geoff Dyer (2005) presents in The Ongoing Moment as 'the simple message that is also there in all photographs: "You are alive"'. This message extends beyond the individuals pictured therein to the Ozas' most enduring subject: Zanzibar itself. Neither elegiac nor recriminatory, then, the series instead transports something of the sense of being a cosmopolitan yet commodifying centre of the mercantile Indian Ocean world into an African present that is being scripted out of the uneven flows of late capitalism. Recalling a time in which Zanzibar was an island metropolis, it refuses the heritage and tourism industry's consignment of this time to the past (see Samuelson, 2016). Rather than the haptic poignancy of being faced with the temporal finitude that Barthes describes, the re-take displays the kinetic force that has drawn the photographer back to the scene and that which releases its subjects into motion across the frame. Bringing into focus the moving quality of photography that has been obscured by Susan Sontag's emphasis on it as a technology that slices through the 'flow of time' (Sontag, 2005: 111), it is evocative for thinking across the revolution even as it may be seen to provide evidence of a rupture.

The term by which Rohit describes his practice of moving frames across time is equally suggestive. Specifically designating these images as 're-takes' draws attention to the conflicting ways in which the camera has been conceived, and thus the conflicting meanings that vie within its frames. In the first volume of her *History of Photography*, Kaja Silverman relates how the verb 'to take' came to replace the verb 'to receive' in ways that have obscured something of the nature of photography. The notion of 'receiving' images dates back, notably, to the eleventh-century Arab scholar Alhazen, who discovered certain optical principles that would inform the camera obscura and which, during the eighteenth century, begins to be thought of in Europe 'as a mechanism for 'taking likenesses', instead of receiving them' (Silverman, 2015). Rohit's term 're-take' reminds us that the etymology of 'receive' is the Latin recipere: re-'back' + capere 'take' (the noun form of capere in turn provides the root of the word 'capture'). Poised between receiving and capturing, the 're-take' is an evocative practice with which to reflect on Indian Ocean Africa – a region that has been composed through practices of accommodation (of sailors, traders and other travellers) and abduction (of women and slaves).

Capital Art Studio thus presents a significant archive and practice that our project seeks to expose and develop further. Its moving frames and circulating subjects suggest profound ways of thinking through Zanzibar's pivotal position between the African continent and the Indian Ocean, and for theorizing the affordances of the photographic medium itself.

Bibliography

Barthes, R. (1980). *Camera Lucida: Reflections on Photography*. Trans. R. Howard. New York: Hill and Wang.

Berger, J. (1972). *Ways of Seeing*. London: British Broadcasting Corporation and Penguin Books.

Boym, S. (2001). *The Future of Nostalgia*. New York: Basic Books.

Dyer, G. (2005). *The Ongoing Moment: A Book about Photographs*. Edinburgh: Canongate.

Frenz, M. (2014). *Community, Memory, and Migration in a Globalizing World: The Goan Experience, c. 1890–1980*. New Delhi: Oxford University Press.

Gilbert, E. (2004). *Dhows and the Colonial Economy of Zanzibar, 1860–1970*. Oxford: James Currey.

Gupta, P. (2012). 'Monsoon Fever'. *Social Dynamics*, 38 (3), 516–527.

——. (2018a). '(Sensuous) Ways of Seeing in Stone Town, Zanzibar: Patina, Pose, Punctum'. *Critical Arts*, 32 (1), 59–74.

——. (2018b). 'Being (Goan) Modern in Zanzibar: Mobility, Relationality, and the Stitching of Race'. In W. Anderson, R. Roque and R. Santos (eds), *Luso-tropicalism and Its Discontents: The Making and Unmaking of Racial Exceptionalism in the Portuguese-speaking World*. New York: Berghahn Books.

——. (2019). 'Balcony, Door, Shutter: Baroque Heritage as Materiality and Biography in Stone Town, Zanzibar'. In *Vienna Working Papers in Ethnography* (VWPE no. 9). Department of Social and Cultural Anthropology, University of Vienna. <www.ksa.univie.ac./vwpe09>

——. (2020a). 'Of Sky, Water and Skin: Photographs from a Zanzibari Darkroom'. *Kronos*, 46 (1), 266–280.

——. (2020b). 'Moving Still: Bicycles in Ranchhod Oza's Photographs of 1950's Stone Town (Zanzibar)'. *Journal of African Cinemas*, 12 (2–3), 191–211.

Gurnah, A. (2001). *By the Sea*. London: Bloomsbury.

Haney, E. (2010). *Photography and Africa*. London: Reaktion Books.

Metcalf, T. (2007). *Imperial Connections: India in the Indian Ocean Arena, 1860–1920*. Berkeley: University of California Press.

Prestholdt, J. (2008). *Domesticating the World: African Consumerism and the Genealogies of Globalization*. Berkeley: University of California Press.

Samuelson, M. (2012). 'Abdulrazak Gurnah's Fictions of the Swahili Coast: Littoral Locations and Amphibian Aesthetics'. *Social Dynamics*, 38 (3), 499–515.

——. (2013). 'Narrative Cartographies, "beautiful things" and Littoral States in Abdulrazak Gurnah's By the Sea'. *English Studies in Africa*, 56 (1), 78–90.

——. (2016). 'Producing a World of Remains in Indian Ocean Africa: Discrepant Time, Melancholy Affect and the Subject of Transport in Capital Art Studio, Stone Town, Zanzibar'. *African Studies*, 75 (2), 233–256.

——. (2018). '"You'll never forget what your camera remembers': Image-things and Changing Times in Capital Art Studio, Zanzibar'. *Critical Arts: South-North Cultural and Media Studies*, 32 (1), 75–91.

Sheriff, A. (2001). 'Afro-Arab Interaction in the Indian Ocean: Social Consequences of the Dhow Trade', *Occasional Paper 13*. Cape Town: Centre for Advanced Studies of African Society.

——. (2008). *Zanzibar Stone Town: An Architectural Exploration*, 2nd edn. Stone Town: Gallery Publications.

——. (2010). *Dhow Cultures of the Indian Ocean: Cosmopolitanism, Commerce and Islam*. London: Hurst, with Zanzibar Indian Ocean Research Institute.

Silverman, K. (2015). *The Miracle of Analogy; or, the History of Photography, Part 1*. Stanford: Stanford University Press.

Sontag, S. (2005 [1977]). *On Photography*. New York: Rosetta.

Steiner, T. (2013). 'A Conversation with Abdulrazak Gurnah'. *English Studies in Africa*, 56 (1), 157–167.

CLARISSA VIERKE

11 Imaginaries of far-away places? Swahili poetry and its changing worlds in the Western Indian Ocean[1]

Mas não é só dos grandes cruzamentos oceânicos que se vive, quando se vive junto à costa. Há características próprias que se instalam nesta faixa humedecida por tudo isto. Cumplicidades entre quem partilha essa mesma condição, contrastes com os do interior. Exemplo do primeiro aspecto é a cultura e a língua swahili, uma mancha costeira que se alastrou por toda a costa desde o sul da Somália até ao norte de Moçambique, e que as actuais fronteiras políticas não conseguiram conter. Como se em todo esse espaço as pessoas partilhassem um segredo. (João Paulo Borges Coelho 2019: 21)

[But living on the coast does not mean just living from big ocean crossings. There are particular characteristics which have become established in this wet strip. Complicities between those who share the same condition, contrasts with those in the interior. An example of this is the Swahili culture and language, a coastal belt which extends all along the coast from the south of Somalia to the north of Mozambique, and for which the current political boundaries do not count. As if in the whole of that space people share a secret.]

1 My heartfelt thanks go to Ana Mafalda Leite, Jessica Falconi and Elena Brugioni, the organizers of the inspiring conference 'Colóquio Internacional Oceano Índico', for allowing me to join the wonderful 'last minute' group. This article is based on research carried out in Mozambique in 2015 with a grant from the Thyssen foundation, and in Zanzibar in 2019 within the framework of the research project 'Multiple Artworks – Multiple Indian Oceans' which is part of the Cluster of Excellence 'Africa Multiple, Reconfiguring African Studies' at Bayreuth University funded by the Deutsche Forschungsgemeinschaft (DFG, German Research Foundation) under Germany's Excellence Strategy – EXC 2052/1 – 390713894.

My gratitude also goes to my 'partner in crime', Chapane Mutiua, to Nassor Hilal Kharusi and Chale Mussa for their generosity and to Ruth Schubert for proof-reading the English text.

Point of departure: The 'Shared Secret' of the East African coast

In the above quotation, the Mozambican writer João Paulo Borges Coelho refers to common characteristics of the East African coast, 'the Swahili culture and language' as a 'shared secret'. The notion of the 'shared secret' guides me in my exploration of Swahili poetry in this article in two ways: Firstly, I will work within the geographical area suggested by João Paulo Borges Coelho's view, as it ignores the political boundaries of present-day nation-states and their language policies involving the former colonial languages, English and Portuguese, which tend to bury all other historically grown relations and languages under them so that these become 'secrets'.

Secondly, I will explore his notion of the 'secret' by presenting an aesthetic perspective on the Indian Ocean. By aesthetic perspective I mean aesthetic practices – here foremost poetry – which have played a crucial role in the coastal communities. The poems offer a perspective on common coastal practices, which, given that they are embodied knowledge, often escape our attention and remain secret. Furthermore, all along the coast, poets have exchanged poetry for centuries, so that the circulating poems add to and describe a broad Swahili coastal map. Lastly, to explore the coastal 'secret', I find it productive to adopt an aesthetic approach in the etymological sense of sensual perception, and to make philosophical inquiries about aesthetic experiences as a way of understanding that goes beyond logical arguments. An aesthetic perspective leads us to consider the Indian Ocean not merely as constructed in narratives of manifold relations that crisscross the ocean, but as evoked sensually through shared scents, tastes, sounds, visions and practices that produce an experience of a shared Indian Ocean world. They reveal a common coastal and oceanic world, create manifold associations and bring a wider world to life, but escape any straightforward translation into ordinary words – and hence remain a secret: how can the experience of a scent or a rhythm and the manifold associations they trigger be rendered in common language?

Outside Western space and time: Recent approaches

Alternative geographies and timelines have been a primary concern in Indian Ocean studies. The Indian Ocean came into the focus of scholarly attention rather late – the last in the list, after the Mediterranean, the Atlantic and the Pacific. In comparison to the other oceans, the Indian Ocean has been characterized as a most culturally, linguistically and politically diverse oceanic space (Pearson, 2015: 3), where the European presence was of rather marginal importance (see also Brugioni, 2020: 88): '(F)or most of its history, the Indian Ocean was crossed and used by people from its littorals, not Europeans' (Pearson, 2010: xv). Along the Atlantic coasts, Christian and Western missionary efforts led to institutionalized churches and schools, and to a 'uniform domination of Euro-American Christian culture' (Alpers, 2000: 86), which shaped important (postcolonial) critical thinkers of African descent, like, for instance, Aimé Césaire or Frantz Fanon. In the Indian Ocean, this dominance does not exist in the same way: there is far more cultural and linguistic variety between the Swahili Coast, southern Somalia and southern Arabia, Oman, the Persian Gulf, Sind and Gujarat, where Islam rather than Christianity plays a key role (Pearson, 2010: Brugioni, 2010). As Alpers argues, reflecting upon the African diaspora in the Indian Ocean, it is rather the lack of a Western-European framework which characterizes the Indian Ocean. This can be considered as an important reason for 'the significant absence of an educated African elite in the Indian Ocean world' who would be the spokespersons of a particular Indian Ocean identity. Or, as I will argue below, the over-emphasis on Euro-American discourses in Western languages makes Indian Ocean research blind to other languages and practices, often embodied and tacit, which, however, also evoke a translocal map of the Indian Ocean. Thus, the Indian Ocean bears the potential of provincializing Europe, questioning existing geographies and the concentration on European colonial powers as the only enduring presence (see also Brugioni, 2019), and requiring us to shift our perspectives.

Despite the emphasis placed by some scholars on their potential for creating an epistemological shift that would marginalize Europe and the former colonial powers, non-Western languages and literary practices, and archives of texts, have been sidelined. This has been pointed out by historians and Islamic scholars, who have sketched out alternative views. For the Western Indian Ocean, for instance, Pier Larson (2009) has shown how Malgache (besides French) was a language of correspondence across the Ocean, linking diasporas on Madagascar and Mauritius. Anne Bang has analysed the network of circulating Arabic books and manuscripts, echoing transoceanic links between Sufi scholars in the Western Indian Ocean, on Zanzibar, the Ilha de Moçambique and Cape Town, with contacts in southern Arabia, in the nineteenth and twentieth centuries (Bang, 2003).

Outside Indian Ocean studies, other scholars have increasingly challenged a unique geography. Their studies can inspire a new view of literature circulating in the Indian Ocean. For instance, Muhsin J. al-Musawi has written a book called *The Medieval Islamic Republic of Letters* (2015), in which he maps a vast Islamic space of discussion and disputation that stretched from India to North Africa, in response to unifying debates on world literature and postcolonial perspectives: he challenges Casanova's *La République mondiale de Lettres* (1999), where the Greenwich Meridian is the centre of 'Eurochronology' (Prendergast, 2001), a single Western, seemingly universal space-time axis. In a productive way, the Indologist Francesca Orsini (2014) adopts an inspiring relational and multilingual perspective in her research on literature in northern India, where Hindi, Sanskrit, Persian and Arabic (and later English) were coexistent and created different, but overlapping arenas of discussion and circulation of texts for centuries. She considers various, often overlapping 'significant geographies' that a text and a (multilingual) author can be part of. Languages map out different spaces and histories and are linked to different practices and materialities.

I take inspiration from these approaches that suggest other maps and timelines, and will look for the geographies that a Swahili literary work evokes and is part of. So far, Swahili literature – including poetry – has not played a role in thinking about the literary imaginary of the Indian Ocean. Rather, Anglophone, Francophone and Lusophone writers, from

Kenya, Tanzania, the Comoros, Mauritius and Mozambique, have gained attention with their alternative Indian Ocean imaginary that cuts through national and continental boundaries (see Brugioni, 2019: 87ff.). And yet, ironically, their use of the former colonial languages endorses the national and linguistic boundaries created by colonialism. While Portuguese links Mozambique with the former colonial power Portugal and its colonial map of discovery and conquest in the Indian Ocean (and beyond), Swahili requires a map that not only includes parts of Mozambique, Kenya and Tanzania but also provides a view of the East African coast as a whole: the term 'Swahili' is derived from the Arabic term for coastal inhabitants. It is a map that shows port cities on the coast and the nearby islands, roughly from Mogadishu to Sofala, the homes of Muslim sailors and merchants, who had far-reaching connections along the coast and across the Indian Ocean, following the winds of the Monsoon. Part of the 'shared secret' of the so-called Swahili coast is not only the language but also a shared material culture: different types of boats (from big dhows to outrigger canoes, ngalawa), and coral stone houses embellished with imported goods, like porcelain and glass, which archaeologists have excavated all along the coast. Shared practices of Islam, and poetry, which will be in the focus here, fundamentally add to the coast's 'secret'.

Contemporary smaller worlds and double affinities on Zanzibar

Francesca Orsini's (2015) compelling argument for overlapping maps is a reminder not to explore the Swahili coast as a neatly definable bounded area of study, fixed in history and defined by one, unchanging language. The city-states along the coast have their different histories echoed in the language. Swahili is a historical continuum of distinct language varieties, with which people proudly proclaim their regional identity as a mwunguja 'Zanzibari', a mwamu 'Lamuan' or a mmvita 'Mombasan'. Speakers from the extreme geographical ends of the coastal map recognize similarities in their dialects but find it difficult to understand each other. Swahili is a Bantu language that has been influenced by Arabic and other languages of the Indian Ocean, as well as the colonial languages, Portuguese, German and English, and, even more importantly, other African languages. These influences depend on the region and the historical period: the coastal map is not homogeneous and is subject to change. While in northern Mozambique Swahili is now only existent in niches, it is a national language in Tanzania and Kenya, and in the Democratic Republic of Congo. It has been a first language to many, and a second or third language to even more people on the coast since at least the end of the first millennium. Particularly from the eighteenth century onwards, when the Omani Sultanate moved to Zanzibar, it expanded widely as a lingua franca in transregional trade along the coast, but also across the Ocean to southern Arabia, Madagascar, the Comoros and Mozambique, and – via caravan trade – far into the East African interior, setting the stage for its later colonial use and use as a national language in Tanzania and Kenya. In the twentieth century, political ideologies which marginalized the Muslim coastal inhabitants turned Swahili from a language of the coast and the islands to a language of the mainland.

A poem by the contemporary Zanzibari poet Nassor Hilal Kharusi, Ingalikuwa hiari 'If I had a choice', which he self-published in his anthology Siku Zapishana ('Days are passing by') speaks of the double affinity of the narrator to the mainland and the sea – which is not without conflict. His emphasis on travelling as an essential human characteristic seems to highlight the dynamic geography I have attempted to describe:

> Tunayo waja safari kwa hakika na yakini
> Iko wazi siyo siri asiyejuwa ni nani?
> Nchi kavu na bahari ni tangu toka zamani
> Safari katu haina hiari (Hilal n.d.: 6, stz.)
>
> [We human beings have been travelling – this is clear,
> it is not a secret, who does not know it? –
> On the mainland and the ocean.
> Travelling is a necessity.]
> (my emphasis)

Nassor Hilal Kharusi's own background reflects the layers of overlapping maps of Zanzibar. Born in 1978, he is of Omani decent, in the sixth generation on Zanzibar, and has a kiosk-like shop in a busy area of the market selling shiny scarves (mostly from China) and dawa ('traditional herbal medicines') – reflecting both a contemporary Indian Ocean world, flooded by products from China, and a historical one of circulating Arabic herbal medicines. When there is no customer around, he composes poetry behind his counter. He refers to Oman, sometimes with the melancholic vocabulary of the diaspora, but more often he underlines his Tanzanian identity. Zanzibar is his home, where he spent all his life, Swahili his only language, and yet he is regarded as an 'Arab' in Dar es Salaam, where he has to pay double the price in the market. The resentment against Arabs, recently reinforced again, is a heritage of the bloody revolution on the island which put an end to the Omani sultanate in the 1960s. In the context of socialist Tanzania, Zanzibar, previously a cosmopolitan hub of trade, whose multicultural flair attracted poets, musicians, traders and scholars from Arabia, the West, India and other parts of Africa, became geographically, politically and ideologically marginal and a troublesome appendix to the state.

This changing geography is also reflected in poetry. On the one hand, Swahili poetry is still the most important and widespread aesthetic practice in Zanzibar, building on a longstanding coastal tradition. There is no wedding without a poetry recitation. Topical poetry is published daily in newspapers, as well as on Facebook, in WhatsApp groups and in other social media. Hard-hitting verbal duels carried out in veiled language between poets are a much commented on part of everyday life – as in other towns along the Kenyan and Tanzanian coast, like Tanga, Mombasa or Lamu. Poetry travels along the coast, overriding national boundaries, to poetic rivals or relatives in Mombasa, but also, for instance, to Oman, evoking older geographies. However, the coast is not the only realm in which Swahili poetry circulates: Nassor also exchanges poetry with poets from the mainland, within the confines of the Tanzanian nation-state, whose imaginary has heavily relied on Swahili as a national language, and on a national canon of modern Swahili literature, created and promoted by Julius Nyerere in the 1960s. Canonic writers like Shaaban Robert or Matthias Mnyampala, who originally worked to support Julius Nyerere's policy of a Swahili-speaking national state of Tanzania with a fresh poetic imaginary are authors that Nassor Hilal Kharusi refers to as his role models, whose works he read in school on Zanzibar. His poetry shows this 'double affinity', sitting at the crossroads of local Zanzibari dialogue poetry and topical verse attentive 'towards the people of his city, the dramas of life and love both as they unfold in private and can be sensed in public and a fascination with the natural world' (Koenings, 2018: 186), and the (initially socialist) concept of the writer as socially engaged, serving the national agenda of development and progress. Like many other poets on Zanzibar, he uses his poetry to openly criticize social ills, like the persecution of albinos or a quickly changing world of 'dramatic tensions' (see Koenings, 2018: 187), but in an increasingly repressive context, his poetry takes on more subtle and allegorical forms – often in line with his inclination towards philosophical musing on human existence.

Apart from the small oceanic reference in the poem quoted above, the Indian Ocean does not play any role in the imaginary of his poetry or that of other poets on Zanzibar, sometimes locally published in anthologies of Zanzibari poetry. References to a wider Indian Ocean beyond the immediate horizon are almost absent. The poets seem to have turned their backs on the sea, concerned in their poetry with their immediate community on the island, and often looking to the mainland in praise or, more often, in protest. Nowadays, Tanzania's Indian Ocean map figures prominently in East African Anglophone novels. These include the acclaimed works by Abdulrazak Gurnah, who fled from Zanzibar shortly after the revolution, like the protagonist of *By the Sea* (2001), or the novels of Moyez Vassanji, a reputed writer of South Asian ancestry, and author, for instance, of Uhuru Road (1991). His stories navigate the lifeworld of the South Asian diaspora on the East African coast. Some novels in Swahili, such as *Vuta Nkuvute* by the Zanzibari writer Shafi Adam Shafi (1999), which has not been taken into consideration by comparative literary scholarship on the Indian Ocean, evoke the manifold oceanic and coastal relations, as well as the fault lines and ruptures, focusing on the dark sides of slavery and social segregation, as well as splendid cosmopolitan moments of cultural diversity. In addition, the imaginary of the Indian Ocean has been explored by the Zanzibar film and jazz festivals, funded by global cultural institutions, which attract a mobile and cosmopolitan audience, mostly from the West – a parallel, inaccessible or insignificant world to most Zanzibari Swahili poets. Their inward-looking poetry does not seem to fit the notion of cosmopolitanism and fluid identities cherished in Indian Ocean studies, which, par contre, has tended to ignore literary production in the Indian Ocean which does not match its concepts and has a narrow perspective on genre and time.

The imaginary map of Zanzibari poets has been subject to much change, and traces of a longer coastal tradition can be dug out in the practices – not the discourses – of poetry composition. The vitality of poetic production can be linked to Nyerere's emphasis on art in the service of the nation, but can also be traced back to an older coastal notion of the habitus of a respected member of society (mungwana), associated with good manners and appearance, and refined rhetoric, best expressed in poetry. Patterns of strict verse prosody, much attacked by modernist poets in the 1960s, are often used to highlight a specific coastal identity. These go back over the Ocean to Southern Arabian models adapted into Swahili in the eighteenth century. Corresponding recitation practices are held in much esteem; these precede but also have outlived, national boundaries along the coast of the Indian Ocean.

Mwana Kombo, another Zanzibari poet, wrote a poem on commission to congratulate the Zanzibari government on building a new road, in line with the modern agenda of the state. Originally from the neighbouring island of Pemba, she teaches the national canon of Swahili literature at the local Haile Selasi secondary school. She also recites poetry to pray for blessings for the bride and the groom in an adapted verse form of the Arabic dua, which in its Swahili form is widespread along the East African coast, as well as the Comoros and Kenya. In its Arabic form, it is used all over the Indian Ocean Muslim world. In the verse forms, the voices and the imagery, one finds traces of older cultural Indian Ocean connections and common poetic practices. Many of them, genealogically bound to the past, have become a 'shared secret'. It is a secret that slowly seems to be fading out of people's cultural memory but does not easily leave the hands of poets, who write in well-balanced lines echoing earlier Swahili manuscripts in Arabic letters, their ears and tongues hearing or reciting verse metrically bound in a modulating voice with an emphasis on pronouncing Arabic loanwords in a more Arabic style than Swahili usually does. In Mwana Kombo's poem about the new road, it is the form of the poem which reflects the wider geography of the Indian Ocean.

The present poetry of a smaller world forms a contrast to the poetry of previous centuries: the Indian Ocean figures prominently in song poetry dating back to the eighteenth century, as I will show. Thus, a transregional perspective is not a product of contemporary ideas questioning the confines of the nation-state, but, in fact, predates the nation-state with its narrow perspectives.

Sensuous oceanic geographies in ancient poetry from Lamu

The earliest Swahili poetry of which we have evidence comes from the Lamu archipelago, geographically situated opposite the coast of what is now northern Kenya. It is a cycle of loosely related song poems, which has been attributed to the heroic warrior-poet and master singer Fumo Liyongo. He is not only the author of songs but also the subject of many narrative oral traditions on the Lamu archipelago and the coastline directly to the west in the delta of the Tana River (Werner, 1926). While the northern Swahili versions were committed to writing using Arabic script in the nineteenth century, when local scholars felt the need to preserve them, the Liyongo songs and stories had been passed on orally for a long time before this, which makes dating a difficult issue (see Steere, 1870 for a discussion). While a few scholars relying on oral sources and local Swahili and Arabic chronicles have dated Liyongo back to a period 'before the 13th century' (Shariff, 1988: 53, my translation), a number of historians situate him in the sixteenth and seventeenth centuries (for an overview, see Pouwels, 2002: 407).

While the Liyongo songs and tales were forged into more coherent, plot-driven narratives by local bards in the nineteenth century (see Meinhof, 1924/1925), oral traditions (some still existing) and manuscripts in Arabic script recount and preserve single episodes and only thematically loosely related songs. There is, for instance, the account of Fumo Liyongo's spectacular escape from the dungeons where his half-brother and enemy Mringwari, had imprisoned him (see Werner, 1926: 249ff.). Mringwari had become the ruler of the important city-state of Pate on the Lamu archipelago, instead of Liyongo, and he sought to trap and kill Liyongo – but in vain: the clever Fumo Liyongo made his way out of prison with the help of a file which his mother had baked into a loaf of bread.

For Pouwels (2002), the stories and songs reflect a fundamental transition of power in the region: power moves from clans in the Tana delta, as, for instance, Liyongo's Bauri clan, to new elites on the islands of Pate, and later Lamu. While the former – perfectly embodied by Fumo Liyongo – adhered to a specific set of cultural practices, influenced by early Indian Ocean connections (Vernet, 2015), but mostly from the African mainland, like beliefs in spirits, cognatic patterns of inheritance, shifting agriculture and oral poetry traditions like the ones described here, the latter – embodied by Liyongo's half-brother Mringwari – introduced more 'literate Shari'a based rules favouring patrilineage' (Pouwels, 2002: 407). Thus, according to Pouwels, the poems echo the beginning of what later becomes a wave of Arabization reaching the coast from across the Indian Ocean, changing older rituals, legal practices and forms of knowledge production. It becomes most manifest in later Islamic poetry of the 18th and 19th centuries, which I will turn to below.

The Liyongo poems are still composed in an ancient metrical long form, the genre of the utumbuizo, where the verse meter depends on the singer's breath and not on abstract syllable count, which accounts for a prosodic vigour in line with the moral strictness of the eighteenth century. The tumbuizo are far more serene and festive, performed 'on solemn occasions' ('dans une circonstance solennelle', Sacleux, 1939: s. v.), such as weddings. In some contexts, they seem to have been linked to dance, the gungu, of which we only have a vague idea. The gungu used to be widespread all along the coast, from what is now northern Kenya to northern Mozambique – another shared secret of the coast. In Mombasa linked to the New Year's celebration that followed the Persian calendar (Sacleux, 1939: s.v., Knappert, 1979: 54), Steere describes it as follows for Zanzibar: 'It is the custom to meet at about ten or eleven at night and dance until day break. The men and slave women dance, the ladies sit a little retired and look on. (…) The first figure is danced by a single couple, the second by two couples' (Steere, 1870: xi).

Differently from the religiously inspired poetry of the eighteenth and nineteenth centuries, the dance poems provide us with sensuous tableaux of the East African coast. They evoke the Indian Ocean connections through precious goods from afar, fruits, scents and clothes, creating carefully cut-out pregnant moments, which completely disappear in later Swahili poetry. For instance, the poem of which a few lines are given below, the Utumbuizo wa Liyongo Harusini ('The Poem of Liyongo at a Wedding'), has a rhythmic force, taking the wedding guests' preparation as an occasion to present their fine attire and the scents of an elegant and cosmopolitan gathering, as if the poet was painting a multisensorial picture:

> Mujipake na twibu hiyari yalotuwa kwa zema ziungo
> Choshi ni maambari na udi fukizani nguo ziso ongo
> Fukizani nguo za hariri na zisutu zisizo zitango
> Pachori na zafarani na zabadi twahara ya fungo
> Na zito za karafuu tuliyani musitiwe t'ungo
> Na itiri na kafuri haya ndiyo mwiso wa changu kiungo
> Lines 10–15 taken from 'Liyongo Songs' (Miehe et al. 2004: 40)
>
> Put on choice scent made of the best ingredients,
> treat your spotless clothes with the **incense** of **ambergris** and **aloe-wood**,
> treat your **silken garments** and finely made ***zisutu*** **clothes**
> with **patchouli** and **saffron** and pure **musk of the civet**,
> and ground them with **clove-buds**, lest people criticise you,
> and (finally) add **perfume** and **fine camphor** to complete the ingredients.
> (my emphasis)

In this poem, according to Jan Knappert (1991: 214) 'the oldest known dance song in Swahili and perhaps any Bantu language', the wedding guests – Liyongo's sister is getting married – are urged to put on elegant clothes, silken garments and zisutu, cotton cloths with small red and blue patterns imported from India (see Ivanov, 2017: 373 for a picture). Textiles were so highly valued that they even served as payment in the founding myths of Kilwa and Angoche (Schadeberg, 1999). In the form of turbans and robes, they became important points of distinction for town dwellers all along the coast. Fragrances are omnipresent in the poem, reflecting the high cultural value put on scents. Creating an 'olfactory rich environment' was not only important in rituals, but was a 'sign of sophistication and refinement' (Hillewaert, 2020: 51). Exquisite perfumes of patchouli (pachori) 'importé de l'inde' (Sacleux, 1939: 722 s.v.), saffron (zafarani), made out of the yellow crocus (Knappert, 1991: 221), and the expensive musk of the civet, zabadi, as well as cloves, karafuu, were widely traded spices in the Indian Ocean world. While civet became an important export good from the Swahili Coast in the Indian Ocean trade (Horton & Middleton, 2000: 13), cloves (like cinnamon, nutmeg and pepper), originally imported from Asia, became such sought-after spices in the Indian Ocean economy, that they were also increasingly grown on huge and (slave) labour-intensive plantations on Zanzibar and the neighbouring green island of Pemba from the eighteenth century onwards.

These traded scents, textiles and spices were part of what the historian Thomas Vernet characterizes as the 'premodern phase' of Swahili Indian Ocean trade. As he shows, relying mostly on Portuguese, French and British sources, between the fifteenth and eighteenth centuries, Swahili seafarers, sailors, captains and boat-owners from Pate island, where the Liyongo songs come from, travelled along the 'Swahili corridor' to Madagascar and the Comoros, and not only received goods from Arab and Indian traders but also – though only to a moderate extent – sent their mitepe, typical East African sewn boats, to southern Arabia and western India. The coastal inhabitants of Eastern Africa exported ivory, tortoise shell and timber, and increasingly foodstuffs, like millet and dried fish. They imported textiles, like cotton and silk, glass, jewellery, fragrances, sandalwood, kohl, spices, pottery and porcelain, and brass and bronze ware (see also Horton & Middleton, 2000: 13).

Not only the actual spices and scents evoke 'the coastal residents' historical transoceanic connections with Arabia' (Hillewaert, 2020: 51) but also the words (which sound foreign to a Swahili audience) referring to the spices and perfumes in the poem, which entered Swahili vocabulary before the eighteenth century, keep alive a memory of their distant origin and evoke a wider oceanic imaginary (see Schadeberg, 2009). Originating from different Indian Ocean languages, they mostly made their entry into

Swahili through a transregionally shared Indian Ocean Arabic (Schadeberg, 2009: 85): For instance, the word kafuri for camphor, an aromatic smell, was borrowed from Arabic (kafūr) (Sacleux, 1939: 317), but originally found its way into Arabic through Indian languages (Sanskrit karpura), and might be of Austronesian origin. To give another example, the word 'pacholi' or 'pachuli', a plant that is 'related to mint', is initially of Tamil origin (Knappert, 1991: 221).

The Utumbuizo wa Mwana Mnazi ('The Utumbuizo of Lady Coconut'), another poem attributed to Liyongo, also presents itself as a miniature portrayal of the rich cosmopolitan, coastal material culture: a lady receives her long-awaited husband (probably Liyongo) at home and urges the servant, Time, to attend to his needs:

Nitiyani hoyo Time aye, ¨ naanguse, ate masindizi.
Apakuwe pilau ya Hindi, ¨ mzababu isiyo mtuzi.
Ete kiti chema cha Ulaya ¨ na siniya njema ya Shirazi.
Mnakashi inakishiweyo ¨ na sahani hung'ara ja mwezi.
Kaamuru hudama na waja, ¨ ai, ninyi, hamtumbuizi?
Basi, hapo akamwandikiya, ¨ naye Time yushishiye kuzi,
Yapijeto mayi ya inabu ¨ kimnosha kama mnwa shizi.
Kimlisha akimpa hiba ¨ kimuonya yakeye maozi,
Kimlisha tambuu ya Siyu ¨ yi laini laini ya Ozi.
Kiukuta k'ono kimwakiza, ¨ kwa iliki pamwe na jauzi.
Lines 22–31 taken from 'Liyongo Songs' (Miehe et al., 2004)

[Call Time, let her come ¨ quickly and let her not tarry.
Let her serve up a dish of Indian pilau ¨ cooked with raisins without curry.
Bring a fine European chair ¨ and a good Persian tray
Engraved with fine designs ¨ and a plate shining like the moon.'
And she ordered the servants and slaves ¨ 'You, why are you not entertaining (us)?'
Immediately she served him, ¨ while Time held the jug
Full of grape juice. ¨ She made him drink it like a drinker of wine.
She fed him lovingly ¨ and gave him a meaningful gaze,
Offering him the special tambuu of Siyu, ¨ as soft as the one from Ozi.
She folded it and put it into his mouth ¨ together with cardamom and betel nut.]

The poem speaks of the lady's ardent desire to please her husband: the servant has to bring him a 'European chair' (kiti cha Ulaya), also called a kiti cha enzi, a commonly shared Indian Ocean piece of furniture from the epoch of Portuguese rule (Ivanov, 2017: 372). More like a throne than a chair, due to its rectangular design and its tall back, with what was later called an 'Indo-Portuguese design' of 'flat pieces of plainsawn hardwood' (Meier, 2016: 363) and an inlay of mother of pearl and ivory, the chair, found all over the Indian Ocean, symbolizes the affluence, power and (specifically Indian Ocean) cosmopolitan sophistication of its owner. The food served to the husband adds to the imaginary of far-away places: an Indian rice dish (pilau), together with precious grape juice (a foreign drink from a Swahili perspective), on a tray with fine engravings from Shiraz (Persia). At the end of the poem, she puts a betel bite (tambuu) into his mouth. This is a stimulating drug, originally from South East Asia, used also as an aphrodisiac. Chewing it became a widespread practice among affluent merchants keen on cultivating a bohemian lifestyle along the East African coast before the nineteenth century. Other Swahili poetry from Lamu of the time sings its praises (Vierke, 2019); as Burton (1872: 430) notes in his nineteenth-century Zanzibar travelogue, the 'Waswahili may be said to exist upon manioc and betel'. Travelling to the Ilha de Moçambique in the nineteenth century, Carlos José Caldeira is seemingly impressed when he catches sight of an elegant, lasciviously dancing woman, who finally offers him a betel bite as the sign of a cordial welcome ('um sinal de cordialidade') (quoted after Saute and Sopa, 1992: 135). Hardly in use anymore today, the ingredients of the betel bite, areca nut, tobacco and spices, were planted locally: as the poem underlines, the betel leaf came from Siyu, on the island of Pate, and Ozi, the home of Fumo Liyongo.

The betel bite she puts into his mouth at the end of the poem, together with her telling look, foreshadow a love scene which the poem leaves the audience to imagine: the borders between the sensuous and the sensual are often fluid in the Liyongo poems. The senses are encouraged to wander on imaginary journeys, inspired by the dense descriptions in the poem. Besides its usage as a stimulant, the betel bite became a poetic leitmotif associated with seduction in poetry of the time, comparable to champagne in a Western context. In the Kamasutra, the betel bite is a recurring topos, placed by lovers into each other's mouths (Doniger, 2016: 30). Thus, for the erudite high class Swahili audience of the time, learned in textual traditions from India and Arabia, the betel bite was a reference to the Indian text, which added to the Swahili poem's erotic reading.

Thus, one would hardly do justice to the poems by merely considering them as the oldest historical documents witnessing oceanic trade in material goods like fruits, scents and furniture in an African language (which in fact they are!). The above poem creates a carefully painted East African tableau, a meal and love scene in a patrician mansion, where the food, the scents, the attire and the furniture celebrate the exquisite lifestyle of the mansion's inhabitants. In her consideration of the particular mundane interior decoration of nineteenth-century Zanzibar, Prita Meier (2017: 364) highlights the intended 'intimate, multisensorial experiences of a range of "elsewheres"', which the much earlier Liyongo poems also seek to evoke. They recreate the larger world of the Indian Ocean not by referential links, but by evoking tastes, scents and the haptic experience and sight of the interior and people's attire. There is a sensuous immediacy of a common Indian Ocean space, with its particular tastes, scents and other sensuous qualities, which invites the audience to indulge in further, unlimited imaginations. The Liyongo poems are not narratives of oceanic connections, tracing an imaginary map like the Sindbad story, but render the Indian Ocean palpable in minutely cut-out scenes, so unique in Swahili poetry. The Indian Ocean here figures as what Ivanov has characterized as an 'aesthetic seascape' (Ivanov, 2017: 369), a shared world that emerges through shared materiality and shared sensual perceptions.

I will draw here briefly on recent debates in respect of the knowledge-producing capacity of artworks, which has a long tradition in aesthetic philosophy and has been re-explored recently in philosophy, the social sciences and the arts. As Durs Grünbein (2007) has put it, the poet does not come up with an abstracting argument, but sticks to the 'abundance of characteristics' of the material world (Grünbein, 2007: 89). And in doing so, the poem stimulates much thought and imagination which cannot be reduced to a neatly expressed argument, for this would abstract from the multitude of impressions. For Grünbein, this is the paradox of an insurgent 'palpability', on the one hand, which the poem creates through its imagery and form, and a 'blurred view', on the other, meaning that it cannot be easily translated into conclusions ('Anschaulichkeit und Unschärfe') (Grünbein,

2007: 84–94). In the ancient Swahili poems, the image of the Indian Ocean certainly remains blurred, a 'shared secret', since it figures only in allusions to the senses – which, however, also accounts for its appeal. In this sense, the poem presents a kind of understanding before there are words for it. Adorno (1997: 126) speaks of the artwork as revealing a truth that is incommensurable with discursive knowledge. It essentially presents itself as a sudden apparition that escapes the logic of a single argument. He compares artworks to elves in a fairy tale, since they stir much thought, yet, since they cannot be pinned to one idea, they appear as riddles (1997: 126).

Changing times and maps: Al-Inkishafi

As hinted at before, the eighteenth and particularly the nineteenth century saw the emergence of a whole new body of Islamic poetry in Swahili, authored by an elite of Islamic scholars who had wide connections across the Indian Ocean, and were well-versed in Arabic, but deliberately turned to Swahili for composing poetry. They belonged to Sufi brotherhoods, foremost the Alawiyya, which had its beginning in Hadramaut, southern Yemen, in the twelfth century. From the sixteenth century onwards, it spread with the trade network, spanning a large part of the Indian Ocean, and linking communities in Arabia, Indonesia, India and East Africa. Adherents of the Alawiyya, who considered themselves Sayyids, that is, descendants of the Prophet Muhammad and accordingly gifted with *baraka* (blessings), arrived on the Swahili coast at that time. But it was in the eighteenth and nineteenth centuries that Lamu and Pate saw a new wave of migration (Bang, 2003: 25). They became centres of a Sufi reform movement which spread along the coast, spearheaded by the highly educated Hadhramis. For them not only asceticism, mystical practice and education in classical Arabic were of importance: composing poetry in Swahili also became a primary concern, with the aim of starting an 'Islamic revival and counterattack against the Portuguese' (Pouwels, 1992: 270; see also Pouwels, 1987: 38–42), and later against the British. Challenging the local establishment, their aim was to pass on Islamic knowledge to a broad section of society. Many people had previously had no access to religious knowledge, which was exclusively taught and discussed by a restricted elite in Arabic. Literacy in both Arabic and Swahili gained in prominence, as well as new ritual practices, supplementing earlier oral institutions and modes of learning (Pouwels, 1992: 92; see also Bang, 2014: 111ff.; Bang, 2003: 31).

The tone, form and imaginary maps of their poetry are of a fundamentally different nature from those of older poems. Lavish dance poetry gave way to poetry that followed an ethos of asceticism, frowned upon earthly pleasures, and overflowed with religious zeal for the hereafter. No other text exemplifies this shift in epistemological perspective better than the poetic masterpiece Al-Inkishafi, written by Sayyid Abdallah bin Nasir (1720–1820), which was held in high esteem for its profound reflections and layers of meaning. The poem depicts the rich, 'lantern-lighted' mansions, which we encountered in the Liyongo poems: the banquet table is set with porcelain, fine goblets and crystal pitchers:

Nyumba zao mbake zikinawiri
kwa taa za k'owa na za sufuri;
Masiku yakele kama nahari,
haiba na jaha iwazingiye.

Their mansions, lantern-lighted, glittered bright
with **brazen lamps** and **lamps with crystal dight**;
Till light as day became their hours of night,
and in their Halls walked Fame and Honour twain.

Wapambiye sini ya kuteuwa,
na kula kikombe kinakishiwa;
Kati watiziye k'uzi za k'owa
katika mapambo yanawiriye

Their banquetings with **ware of T'sin** were spread,
each **goblet with fine graving overlaid**;
And set in midst, the **crystal pitchers** made
bright glitter-glow upon the nappery lain.

And set in midst, the crystal pitchers made bright glitter-glow upon the napery lain.

Taken from Hichens (1939, stz. 37 and 38, my emphasis)]

In the Inkishafi, however, the affluent houses full of music and laughter remain only in the memory of the lyrical I. A few stanzas later, the narrator is walking through houses that lie in ruins, full of empty echoes and scary animal sounds. Instead of the houses' owners and their distinguished guests, bats, spiders and an owl have taken possession of the house, destroying the achievements of civilization: 'Where once in wall-niches, the porcelain stood, now wildling birds nestle the fledgling brood' (stz. 50; Hichens, 1939: 81). The powerful owners no longer reside in their mansions, but lie in 'houses of sand' (nyumba za tanga-tanga, stz. 60), that is, their graves. The scenery is constructed very much in antithetical terms. The ruins are not just empty, but appear as a distorted picture of the previous mansions:

Nyumba zao mbake ziwele t'ame;
makinda ya p'opo iyu wengeme.
Husikii hisi wala ukeme;
Zitanda matandu walitandiye.

[Their mansions bright, with empty echoes ring.
High in the halls the fluttering night-bats cling.
Thou hear'st no outcry; no sweet murmuring.
The spiders, o'er the couches, spin their skein.
(stz. 49; Hichens, 1939: 80]

On the one hand, the poem is a reflection on the brevity and vanity of life, inspired by a new austerity and religious sentiment. The futility of the here and now is drastically underlined by the last stanzas of the Inkishafi, dedicated to the depiction of the circles of hell awaiting the sinner. On the other hand, the poem has also been read as having a historical meaning, depicting the fall of the mighty city-state of Pate, from a major, cosmopolitan port city to an isolated place on the coast, on the periphery (Vierke, 2016). The poem spells out the experience of a fundamental shift in the coastal map of power constellations, a historical rupture related to Zanzibar's rise. It is Pate's history of decay, which in real time was a gradual process following a series of political confrontations and fights over succession to the throne. There was increasing competition over economic hegemony in changing Indian Ocean trade constellations, which favoured Zanzibar as well as its ally, the neighbouring Lamu, which demanded heavy taxes from Pate (see also Hichens, 1939). With a tone of nostalgia that is also present in the poem (besides or parallel to the contempt for the here and now), the lyrical I recalls Pate's splendid days. The narrator evokes the rich material goods only in retrospect, making them palpable to the audience. He highlights the 'cosmopolitanism' of the translocal material culture of the Indian Ocean that characterizes the decoration and furniture of the house (Ho Seng, 2006), in a way that reminds us of the Liyongo poems. But it is the transience of this cosmopolitan world on Pate, the changing map, which the poem underlines. The ruins also foreshadow a changing imaginary map of Swahili poetry, where after the Inkishafi, the tableaux of a sensually rich East African urban landscape, in which far-reaching connections appear in the here and now of the poem, never come back.

Instead, many of the poems, composed in the eighteenth and nineteenth centuries and in the first half of the twentieth century, create another transoceanic poetic link that takes their audience across the ocean and back in time to the Arabian peninsula of the Prophet Muhammad, in all his glory. The Utendi wa Tambuka ('The Poem of Tambuka'), for instance, often considered as the earliest written Swahili utendi, adapted by a poet called Bwana Mwengo for the Sultan of Pate in 1728, recounts the battle of the Prophet against the Byzantinian emporer in a suspense-driven colourful epic narrative, a template for many tendi to follow (Knappert, 1958).

The poets, respected Islamic scholars like Sayyid Abdallah bin Nasir, the composer of the Inkishafi, enthusiastically immersed themselves in the task of reproducing Arabic poetry in the newly created Swahili verse prosody, translating Hadhrami verse patterns into Swahili lines of a new metrical strictness – also visible on the manuscript page. Swahili in Arabic script, previously used only occasionally and foremost in letters and inscriptions, became systematically used to write down Swahili poetry at that time, and manuscripts of the most popular poems travelled all along the East African coast. The first known translation is the Hamziyya, a translation into archaic Swahili of the Arabic poem Umm al-Qurā by the thirteenth-century Egyptian poet, Muhammad ibn Sa'īd ul-Būsīrī. The first Swahili manuscript of it is considered to have been written in 1749 by Sayyid 'Idarusi bin Athman, a prominent theologian and scholar of Arabic from Pate, related to the composer of the Inkishafi (see Knappert, 1966; Harries, 1952). Manuscripts of the poem have been found in what is now Somalia, Tanzania and Kenya. Nineteenth-century translations spread from what is now Kenya to northern Mozambique along the scholarly networks of the Sufi brotherhoods (Bang, 2014; Knappert, 1971). These included, for instance, Ya Dura Mandhuma 'Strung Pearls' (Harries, 1953) by Sayyid Umar bin Amin, the chief kadhi in Siyu on Pate; the Maulidi Barzanji (Neuhaus, 1935); and the Tabaraka (Dammann, 1960), an austere poem preaching the brevity of life, by the most reputed Sharifu Mwenye Mansab, whom I will turn to below. Furthermore, a great quantity of epic narrative poetry in the Swahili utendi genre of the kind mentioned above came into being, accounting for a huge body of Swahili manuscripts kept in private archives in Lamu, Mombasa and Tanga, as well as the Ilha de Mozambique and Angoche. This poetry describes legendary battles in the early history of Islam, or provides hagiographies of the Prophet Muhammad, or other prophets, such as Ayubu (Job).

While the production of Swahili manuscripts started on Lamu and Pate, in the nineteenth century Zanzibar attracted many of the most erudite Hadrami scholars. Learned in local jurisprudence, they not only occupied high-ranking positions as kadhi under Omani rule but also brought their poetry with them. It was also under Zanzibari tutelage, and with the help of Comorian networks, that trade connections expanded significantly, but also the Sufi movements, and with them the Swahili language. It was in the nineteenth century that the Swahili coastal map reached its widest extension. In a seemingly paradoxical way, at a time of colonial onslaught and an increasing influx of Arabic, Swahili spread through maritime exchange to Madagascar, the Comoros and Southern Arabia, as well as into the interior along the caravan routes, and all along the coast, from Somalia to Mozambique (see Alpers, 2000; Ingrams, 1967: 6). Below, I will take a closer look at one of the most widespread poems, the Maulidi Barzanji.

The coast as a transregional Swahili literary cosmopolis

Wewe ndiwe yua, ndiwe mwezi utimiyiwo wewe ndiwe nuru weye nuru u yiu lao

[You are the sun, you are the full moon, you are the light, you shine above them]

(Neuhaus 1935: 159, stz. 24d with a slight adaptation of the Swahili, my translation)

In the nineteenth century, the most widespread and best-known text all along the East African coast, from Barawa in Somalia to northern Mozambique, and on the islands including the Comoros, was a praise poem, in Swahili maulidi, recounting the Prophet's birth and life, interspersed with ecstatic devotional praises of his outstanding personality (Bang, 2014: 24). In the Sunni-Islamic world, mawlid celebrations with various forms of recitation of texts (prose and poetry) reach back to the thirteenth century, while the globally most widespread mawlid text was composed in Arabic in the eighteenth century by Ja'far b. Hasan al-Barzanjī (1690–1765) (Bang, 2003: 149). The Maulidi Barzanji was translated into various other languages in the Muslim world, including Turkish (see Knappert, 1971) and Acehnese in Indonesia (see Harries, 1958), as well as Swahili. In the late nineteenth century, recitation of the poem in Swahili played a central role for the Sufi brotherhood of the Alawiyya on Lamu, whose members publicly displayed their ritual affection and love for the Prophet in front of larger audiences. The recitation practice of the Maulidi quickly spread to Bagamoyo, as noted by Mtoro bin Mwinyi Bakari in his early twentieth-century account of Swahili customs (Velten, 1903: 11), Zanzibar (Dale, 1920: 65), and almost all towns along the coast (Bang, 2003: 149).

It is not clear exactly when Arabic versions of the Maulidi Barzanji in prose and poetic form reached East Africa, but they were recited in almost all towns along the Kenyan, Tanzanian and Mozambican coast, and in the Comoros (Bang 2003: 149). One of the earliest, if not the earliest (Harries, 1958: 27), adaptation into Swahili was done by Al-Seyyid Abu Bakr bin Abdul Rahman al-Husseiny (1828–1922), commonly known as Mwenye

Mansab, a reputed and charismatic Islamic scholar and healer from Lamu. Similarly to other Sufi poets mentioned above, he was of Hadhramy origin, came from a family of scholars and poets, and was considered to be a descendant of the Prophet Muhammed endowed with special blessings. He studied law and theology under important scholars in Mecca and the Hadhramaut before becoming kadhi at the Zanzibari court, and later, after the Sultan's death, in Lamu town. He not only translated texts on fiqh (jurisprudence), and gave advice on Bible translations, but was also an authority on Swahili poetry, 'unrivalled in his knowledge of archaic Swahili' (Hichens 1939: 5). He adapted a number of Arabic poetic traditions, like the above-mentioned paranetic Tabaraka (Dammann, 1960), or the mystical Kishamia (Dammann, 1940); he copied Swahili manuscripts, and he knew poems like the Inkishafi by heart. He became reputed for his Swahili version of the Maulidi Barzanji. His well-arranged and beautifully adorned Arabic text with interlinear Swahili translation was recited not only on Lamu and Pate but also in Tanga and Zanzibar (Ahmad, 2002: 13ff.).

In 1891 Mwenye Mansab composed at least two Swahili versions of the Maulidi Barzanji of different length and with different rhyme schemes. The multitude of later Swahili versions seems to echo the variation in the Arabic tradition (see Dammann, 1958: 43–44; see Neuhaus, 1935, Harries, 1958, Knappert, 1971). The Arabic text of the Maulidi Barzanji he based his Swahili adaptations on was an abridged version of the original text by Ja'far b. Hasan al-Barzanjī, as Mwenye Mansab mentions in the introductory stanzas of the poem (the so-called dibaji): indicating one's sources is common practice in this context. Furthermore, Mwenye Mansab also gives a reason, why he decided to translate the Prophet's praises into Kiamu, the dialect of Lamu: on the one hand he wants to become part of a wider Indian Ocean, or even global, community of Muslims praising the Prophet, reaching out beyond the East African coast – an aspect that is key in all poetry at the time. Leaving the small world of the coast and creating a – transoceanic – bridge in time and space to Arabia at the time the Prophet lived there, joining a wider community of worshippers, and meditating on his wonders, his beauty and his perfection, is a typical feature of the poetry at that time which is echoed in Mwenye Mansab's words. On the other hand, to bridge the distance in time, space and language, he brings the Prophet closer to the people in East Africa by using the Swahili dialect of Lamu, and not 'their language': he thus distances himself from Arabic and emphasizes his local identity:

> Taitiya kati ya wasifuwo henda hazipata tatuko zawo
>
> Illa sitayuwa kwa lugha yao tanena Kiyamu mutafamau
>
> [I will include myself among those who praise so that I may follow in their steps
>
> But I will not use their language I will speak Kiamu and you will understand.]
>
> (Harries, 1958: 31, stz. 2)

The poem – though taking on different forms in different versions – has the same core, 'ranconter en psalmodiant, ou plutôt en chantant, la sīra (biographie) du Prophète Muhammad dans la passion et l'allégresse' (Ahmad, 2002: 13). After introductory verses in which the poet tells the audience to prepare the stage for the important event – 'clear the place where the poem is recited/make it look pleasant' (safidi mahala pa kusomeya/ufanye uzuri, Harries, 1958: 31, stz. 5) – recite the opening fatiha and burn incense, the actual poem starts by invoking God's blessings and praising the Prophet as the 'wall of his light' (kuta la nuruwe). The poem uses hymnal language: 'in countenance and conduct' (stz. 35) the Prophet is depicted as surpassing all beings: he is of fair skin colour and medium height (stz. 39), full of grace, and intercedes on behalf of all mankind. Besides the interspersed praises, sometimes recited by a choir, aspects of the Prophet's life are recounted: his ancestry, his birth, foreshadowed by miracles (in the version edited by Neuhaus, wild animals announce his birth (Neuhaus, 1935: 157, stz. 19)), the maidens who breastfed him (stz. 25) and his mysterious ride to the heavens, when he saw 'the Lord with the naked eye' (stz. 34). In Mwenye Mansab's and later versions, the poem ends with duwa, prayer, imploring the Prophet to bestow peace and good health on the audience (stz. 47 and 48) to open their hearts, and make them firm in religion (stz. 49).

The poem is more than just an account of a life; it has a tension-driven plot. At the centre of the poem, the Prophet is born and later meets God in the heavens, underlined by praises. It has an uplifting, vital force, a power that makes people overflow with love for the Prophet, who is not only a moral guide but, on a more mystical level, 'a carrier of the divine, eternal spiritual essence'. The light metaphor is so strong in the poem, which extols him as 'the sun and the moon' (stz. 15), 'the lamp of good hearts' (stz. 15), or 'the full moon', so that 'all (other) moons were extinguished' (stz. 14). Such ecstatic language is rather unusual in East African Sufi poetry, which typically takes a more sober or warning tone (see also Ahmad, 2002). The poem echoes the strong devotional focus of the Alawiyya and other Sufi brotherhoods on the Prophet Muhammad, the centre of mystical practice, created from the divine light (Bang, 2003: 18ff., Schimmel, 2003: 171–212), which Sufi disciples are eager to approach through exercises and ritual practices – much opposed by orthodox Islamic currents at the time and even more so today. Coming back to the notion of the aesthetic experience as an untranslatable sensuous revelation, the poem's aim, particularly in its recitational form, is to go beyond words, to open the hearers' hearts to a transcendent experience, the experience of the Prophet's presence, which constitutes a higher form of 'understanding', since it leaves the limits of human language behind. In present-day recitations of the Maulidi Barzanji, which may be similar to recitations in Mwenye Mansab's time, body movements add to this effect: at the beginning the reciters are seated, swaying their torsos back and forth, but when the poem reaches the point of the Prophet's birth, the performers and the audience have to stand up to pay reverence to him, and start to move rhythmically to the left and right in unison with the poem's rhythm, often enhanced by drums (see also Ahmad, 2002: 75).

In East Africa in the late nineteenth century and early twentieth century, the poem took on slightly different textual and recitational forms, sometimes putting more emphasis on music and performance, sometimes shying away from translation into local languages, and it spread further south to what is now northern Mozambique. This reminds us that though the coast shared texts and recitational practices, the adaptation, interpretation and performance of texts differed according to the hermeneutic practices, religious and poetic influences of the specific location. On Lamu, for instance, the ritual was significantly altered by Mwenye Mansab's most important disciple, the charismatic Salih b. Alawi Jamalil-Lail (1844–1935), of Hadhrami origin, born on the Comoros, and commonly and fondly known as Habib Saleh (Bang, 2003: 149). Not only did he gradually put it on a different textual basis (the so-called Maulidi al-Habshi, named after its Yemeni composer), but the aspect of musical performance became much more important: in the early twentieth century, the Maulidi started to take the form of a cheerful procession through town, accompanied by strongly rhythmic music driven by tambourines (matari) and dances, inviting all members of society, including laymen, to participate in a 'proper' Islamic ritual. Initially met with much opposition from the establishment, from 1910 onwards, the Maulidi celebrations started to attract people from all over East Africa (Ahmad, 2002; Bang, 2014), and still continue to do so today. It became a hallmark and a celebration of translocal East African Islamic identity. The practice of processions spread to the Tanzanian coast, for instance, to Bagamoyo (Velten, 1903), while on the Comoros, the Maulidi was preserved in the form of Arabic recitation, performed either by men in the mosque and in semi-public, or by women in private contexts (Ahmad, 2002).

The Maulidi Barzanji also reached the Ilha de Moçambique, presumably at the end of the nineteenth century, since Maulidi performing groups, which my colleague Chapane Mutiua and I visited in August 2015, reported to us that it has been performed at least since the time of their great-grandparents, and still continues to be performed by men's or women's groups. We also found many people on the island who owned recently produced cheap printed brochures of the Arabic text, which the women in the front row, at the Maulidi performance we were allowed to attend, held in their hands – more for symbolical value, since no one was reading from it. The Maulidi women's group, which is regularly invited to perform the Maulidi to celebrate important events, sang the Arabic poem by heart, throwing in praises of the Prophet in Enaharra, their local variety of Emakhuwa. What surprised us most was that they also interspersed the Arabic lines with Swahili verses and praises, although none of the performers spoke or understood any Swahili, nor did we ever come across any printed or handwritten Maulidi text in Swahili. The coach of the group was Chale Mussa, an Islamic scholar, learned in Arabic, who came originally from Moussuril, and earned his living as a tailor, but also ran a madrassa for girls on the Ilha de Moçambique. He trained the women's group to memorize and sing the Arabic text and explained to us that the Swahili verses have always been part of the performance, as far as he remembers. Chale Mussa was born in the 1930s. He also told us that he remembered that long epic Swahili poems of the utendi type, like the Tambuka and the Kaiza, were regularly recited on the Ilha de Moçambique when he was a child. Recitations at night would attract large numbers of people, who would listen to the story of the First World War depicted in the 'Poem of the Emperor' (Kaiza) for three nights in a row. Nowadays, Swahili poems, of which Chale Mussa and other elders remember some stanzas, are not recited on the Ilha de Moçambique anymore, but, to some extent, in neighbouring Angoche, which, like the Ilha de Moçambique, had a strong connection to the Comoros and Zanzibar in the nineteenth century. There are traces here of an older nineteenth-century huge coastal Swahili map, reaching from Somalia to the Comoros to Mozambique, now buried under layers of new political realities and a turbulent history in the twentieth and twenty-first centuries. At the Maulidi performance we attended, this map seemed to resurface, preserved, like a fossil, in the practices of recitation – the women stood up at the point of the Prophet's birth, swaying their bodies back and forth – and in the interspersed Swahili lines, whose meaning was clear to only a few of the women.

Connections between sultanates along the Mozambican coast and the city-states of what are now Tanzania and Kenya have a long history of maritime trade networks, sharing material culture, political and familial relations, and Islamic practices (see also Newitt, 1995): Islam arrived on the Quirimba archipelago, the Ilha de Moçambique, Angoche and Sofala between the twelfth and fifteenth centuries (Newitt, 1995; Macagno, 2007: 85). At the end of the nineteenth century, the wave of Sufi reform reached Mozambique. The north of the country increasingly came under the influence of Omani Zanzibar, and the Sufi orders which grew under its tutelage (see Macagno, 2007: 86). The brotherhood of the Qadiriyya, which Chale Mussa was linked to, was brought to the Ilha de Moçambique from Zanzibar in 1904 or 1906 by a 'native of Moroni', the Comoros, Issa bin Ahmed. He had done his Quranic studies on the Comoros and later in Zanzibar (Alpers, 2001), and was often mentioned to us as a pioneer on the Ilha de Moçambique. The Comoros played an important role in creating a triangle between Zanzibar and the Ilha de Moçambique, because they had become a centre of Hadhrami migration and home of many scholars. A close connection between Comorian scholars of Hadhrami descent, who had connections with their counterparts on Lamu, Pate and later Mozambique, and the rising Busaidi state of Zanzibar evolved from the mid-1830s onwards (Pouwels, 1987: chap. 6 & 8). The Shadhiliyya, another brotherhood, brought by emissaries from the Comoros, established a foothold on the Ilha de Moçambique in 1896 and soon spread to other coastal towns and the hinterland.

In this context of the rising and well-connected Sufi brotherhoods, which came to northern Mozambique with a programme of Islamic education (da'wa), Swahili became a language of learning and teaching in the newly founded Quranic schools, supplementing Arabic, and reaching large parts of society (Alpers, 2001: 85; Newitt, 1995: 188). This was not the first appearance of Swahili on the Mozambican coast, as it had left its imprint on local languages at several points in history, and a sense of deep-rooted kinship certainly helped to foster Swahili as a medium of instruction at this time. While the larger Swahili world was important in the early history of the coastal towns – Angoche, for instance, derives its origin from Kilwa (in present-day Tanzania) – the arrival of the Portuguese in the fifteenth century and the destruction of Kilwa, made the city-states in northern Mozambique subsequently lose their orientation towards what is now the Tanzanian and Kenyan coast (Newitt, 1995: 10ff.) and turn towards the Mozambican interior. This shift has left traces in some smaller local languages, such as Ekoti spoken in Angoche (Schadeberg & Mucanheia, 2000), or Kimwani spoken in the Quirimba archipelago. A large part of their vocabulary is Swahili (Schadeberg, 1994), but in structure they are closer to Emakhuwa and Shimakonde, respectively, the dominant languages in the area. The Emakhuwa varieties which have been spoken in the area around the Ilha de Moçambique and in the important sultanate of Angoche for centuries (Mutiua, 2015) are not mutually intelligible with Swahili. However, for many on the northern Mozambican coast – including the Ilha de Moçambique – Swahili became a second or third language in the eighteenth century, and particularly in the nineteenth century, in the context of growing Zanzibari trade influence and the impact of the Sufi movements.

Swahili also became a transregional language of wider communication for merchants and Muslims on the Comoros and the west coast of Madagascar opposite the coast of Mozambique, whose languages were very different from the Swahili spoken on the East African coast of today's Kenya and Tanzania. As evidence from letters written in the eighteenth century suggests, Swahili was used in written communication, as the language of power and authority, to the detriment of other local languages. Even the Portuguese colonial administration used it in the eighteenth century, and more regularly in the nineteenth century, for its written correspondence with the coastal city-states of Mozambique and across the Ocean to India (Mutiua, 2015).

In addition, Swahili became a language of Islamic poetic composition and recitation in northern Mozambique in the nineteenth century – as remembered by Chale Mussa. Manuscripts of Swahili poems, composed in other parts of the Swahili coast, reached the Ilha de Moçambique and Angoche, such as the Poem about the Prophet's Death, locally called Nazajina (Vierke and Mutiua, forthcoming), and have been copied and recited in Angoche since then – although recently less and less. At the beginning of the twentieth century, local poets started to compose their own hagiographic Swahili poetry, for instance, the Utenzi wa Nabii Musa ('The Poem of the Prophet Musa') or the above-mentioned Tambuka or Kaiza.

Elder poets born in the 1930s, 1940s and 1950s speak with admiration of glorious and powerful Zanzibar and the Comoros, centres of learning which were very much part of their imaginary map of the coast, while for the younger generation, these places mean very little. For Chale Mussa, or Fundi Haliti, a renowned poet in Angoche who passed away three years ago, the recitation of Swahili poems not only made them part of a larger Muslim community, lifting them out of the narrow confines of their immediate context, but, more specifically, made them part of a vast Swahili coast, creating a bridge to Muslims on the Comoros, Zanzibar and Lamu. The practice of reciting and copying Swahili texts created an imaginary bridge to other distant places along the Swahili coast (Bang, 2014). Traces of the map can also be experienced with the women who perform the

Maulidi Barzanji, the youngest of whom was born in the late 1950s. They speak about Zanzibar with shining eyes, underlining that they perform the ritual the way it is done on Zanzibar. The Swahili verses open a window onto a larger world, although the political map has changed significantly.

Reciting these widely circulating poems was, and still is, a form of participation in a translocal 'literary network' or 'literary cosmopolis' connecting 'Muslims across the boundaries of space and culture', as Ronit Ricci (2011: 2) has said of the circulation and adaptations of Islamic texts in South-East Asia. This literary network does not merely echo existing networks of scholars, but also, according to Ricci (2011: 3), produces a sense of belonging to the same translocal Muslim community, providing people with a set of shared stories, historical and hagiographic accounts – sustaining other networks and nourishing the imagination of a shared world. The concept of the literary cosmopolis, originally explored by the Sanskritist Sheldon Pollock (2000), is defined by Alexander Beecroft (2015: 105), who developed it further, as 'a vast translingual and transpolitical space', both echoed and created through literary imagination. For Pollock, the use of Sanskrit and the circulation of Sanskrit texts in a vast space from Afghanistan to Java in the first millennium CE mapped out a broad transcultural and translingual polity, just as Swahili language and poetry gave a sense of belonging to speakers of different languages in Mozambique, the Comoros, Madagascar, Kenya, Tanzania and southern Somalia. More specifically than Arabic, the texts enabled people to identify themselves with a specific Swahili East African coastal community: while the city-states were linked to each other by a sense of Swahili kinship and by Swahili as a transregionally used language, the transregional East African Muslim community also projected itself into shared narratives, of which the Swahili formulaic verses are traces. As Beecroft (2015: 110) points out, textual and embodied practices that span a large cosmopolitan space often persist, even after the factors which have led to the spread of these practices – like trade, colonization or conquest – have changed: long after the fall of the Roman Empire, Latin continued to be the accepted literary language in Europe, and Greek theatre practices in Afghanistan continued to echo the huge extension of the Greek empire long after the conquests of Alexander the Great.

Beecroft's examples speak of the enduring nature of language and embodied practices. Once they have become established, they outlive discourses and meaning, as in the case of the poetry of the Zanzibari poets. In northern Mozambique, Swahili persists in niche practices and texts, although links to the rest of the Swahili coast have been disrupted, not only by the struggle for independence and the Mozambican civil war but also by national ideologies, inward orientation and language policies. It is not in the context of colonialism, when a transcoastal Muslim cosmopolis was thriving (in opposition to colonialism), but in the context of the nation-states of Kenya, Tanzania and Mozambique that the Swahili coast has largely lost its former coastal map, that is, cultural and religious links beyond national borders.

Conclusion

My paper has been a journey guided by changing maps, along the coast and back and forth in time, like an archaeological excavation digging through different layers of soil – and, importantly, deeper and wider than the period of colonization or the nation-state. Starting with contemporary, inward-looking Zanzibari poetry, whose form hints at multiple far-reaching coastal connections, I turned to the Lamu archipelago and the oldest surviving body of Swahili poetry, which opens up a wider oceanic perspective from the distinctively coastal standpoint of the merchant's mansion. The festive Liyongo poems bring the Indian Ocean to life as an 'aesthetic seascape' of 'shared material and aesthetic expression' (Ivanov, 2017: 369). This is echoed in the rich interior of the houses, which the Al-Inkishafi, composed in the early nineteenth century, symbolically destroyed, reflecting a shift in worldview as well as changing political maps of the coast. Hymnal Muslim poetry was composed eagerly and in large quantities in the nineteenth century, reflecting the religious zeal of the Sufi scholars and adepts. This created a vast Swahili coastal Muslim literary cosmopolis, sharing poetry and ritual practices, and creating an even bigger bridge across the Ocean to other Muslims in far-away places, enhanced by the distant literary landscapes of Arabia. Finally, digging for remnants of the earlier Swahili cosmopolis, I have shown how a panegyric poem, the Maulidi Barzanji, travelled all the way along the coast from nineteenth-century Lamu to the Ilha de Moçambique in the twenty-first century. Here, in a context where Swahili was to a large extent a foreign or forgotten language at the end of the nineteenth century, Swahili poetic practice is fundamental in creating a sense of belonging to the wider East African coast, vestiges of which survive in poetic form, often unnoticed, like a 'shared secret', or – to remain with the Indian Ocean vocabulary – a coral reef below the surface.

In my account, I hope to have raised some far-reaching issues to guide further research on literature and its relation to the Indian Ocean: Firstly, in the Indian Ocean, multilingualism is the norm rather than the exception. The colonial languages add to the existing multilingualism rather than doing away with it. A focus on the simultaneity of languages requires research across languages (and across the academic disciplines ordered by national philologies), with a view to studying multilingual contexts and individuals, with their multiple affiliations, since each language is part of a different cultural geography, speaks to different audiences, and comes with different genres and traditions – a huge challenge for literary scholarship. Secondly, a focus on Indian Ocean literature should not value widely travelling poetry more than poetry that is more narrowly confined in imaginary and reach, since this blocks from view the dynamic nature of changing literary maps. We need to see moments of stasis, forgetting or changing direction, as I hope to have shown. This can produce a more refined historical perspective, which takes the nation-state into account, but also looks beyond it. Thirdly, we need not only a more subtle view of history but also of literature. We need research not only on imaginaries but also on travelling literary forms, like prosody, writing, performative practices and imagery, as well as travelling texts. There are two reasons for this: firstly, scholars such as Edward Alpers (2000) have argued that in the context of the African diaspora in the Indian Ocean, not testimonies or essays, but embodied practices, like music, folklore and religious rituals, and formulaic language have persisted over centuries, but have gained far too little scholarly attention. This implies, secondly, broadening our perspective on the Indian Ocean, seeing it not only as a discursive and narrative space but also as an aesthetic space, defined by shared materiality and shared sensuous experiences evoked by the immediacy of sound, taste and scent and the practices of verbal art. This also implies to question the very notion of literature, since in many cultural contexts of the Indian Ocean this term does either not exist or is of recent and unnecessarily narrow reference, excluding performances, music and ritual, which can often hardly be separated from forms of verbal art.

Bibliography

Adorno, T. (1997 [1970]). *Aesthetic Theory*. London: The Athlone Press.

Ahmad, A. C. (2002). *Ngoma et Mission Islamique (Dacwa) aux Comores et en Afrique orientale*. Paris: Harmattan.

Al-Musawi, Muhsin J. (2015). *The Medieval Islamic Republic of Letters: Arabic Knowledge Construction*. Notre Dame: University of Notre Dame Press.

Alpers, E. (2001). 'A Complex Relationship: Mozambique and the Comoro Islands in the Nineteenth and Twentieth centuries'. *Cahiers d'Etudes Africaines*, 41 (161), 73–95.

——. (2002). 'Recollecting Africa: Diasporic Memory in the Indian Ocean World'. *African Studies Review*, 43 (1), 83–99.

——. (2009). *East Africa and the Indian Ocean*. Princeton, NJ: Markus Wiener.

Arnold Koenings, N. (2018). 'Seeing the World with Zanzibari Poet Nassor Hilal Kharusi'. *Swahili Forum*, 25, 186–195.

Bang, A. (2003). *Sufis and Scholars of the Sea: Family Networks in East Africa c. 1860–1925*. London: Routledge.

——. (2014). *Islamic Sufi Networks in the Western Indian Ocean (c. 1880–1940): Ripples of Reform*. Leiden: Brill.

Beecroft, A. (2015). *An Ecology of World Literature: From Antiquity to the Present Day*. New York: Verso.

Bonate, L. (2010). Documents in Arabic Script at the Mozambique Historical Archives. *Islamic Africa*, 1 (2), 253–257.

Borges Coelho, J. P. (2020). 'O Índico como Lugar'. In A. M. Leite, E. Brugioni and J. Falconi (eds), *Estudos sobre o Oceano Índico. Antologia de Textos Teóricos*, pp. 13–28. Lisbon: Ediçoes Colibri.

Brugioni, E. (2019). 'Literaturas Africanas e o Oceano Índico'. In E. Brugioni, *Literaturas Africanas Comparadas. Paradigmas Críticos e Representaçoes em Contrapunto*, pp. 87–112. Campinas: Editora da Unicamp.

Burton, R. F. (1872). *Zanzibar, City, Island and Coast*. London: Tinsley Brothers.

Casanova, P. (1999). *La République mondiale des lettres*. Paris: Seuil.

Dales, G. (1920). *The Peoples of Zanzibar: Their Customs and Religious Beliefs*. London: Universities' Mission to Central Africa.

Dammann, E. (1940). *Dichtungen in der Lamu-Mundart des Suaheli*. Hamburg: Friederichsen, De Gruyter.

——. (1958). 'Die Überlieferung der islamischen Suahelidichtung'. *ZDMG*, 108 (33), 41–53.

——. (1960). 'Die paränetische Suaheli-Dichtung Tabaraka'. *Mitteilungen des Instituts für Orientforschung*, 7 (3), 411–432.

Doniger, W. (2016). *Redeeming the Kamasutra*. Oxford: Oxford UP.

Grünbein, D. (2007). *Gedicht und Geheimnis. Aufsätze 1990–2006*. Frankfurt am Main: Suhrkamp.

Gueunier, N. (1983). 'Les Poèmes de Maulidi Manganja. Poèmes swahilis recueillis à Nosy Be'. *Bulletin des Etudes Africaines de l'Institut National des Langues et Civilisations Orientales*, 3 (6), 7–76.

Gurnah, A. (2001). *By the Sea*. London: Bloomsbury.

Harries, L. (1952). 'A Swahili Takhmis from the Swahili Arabic Text'. *African Studies*, 11 (2), 59–67.

——. (1953). 'Strung Pearls: A Poem from the Swahili-Arabic Text'. *Bulletin of the School of Oriental and African Studies*, 15, 146–156.

——. (1958). 'Maulidi Barzanji: The Swahili Abridgement of Seyyid Mansab'. *Africa und Übersee*, XLII, 27–40.

Hichens, W. (1939). *Al-Inkishafi: The Soul's Awakening*. London: Sheldon.

Hilal, N. (n.d.). *Siku Zapishana*. Zanzibar.

Hillewaert, S. (2020). *Morality at the Margins: Youth, Language and Islam in Coastal Kenya*. New York: Fordham University Press.

Ho, E. (2006). *The Graves of Tarim: Genealogy and Mobility across the Indian Ocean*. Berkeley: University of California Press.

Horton, M. and Middleton, J. (2000). *The Swahili: The Social Landscape of a Mercantile Society*. Oxford: Blackwell.

Ingrams, W. H. (1967 [1931]). *Zanzibar: Its History and Its People*. London: Frank Cass.

Ivanov, P. (2017). 'The Aesthetic Constitution of Space: Mimetic Appropriation of Foreign 'Styles' and the Creation of Transoceanic Connections on the Swahili Coast'. *Comparative Studies of South Asia, Africa and the Middle East*, 37 (2), 368–390.

Knappert, J. (1958). *Het epos van Heraklios: een proeve van Swahili poëzie*, PhD Dissertation, Leiden.
——. (1966). 'The Hamziya Deciphered'. *African Language Studies*, 7, 52–81.
——. (1971). *Islamic Swahili Poetry*, vol. III. Leiden: Brill.
——. (1991). 'Liongo's Wedding in the Gungu Meter'. *Annales Aequatoria*, 12, 213–226.
Larson, P. (2009). *Ocean of Letters: Language and Creolization in an Indian Ocean Diaspora*. Cambridge, UK: Cambridge University Press.
Macagno, L. (2007). 'Islã, transe e liminaridade'. *Revista de Antropologia*, 50 (1), 85–123.
Meier, P. (2017). 'Unmoored: On Oceanic Objects in Coastal Eastern Africa, circa 1700–1900'. *Comparative Studies of South Asia, Africa and the Middle East*, 37 (2), 355–367.
Meinhof, C. (1924/1925). 'Das Lied des Liongo'. *Zeitschrift für Eingeborenensprachen*, 15, 241–265.
Miehe, G. et al. (eds). (2004). *Liyongo Songs: Poems Attributed to Fumo Liyongo*. Cologne: Köppe.
Mutiua, C. (2015). 'O Islão e o Processo de Literacia no Norte de Moçambique entre os finais do século XIX e princípios do século XX'. In T. Cruz e Silva and I. M. Casimiro (eds), *A Ciência ao Serviço do Desenvolvimento? Experiência de Países Africanos Falantes da Língua Oficial Portuguesa*, pp. 205–219. Dakar: CODESRIA.
Neuhaus, G. (1935). 'Kitabu Mauludi. Buch der Geburt Muhammed's. Suaheli-Gedicht des Lamu-Mannes Scharifu Mansabu bin Scharifu Abdurrahmani al-Hussaini, Manuskript in alabaster Schrift, wiedergegeben in Autotypien, übersetzt und erläutert'. *Mitteilungen des Seminars für Orientalische Sprachen*, 38 (3), 145–201.
Newitt, M. (1995). *A History of Mozambique*. Johannesburg: Witwatersrand University Press.
Orsini, F. (2014). 'Traces of a Multilingual World: Hindavi in Persian Texts'. In F. Orsini and S. Sheikh (eds), *After Timur Left: Culture and Circulation in Fifteenth-Century North India*, pp. 403–436. Oxford University Press.
——. (2015). 'The Multilingual Local in World Literature'. *Comparative Literature*, 6/7 (4), 345–374.
Owuor, Y. A. (2019). *The Dragonfly Sea*. New York: Knopf.
Pearson, M. (2010). 'Preface'. In S. Moorthy and A. Jamal (eds), *Indian Ocean Studies: Cultural, Social and Political Perspectives*, pp. xv–xvii. New York: Routledge.

——. (2015). 'Introduction: Maritime History and the Indian Ocean World'. In M. Pearson (ed.), *Trade, Circulation, and Flow in the Indian Ocean World*, pp. 1–14. New York: Palgrave Macmillan.

Pollock, S. (2000). 'Cosmopolitan and Vernacular in History'. *Public Culture*, 12 (3), 591–625.

Pouwels, R. (1987). *Horn and Crescent: Cultural Change and Traditional Islam on the East African Coast, 800–1900*. Cambridge: Cambridge University Press.

——. (1992). 'Swahili Literature and History in the Post-structuralist Era'. *The International Journal of African Historical Studies*, 25 (2), 261–283.

——. (2002). 'Eastern Africa and the Indian Ocean to 1800: Reviewing Relations in Historical Perspective'. *The International Journal of African Historical Studies*, 35 (2/3), 285–425.

Prendergast, C. (2001). 'Negotiating World Literature'. *New Left Review*, 8, 100–121.

Ricci, R. (2011). *Islam Translated: Literature, Conversion, and the Arabic Cosmopolis of South and Southeast Asia*. Chicago: University of Chicago Press.

Sacleux, C. (1939). *Dictionnaire Swahili-Français*. Paris: Institut d'Ethnologie.

Saúte, N. and Sopa, A. (eds). (1992). *A Ilha de Mocambique pela Voz dos Poetas*. Lisbon: Edições 70.

Schadeberg, T. (1994). 'KiMwani at the Southern Fringe of KiSwahili'. In P. Bakker and M. Mous (eds), *Mixed Languages: 15 Case Studies in Language Intertwining*, pp. 239–244. Amsterdam: IFOTT.

——. (1999). 'Nguo-nyingi Mote: Mwanzishaji wa mji wa Ngozi (Angoche)'. *Swahili Forum*, 6, 121–130. <https://nbn-resolving.org/urn:nbn:de:bsz:15-qucosa-97766>.

——. (2009). 'Loanwords in Swahili'. In M. Haspelmath and U. Tadmor (eds), *Loanwords in the World's Languages*, pp. 76–102. Berlin: De Gruyter.

Schadeberg, T. and Mucanheia, F. (2000). *EKoti: The Maka or Swahili Language of Angoche*. Cologne: Köppe.

Schimmel, A. (2003). *As through a Veil: Mystical Poetry in Islam*. Oxford: Oneworld.

Shafi, A. S. (1999). *Vuta N'kuvute*. Dar es Salaam: Mkuki na Nyota.

Shariff, I. N. (1988). *Tungo Zetu*. Trenton: The Red Sea Press.

Steere, E. (1870). *Swahili Tales as Told by the Natives of Zanzibar*. London: Bell & Daldy.

Vassanji, M. G. (1991). *Uhuru Street*. Oxford: Heinemann.

Velten, Carl. (1903). *Desturi za Wasuaheli*. Göttingen: Vandenhoeck.

Vernet, T. (2015). 'East African Travelers and Traders in the Indian Ocean: Swahili Ships, Swahili Mobilities ca. 1500–1800'. In M. Pearson (ed.), *Trade, Circulation, and Flow in the Indian Ocean World*, pp. 167–202. New York: Palgrave Macmillan.

Vierke, C. (2016). 'From across the Ocean: Considering Travelling Literary Figurations as Part of Swahili Intellectual History'. *Journal of African Cultural Studies*, 28, 225–240.

——. (2019). 'Frau Betelpfeffer und die lustvollen Stunden. Die Inszenierung sinnlicher Erfahrung, Erinnerung und Erwartung in früher Swahili-Dichtung'. In L. Henningsen, K. Wiegandt and C. Battegay (eds), *Gegessen? Essen und Erinnerung in den Literaturen der Welt*, pp. 125–148. Berlin: Neofelis.

Vierke, C. and Mutiua, C. (forthcoming). 'The Poem about the Prophet's Death in Mozambique – Swahili as a Transregional Language of Islamic Poetry'. *Islamic Africa* (Special Issue edited by Andrea Brigaglia).

Werner, A. (1926). 'The Swahili Saga of Liongo Fumo'. *Bulletin of the School of Oriental Studies*, 4 (2), 247–255.

Wood, M., Dussabieux, L. and Robertshaw, P. (2012). 'The Glass of Chibuene, Mozambique: New Insights into Early Indian Ocean Trade'. *South African Archaelogical Bulletin*, 67, 59–74.

PART III

Literatures across the Ocean

ANA MAFALDA LEITE

12 Foreword

This section of the book is a foretaste of how the Indian Ocean is imagined and (re)written by poets and novelists from different spaces, such as Mauritius, Mozambique and Goa, and frames the essayistic and literary meditations produced by different writers from the Indian Ocean on the cultural and historical and, above all, creative specificities of this sea. It begins with two essays, the first by the Mauritian poet Khal Torabully and the second by the Mozambican historian and novelist, João Paulo Borges Coelho.

Torabully elaborates on the concept and theory of 'coolitude', which was inspired by the transnational system of indentured labour, known as the coolie trade. He historically redefines the Indian-oceanic space by placing the centrality of oceanic travel in the poetics of the coolie and by breaking with a narrative centred on the land. According to the author, this is the first specific aspect of the concept, which the poet later characterizes as an open, plural and structuring poetics, in which the Indian Ocean is the fundamental binding element. The author's lyrical work is thus articulated with his essays by re-situating the representations of peripheral circulations in the Indian Ocean.

In an equally unique way, the novelist and historian João Paulo Borges Coelho discusses the potential of the Indian Ocean as a narrative space. Conscious of its antiquity and historical, geopolitical and economic diversity, which contributes to the formation of a Mozambican literary Indian Ocean, Borges Coelho, along with other Mozambican writers, constructs it as *a home of fiction and poetry*. This is what we read in the four poems by Luís Carlos Patraquim, which evoke the emotional and historical interiority of different scenarios and motifs in the Indian Ocean.

This space of poetic and narratable 'indianoceanity' also appears in the beautiful lyrical prose by the Goan writer Jessica Faleiro, *Merman*. In turn, Torabully's four poems from *Voices at the Aapravsi Ghat-Indentured Imaginaries*, are another embodiment of Indian Ocean poetics, fitting into the coralline poetics of coolitude.

KHAL TORABULLY

13 *Coolitude*, when indentured labour rewrites the (Indian) Ocean and history in unceasing resistance and creativity[1]

Preamble to a celebration

Ladies and gentlemen, thank you for giving me the opportunity to commemorate thirty years of writing about coolitude, begun in 1989, when we were commemorating the bicentennial of the French Revolution.

1989 was also the year that saw the fall of the Berlin wall, a planetary wall that fell and, regarding representations and circulations, that constitute an essential landmark, especially in respect of the new walls that are now going up geographically, culturally and mentally in our world.

In addition to this, the date of this colloquium could not be better since it gives me the opportunity to commemorate thirty years of that oceanic vision of coolitude that I formulated in 1989. Oscar Wilde said that when a woman tells you her age she is willing to talk about anything; in this case, it is the poet who is telling his age in relation to the conceptualization of coolitude. I hope to prove that Wilde is wrong, in light of the observations I shall make in this opening speech.

I was 33 years old at the time and I had a utopian vision of a totally connected world. I articulated memories, (untold) stories and archives from different places, in a desire to be inclusive, believing the ocean both dissolves and binds. It was this first attempt to place the ocean at the centre

1 Opening speech by Khal Torabully at the International Indian Ocean Conference: Circulations and Representations, Lisbon, December 2019, NILUS Project.

of the literature of the coolie, commerce or servitude, that was theorized as a poetics of coolitude.

Yes, thirty-three years have elapsed …

Seen in this light, it strikes me as entirely natural that all these years after my crossing of the signs, I should find myself with you, here in Lisbon, a place open to maritime adventures. I must add that there is part of my own memory and poetics in Lisbon, to which I am returning for at least the tenth time.

Here we remember Camões and the Lusíadas, as well as Vasco da Gama and Prince Henry the Navigator, who led the Portuguese people to India. Royalty was also centreed on the ocean, coming from Europe to travel to India whereas coolitude travels from India to the world. To reciprocate the Navigator's courtesy, I began to write the Indiades, here in Lisbon, in the 1990s, as an answer from the South, the seas of India, in dialogue with Camões' Lusíadas. The manuscript of the inverted poetics of the Indiades already exists and awaits your inverted gazes in the antipodes of our imagination.

Camões shows how a poet is needed to write the crossing of the seas, bound for civilizations rich in their diversity, upstream from the Lusitanian epics. For dialogical reasons, this foundational text by Camões is an extension of my coolitude in the Portuguese imaginary, even if I speak more about coralline relations than discoveries or conquests.

In the extension of these clashes of ships, in capes of storms, separating or connecting my Indian sea and the Atlantic, I consume vindaye, a dish brought by the Portuguese to Goa, which travelled to the island of Mauritius. Like the Maria biscuits, which an artisan made in Port-Louis, not very far from the home of my birth. Not to mention the sardines and the salted fish, that we also imported from the near yet distant land of Portugal.

This is all by way of telling you how honoured and moved I feel, finding myself at this important moment in my work, among researchers, authors and thinkers who are known for their contribution to the field of Indian Ocean Studies, and who have reflected on common paradigms of the ocean.

Obrigado, obrigada. Danke. Merci. Thank you. Gracias. Shoukrane. Shoukria.

Coolitude and thalassography

In the 1990s, my theory of coolitude was inspired by the transnational indenture system known as the coolie trade. My first idea was to write about the (Indian) Ocean from a redefinition/reappropriation of the ocean space. What this accomplished was a kind of 'oceanography' which placed ocean voyages at the heart of the poetics of the indentured worker or coolie, breaking with a narrative centred on land. That is the first specificity of its poetics, investing in a maritime transnational space, connecting the memories, stories, cultures and imaginaries of the world.

Later, I developed a field and a coralline vision, experiencing the semiotic/symbiotic crossing of the indentured worker or coolie as the possibility of an open, plural, structuring poetics in which the (Indian) ocean is the fundamental binding element. This latter, breaking with the geographical, anthropological and memorial marginalization of representations and circulations, defines a History of the peripheries in relation to the sea reappropriated by the margin, this ocean sea that is too often written from a dominant centre .

Also, in this respect, it is not surprising that the ocean should become a space of deconstruction. The sea is conceived as binding and diluting memory and history, as a fundamental paradigm of cultural imaginaries and interactions, this time, from a polycentric, no longer eurocentric point of view.

In this extended area of water, writing (on) the sea is the equivalent of holding up a mirror to the writing of history, of hi/stories. This is the core of my reflections on the subject that concerns us here, namely the representations and the circulations which took place in my ocean of the Indias.

Places, texts and Gandhi, who was an emblematic figure in respect of indenture, will be evoked in my talk, because they sketch a mental and historical geography resulting from the way in which circulations, representations and controls of migratory flows were implemented, first in the Indian Ocean and elsewhere. After slavery, it was effectively the coolie trade that was established in the vast field of nineteenth-century maritime empires.

Also, opening the sea with a different key, provided by the peripheries, coolitude allows us to grasp how a story transmitted by the voiceless of History (the coolies), has allowed a translation of worlds into a maritime perspective, operating as the matrix of their transformation and their emergence as subjects with the ability to create their own representations, with creativity born from the coolie trade. Even if these workers have often been limited in their physical and geographical circulation.

Also, in this articulation between circulations and representations, the Indian Ocean, the ocean of the coolie trade, is approached as a space of transformations and creativity. And what we will read there is a coralline poetics, developing an essentially marine ecology in which diversities are interdependent, placed in relation to one another.

In line with the framework of this colloquium, I shall start by exploring two paradigms suggested above in my opening exposition, 'circulation' and 'representation'.

These will be the recurrent threads of my lecture.

Defining the two terms of the colloquium

Circulation may be defined as travelling rapidly, like a free and continuous movement within a system or closed area. It also includes the notion of communication going from one group of speakers to another; it connotes knowledge which is disseminated. We might say, for example, that someone's thoughts circulated widely. Circulation also refers to the number of copies of a newspaper or magazine that are sold. Here we touch on the reading public, naming a movement of the thing that circulates, moves or spreads.

In my observations, I shall define the term circulation principally from the standpoint of people or groups of indentured people who moved, subject to organized systems of promoting and controlling migratory flows, in particular in the ocean space, with relays linked to the country of departure and the country of arrival.

In respect of 'representation', I refer primarily to mental representation, an action, the fact of representing something for oneself, or the way someone represents something in a speculative way, to cite Graw (1981). Lal (1968) says that it is the content of an act of thought, the reproduction of a previous perception.

Writing the Indian Ocean between these two definitions, according to coolitude and the coralline field, leads us into a kind of thalassography, a marine writing in motion. This marine writing explores the crossing of a semiotic/symbiotic text, which structures this writing in terms of a detailed oceanic matrix that defines a very particular history, one which saw migration become linked to indenture, and which changed the history and the cultural anthropology of the Indian Ocean perimeter and beyond.

It is obvious, as we shall see below, that this thalassography is not disconnected from oceanography or marine ecology, namely, the poetics of coral, deriving from this writing with strong marine connotations.

Here, my poetics poses a question: how is it that writing a history, carried by the voiceless, has allowed a translation of worlds onto that liquid material that is the ocean, which is also the matrix of their own transformation? In my readings of circulation and representation, I shall also include the Indian Ocean as a space where representations were transformed, namely through the creativity that was linked to indenture. I pay particularly close attention to the movements of Gandhi in this respect, because the experience of indenture also shaped the Mahatma.

In view of this conceptual choice, let us say from the outset: my vision of thalassography mixes archives, movements, fusions and histories, what is not said and also what is prospective. It is a narrative that forges its own representations by disrupting the fixed frontiers of colonial historiography and geography.

So, for me, the period from 1989 to 2019 has also been thirty years of writing in motion, exploring the representations inspired by indenture, disrupting the writing that has bound it to a truncated, partial representation and establishing an idea of specific circulations and transformations.

Why such a long crossing? Because the coolie trade lacked its own creative text.

Writing indenture, with the sea circulating in its representations and fluidifying its topoi, means rejecting the margin to which the dominant History confines you.

Rewriting the narrative of the sea

Very early on, I became aware of two things in my writing that led to my book Cale d'Etoiles-Coolitude in 1989, the pioneering text in studies of the coolie trade and a (re)foundational moment for indenture.

1. An archipelago consciousness, because, even as a child, I knew we were connected to Rodrigues, Saint Brandon, Tromelin and Agalega. Born on this sprinkling of British Empire islands, I learned very early on that Britannia rules the waves. In my child's understanding, the sea was almost a country that divided and connected, that fragmented but opened up to the world.
2. The second thing that struck me as an adolescent, was the knowledge of that famous excision that was made in our archipelago, our excised Diego Garcia (I say in one text that he is the only 'man' who was not castrated but rather excised, because island is a feminine word). This dissection undertaken by the English, which we could also call a dismemberment if you like, of the Chagos archipelago, of which Diego Garcia is an island.

Of course, the words Diego Garcia remind us that the Portuguese reached our corner of the world on their way to the India of spices and other fabulous riches.

They gave our island the name of Cirné or island of swans – ilha dos cisnes – (the dodo inspired the Portuguese to choose this name) that I have often rewritten as island of signs (île aux signes). They also named our neighbouring island, little Rodrigues, and even this group of islands where we live, the Mascarene Islands. This was in memory of Pedro Mascarenhas, who discovered Diego Garcia in 1521. He named it 'Dom Garcia' in homage to his patron, Dom Garcia de Noronha.

For me, as a child of the islands and despite my fear of the unknown, there was something comforting about the Indian Ocean, the only one that bore the name of a country or a nationality, giving me a sense of continuity with the Great Peninsula where my grandparents had come from, Bengal in the case of my mother, and my father, born in Trinidad and Tobago, from Kokane on the west coast of India. In addition to this, the ocean filled me with a 'sense of historicity', the idea that we lived on the outskirts of a plural world where Arabs, Portuguese, Dutch, French and English had dropped anchor. Some people say that the Phoenicians arrived before them in Dina Mozare, the Arabic name for the Island of Mauritius, as it is transcribed in the Cantino Planisphere, housed in the Maritime Museum in Lisbon, not far from here. I went there, but the museum was closed. But I will go back.

Since everyone had come to us, to the outskirts of the world, we developed an inflated pride, displayed in the motto Clavis stella mare indici, Star and Key of the Indian Ocean. Inflated because we were not the centre of the Indian Ocean. To go to India, the Portuguese and the other voyagers did not come first to the city of my birth, Port-Louis. They set out from the east coast of Africa, Mombasa or Melinde, or some other place, to sail towards Cochi and Calicut.

Our coat of arms, on which the lance-shaped leaves of sugarcane are artistically traced like fine lace fluttering above our emblematic animals, the deer and the dodo, reinforces for us the power of attraction associated with Mauritius, its representation as a paradise to be attained. Except you will understand very readily that sugar is not that sweet, actually. And this product, that was bitter for many people, brings me to the coolie trade and to coolitude. Forgive me if this link seems a little too simple, but that is how the power of representation works on the imagination to produce movement and circulation.

I began with an important idea: he who writes the sea, writes History …

The coolie trade was born in the Indian Ocean, although the indentured servant, his ancestor, was born in the sixteenth century in the Atlantic, that other ocean that is better known as the ocean of slavery. Although it is a long way away, the Cape of Good Hope brings them together. Those two oceans are no longer in the forbidding shadow of an Adamastor from the Lusíadas who barred circulation because, as the poet Camões represented it, the place where these oceans meet is a place of horror, like Cerberus guarding the gates of hell. But as we know, that monster-guardian of the Cape of Storms was a ruse invented by mariners and geographers to discourage possible rivals on the path to the riches of India. This is where the absolute Other is watching you …

Sometimes we bathe in the same waters of memory, with our own representations and circulations. And ultimately, we also realize that we must deconstruct the discourses, itineraries and representations of previous discoverers.

Cale d'Etoiles and the centrality of the ocean

When I wrote the first oceanic song of the coolies, I opted for the centrality of the ocean, as I have already stated. That vision permeates Cale d'Etoiles-Coolitude, the foundational text of the coolie trade, which is a triptych.

The book of the sea is central to it, it is not just a fanciful choice, it is a semiotic one. This book may be read in two directions, from beginning to end or from end to beginning. The circular chronology is closer to an oriental poetics than the linear chronology of western historiography. The form matches the substance of the project.

Writing the sea, in this respect, is rewriting history, a history that I learned as a subject of Her Britannic Majesty. But I wanted to go beyond that colonial representation that was established for me, without me, and to assume my own voice.

I could have been vengeful, but I did it without essentialism, without setting identity in the non-fluid medium of stone. The sea bathed my texts with a desire for polycentric universalism.

I also drew on the vocabulary of mariners, Bretons and Scandinavians, in Cale d'Etoiles, because that crossing of the Indian Ocean was also my own crossing of signs in order to awaken other histories and memories, as I indicated above. Linguistic polyphony was naturally imposed upon me, especially as regards transgressing the taboo of the black water or the kala pani.

My poetics is the first to deconstruct this limit to the circulation and representation of the Indian sea and the indenture system, this time not from the perspective of the colonizers, but of the colonized themselves.

I had to go beyond myself twice over, inside and outside, to better represent myself and to circulate.

The text Cale de'Etoiles-Coolitude is the founding act of this open Indian Ocean perspective, freeing up the discourses and the limits imposed by the dominant class. It rethinks my place in history and geography. It offers a proposition according to which the sea is also a matrix and it is not only the maati, the Earth – the sea-earth [la mer-terre] central in Indian cosmogony, which allows regeneration, being reborn here and elsewhere. To unpick this knot, I had to shatter the chains of taboo that surround the sea in the Hindu subconscious.

In order to do this, I re-read the meaning of the sea in the history of India, concluding that the Indian Ocean, the only one that bears the name of a nationality, was made contingently taboo, sometimes to serve the needs of a particular objective. It was the invention of the kala pani, the black water, that established a huge barrier to the circulation of Hindus in the great blue ocean. In effect, ocean voyages by Hindus were prevented for reasons I have already explained in other writings. In essence, it was a way of limiting circulation, sometimes in the interests of either higher castes or colonizers.

Anyway, I allowed myself to snuff out this taboo of the black water that forbade circulation over the oceans. Besides, when I wrote Cale d'Etoiles (the very title of which signals an oceanic perspective), the maritime space offered me the possibility of rebirth, of creating worlds and identities, not just confinement in a murky abyss and the irremediable loss of the migrant's soul. I knew, in order to break this taboo, that 'Britannia rules the waves' and that he who writes the sea rules the world and it is not just the hand that rocks the cradle that rules the world.

From this concept of un-creating/creating the world on the sea, I then wove a creative, hybrid vision of language, and thence of cultures, spiritualities and memories, staging the notion of jahaji bhai, behen or shipboard brothers and sisters. Faced with the dangers of the ocean crossing, it was solidarity that the coolies developed onboard the ships, through a common, unified and ultimately liberating representation of the crossing. From this concept of India being reconstituted on the water, I drifted towards the humanism of indenture's diversity, with a view to reappropriating the sea from the perspective of a marine poetics proper to the coolie. An ontological subversion derived from a different mode of Indian circulation and a representation of the Indian Ocean that is rather less black. But was the Indian Ocean frozen in our imaginaries?

India subverted and the thousand and one nights

In these ties woven in adversity by shipboard brothers and sisters, in this India in motion, we see a total disruption of the indigenous family and its social, cultural, linguistic and religious rules and hierarchies. Faced with the immensity of the sea, people became shipboard brethren over and above anything else, so that they would not fall overboard in the vertiginous experience of the crossing and the unknown that awaited the indentured servant at the end of his voyage. The indentured labourer would bond with Indians he considered to be 'other' in the context of adversity, so a Tamil, for example, would bond with a Muslim.

Here the sea clearly dissolved the rules that segmented human groups. I believe that this prioritizing of solidarity occurred mainly at sea and less on land, except when they regrouped to confront the dangers of caravans or journeys. But I have not found, even though it must have existed, the land equivalent of jahaji bhai (shipboard brother) which could be maati bhai or behen for journeys overland. Perhaps the term travelling companion would be the closest equivalent on land.

Let us also add that in the green sea of sugar canes, far from India, far from the ship, those bonds of solidarity continued in the villages described as Indian.

But, at the same time, there was competition between coolies to get women, food and the master's approval, but also to resist the masters, the emancipated labourers, the unequal laws of the plantations, the repressive authorities and the police.

I think the sea which creates diversity also brings out a particular duality, owing to the limited space onboard the ships: us against them. The diversity of the sea is already there in the us that constitutes the coolies, the human diversities of India from which we depart with the added value of new bonds of solidarity, rising above caste and taboo to express a new definition of India in the open, mobile space of migration.

It is that focus on the Indian Ocean that I affectionately call my sea/mother (mer/mère, mamta/motherly love) of India, that set me on the maritime routes of that ocean of writing, to realize that the Indian Ocean was also a mare nostrum, and a more dynamic one than the Mediterranean, the first matrix of globalization. Here there was a culture of navigation, trade and languages that bathed the shores of East Africa, the Arabian Peninsula, India, South East Asia and China, for millennia.

This 'ocean as connecting space' was also the birthplace of the foundational narrative, 1001 Nights which is deeply rooted in my heritage, as every night my mother recounted the adventures of Sindbad (he who comes from Sind, from India). This gave me the first intertextual link to my ocean.

For me, it is important to remember three things about the 1001 Nights, which are the foundational texts for my representation of the Indian Ocean and my first contact with cross-border literature which flouts policed representation and circulation. Those three elements are:

1. The 1001 Nights introduce an element of serialization into the narrative, in the form of suspense. This left its mark on world literature and on the art of telling stories. And it set me on the path of writing in fragments, tessera, where the parts respond to one another.
2. The 1001 Nights are intrinsically transnational, like archipelagos, and the stories make a mockery of both land and sea borders. Sindbad, at the mercy of shipwrecks, turns up in several different countries, including in the untraceable land of Waqwaq, which

would have been somewhere between Madagascar, Mauritius and Japan. Imaginaries become porous in Sindbad's maritime adventures. Representations go beyond frontiers and circulate, through the ocean voyages of this merchant-seaman. The circulation of these stories has favoured the dissemination of common representations in this ocean, a singular feat in world literature: the 1001 Nights are essentially transnational.

3. The 1001 Nights are the first stories that happily ferment intertextuality on a transnational, oceanic level. They mix Persian, Arabic, Chinese, Indian and other sources. It is the text par excellence of literature without frontiers, which opens up to a globalization of the imaginary. It was the first anchorage point for oceanic literature and poetics that went beyond established representations.

Those stories had a special resonance for me, born as I was into a family of thirteen children. My father was a sailor from Trinidad (so from the Atlantic) and my mother was born in Mauritius, while her parents came from Calcutta, from the east coast of India which was open to the navigation of South East Asia. This web of maritime meaning affords the primary geography underpinning the poetics of coolies, who worked the land obviously, but who in order to do this had first to cross the sea, though we also had coolies of the sea, the lascars who ploughed all the seas on board English or other ships. Lascars served on all the European ships that reached the Indian Ocean.

Writing about the sea as an aid to imagination, memory and histories, mixing the archives, the unsaids and the prospective visions of coolitude, I abandon a telluric perspective, a tellusophy, in favour of a thallasosophy, opening up the sea not as a flat surface but as a palimpsest where layers of meaning overlap.

The sea of coolitude is complex and polysemic. In that respect, it is the sea of peripheries, whilst also being the mother of globalizations, unconsciously opposing the monosemic, unilateral discourse of 'Britannia rules the waves', an exclusive and hierarchical perspective on the History of the World.

Faced with this unilateral discourse on the Indian Ocean, I developed a vision, a coralline poetics that unlocks perceptions and opens into the Mediterranean and the Atlantic. My sea is not mean or exclusive, it connects beyond the Cape of Storms.

Precisely by writing coolitude, as I said at the time, I was claiming my share of the sea, and my representation of the sea was open. And what is more, anyone who writes about the sea writes the world more fluidly and from different angles. That was my poetological desire.

Intermediary seas and spaces

Kamau Braithwaite said that hybridization or mestizaje is not limited to the land, it also begins in maritime spaces and along coasts. The fraternity of the jahaji bhai and behen proves him right because in the depots of recruits near the Indian Ghats, where the authorities would hold the recruits waiting for their ships, this social, cultural and religious upheaval was already happening: the Muslim, for example, was frequently contracted to cook for Tamils and Hindus, all the castes together. They were already adjusting to the ocean voyage to come, where traditional boundaries would be rudely put to the test.

Once they were at sea, the level of mestizaje increased, in light of the near-total void the coolies experienced onboard the ship. I do not mean biological mestizaje here, but rather the mestizaje of cosmogonies, religions, cooking, languages and also anthropomorphic spaces, where Brahmins and lower castes shared the same zone, without any distinctive sign of socio-religious hierarchy or traditional representations of the other.

Here, at the level of self-representation, something crucial takes place. The coolie suffers social anomy as soon as he enters indentured servitude, coolitude. But this opens him up to other representational capacities. Hybridization is also the means by which stories, representations and myths, come into contact and fertilize one another, breaking down barriers that would be considered hermetic on land.

At sea, the country becomes movement, the voyage, an apprenticeship in alterity and only rarely, that certainty of immutable classification that is found on Indian terra firma. Effectively, the Indian Ocean reconstitutes the Indian ontology of being.

Moreover, if I try to think about an ontogenesis of coolitude, I would choose to research not some form of origin, but rather the intersection of origins in their Indian-oceanic matrix, the cradle of the jahaji bhai and behen, that inspired in me the philosophy of the humanism of diversity arising from the indenture, at the heart of coolitude. This is born of observation in the country of India (India is a human, religious and linguistic mosaic, hence the plural signifier that I use in this respect, the Indias), and the transformation of origins in the hold (cale) of the ship, on the gangway, and later in coralline representations of the world.

Obviously, we cannot discount the fact that the coolies' maritime perspective already encompasses the complexity of societies undergoing creolization, adding to this at sea, a further layer of intra-Indian diversity even before they come face to face with the creole hybridization produced by the violent encounter between blacks and whites in the plantocracy, the mines and the railways.

Mare tenebrosum or blue sea?

With regard to those representations of the ocean that are linked to physical circulation, one fairly anodyne question occurs to me: is the sea the same colour for everyone?

I write the blue kala pani, which differentiates/contrasts readings of the sea by placing two marine colours in opposition, thus giving the sea shades that distance it from the negative colouring which invaded the Hindu understanding of the sea in the indenture period.

We should note here that all peoples have already experienced this sea of the unknown; medieval maps frequently showed monsters around it and the designation mare tenebrosum was often used to represent the Atlantic. It was an Atlantic version of the kala pani, the black water.

In Hindu mythology, the Indian Ocean was born from the spittle of Vishnu and for Muslims, both coolies and sailors, the sea brings the benefits of trading and fishing, and makes it possible to undertake the pilgrimage from India to Mecca. The Arabian Sea and the Indian Ocean are both blue. So the Indian Ocean has two colours in India.

This more open vision of the ocean of traders and Muslim pilgrims does not, however, obscure the hardships endured crossing the Arabian Sea, for example, in common with their shipboard brothers and sisters in the coolie trade, even if this part of the Indian Ocean was a kind of inland lake for the people of Gujarat.

What should be highlighted here is that during the crossings of coolie ships, the mare nostre, the ocean of Indian nationality, became a mare tenebrosum in the imaginary and the representations of the Hindus. The southern part of the Indian Ocean is not as familiar as its eastern and western sides, where it was possible to sail along the coast. Add to this the kala pani taboo of the sea, a Hindu fear of the sea, arising from the prohibition of crossing the ocean, for religious reasons. It was inhabited by houglis and monsters that sucked people into the depths of the abyss, cutting short their cycle of reincarnation.

That blue fear, if I can call it that, a kind of sea panic, allied to the tough conditions of the crossing, meant that the connections and representations linking the different Indian groups in circulation and in motion, were more flexible and more porous. It is well known that when nature dominates you, human solidarity is a necessary response. The representation of the 'other' in India, changes in the course of this circulation. At the same time, the representation of the European 'other' starts to take shape.

At this point in my discussion, it is worth defining more precisely the modes of circulation and representation that pertained to coolitude, with specific reference to the emblematic spaces of indentured labour.

Emblematic spaces and images of the coolie trade

In terms of geographical circulation, the emblematic spaces of coolitude are rivers, water courses, the sea, ships, warehouses for ships departing and arriving, the camps and cane fields of the sugar mills, rubber plantations, mines, railways, depots for vagrants and courts of justice, etc. These were the emblematic places in which circulation and representation were articulated.

In an extremely hierarchical Indian society, social mobility was effectively very limited, by taboo and by the specific roles that were allocated to each person. The caste system even enforced geographical confinement on the lower castes. Because it broke with the practice of reserving specific tasks for pariahs, migration for work represented the seismic shift that produced both coolitude itself and the humanism of diversity that the coolie could then peddle, precisely by virtue of the social and geographical mobility that indenture brought with it.

Even if it was often the poorest who made the journey, Brahmin also became part of this movement towards the lands beyond the kala pani.

That circulation was fundamental, because it was very literally circulation, even if it was controlled. It is a topos of the coolie trade, referring to a controlled migratory flow rather than free movement. But, even with this limitation, it represented a social conflagration in the Indian world.

Being bundled together in enclosed spaces or on ships specially equipped and chartered for the coolie trade, undermined representations of self and other, creating new bonds of solidarity in the process of circulation. There are numerous historical references to this.

Take us as an initial example, the 'waiting' depot where the coolie waited. Obviously, the colonial administration and recruiters did not provide separate spaces for castes or, at times, even for men and women, in line with Indian tradition. This forcing together of different ethnic groups, castes and religions, etc. was of seismic consequence for many. The divisions that maintained the socio-economic, religious and geographical separations of Indian social organization had been broken down. The fact of circulation disrupted the representation of the 'other' in Indian groups.

In these reconstructed Indias, the Indians also came into contact with the colonial authorities and the recruiters, the mestris, the agents of those companies which represented the plantation owners' interests, A whole system engaged them in a mode of production that was taylorized ahead of its time.

They needed, therefore, to communicate among themselves beyond traditional barriers. The Indian thinks about his alterity when faced with an alterity that is even greater. He would acquire a (relative) level of knowledge about his contract, which defined the representation of the coolie, the Indian, the descendant of an age-old civilization, as an employee of the Empire and even then, only in its distant possessions. It should be noted that the forerunner of this wave of migrations was the hill coolie, who may be regarded as an Indian very much at odds with traditional Indian society. He was effectively a semi-nomad, frequently working in the indigo fields, who broke with the image of the Indian whose circulation was impeded, by being able to travel beyond the place allocated to him under the caste system. The hill coolie was already in circulation in the Indian interior. Because the hill coolie was an ideal candidate for the exodus, the Indian became the empire's pair of arms, a synecdoche, a metonym in an international system of production. He was contracted for manual labour, the labour of the colonies. So an entire cosmogony was already called into question on land in India, even before the departure of ships, and the Indian gradually came to represent the manual labour initially of India, and then of the colonies overseas.

This was followed by the middle stage of the ship – and I would like to note here that Cale d'Etoiles reiterates these three stages of departure, crossing and arrival – where the loss of reference points on land and the immensity of the sea, along with strange maritime technology, foreign sailors and captains speaking in another language, were the profoundly traumatizing experiences with which this exile began.

There was also, very specifically, as we have already observed, the dystopian kala pani, that black water to be crossed, that fear of the end of the cycle of reincarnation for Hindus, that terrible representation of the sea that the Indian retained in his subconscious. And we know that the English employed a particular ruse to overcome this. Jars of water, supposedly from the Ganges, were sometimes placed on ships, to force the more recalcitrant ones to come abroad, because the purifying waters of the reincarnation were on the black water. This subterfuge kept circulation moving, despite the Hindu representation of the sea …

Coolitude draws the essence of its poetics from the heart of this negative poetics of the sea, going beyond it, establishing it as the constitutive centre of new identities and representations born out of the wound that is the ocean. Coolitude considers the sea to be a matrix from which new visions and representations emerge, because in the very circulation of coolies there is a circulation of identities, a reconstitution of cosmogonies.

I have poeticized that ocean voyage positively, sketching visions of beauty, like phosphorescent trails, the aurora borealis, onto that wide expanse thought to be inhabited by houglis and monsters. This centrality of the ocean also disturbs the fundamental representation of coolie poetics, which is essentially terrestrial, because the indentured worker is frequently associated with working the land, thus eclipsing the marine poetics of his circulation.

On arriving, in the ghat, in the Mauritius Islands or elsewhere, the coolie suffered an injury to his family name, which was written down incorrectly. This would show that the voyage had changed him not only in relation to what he had experienced during the crossing but also in the system that controlled his circulation.

On arriving at the depot, he was listed and allocated to a sugar planter and then had to walk on foot to the old slave quarters. He worked the fields after the slaves had been emancipated. But this did not mean that he was free: his mobility was reduced, and his circulation was limited. He needed to have a 'pass'; otherwise, he would be sent to the depot reserved for vagrants. He had to pay fines and then he would be taken back to the sugar plantation.

Modes of circulation and representations

At this point, I think it is important to return, albeit rapidly, to the modes of circulation and the representation of the coolie, because those are the two main themes of this colloquium.

First, as we have seen, the coolie had a contract. He signed and put his thumb-print on the five-year indenture; this registered him as a human being, unlike the slave, who was an object, a 'moveable asset' (cf. Black Code).

The coolie, with his name (albeit in a mangled form) on the piece of paper, represented a relative advance in the work patterns of people from the colonies. That conviction that the coolie was treated better and enjoyed a more positive image on the human scale, also proved a controversial issue between coolies and slaves, as they represented themselves differently in relation to the roles they were allocated.

Freed slaves represented the coolies as new slaves. There was no solidarity between these two groups of the oppressed. Some used the term coolie as an insult, which explains why in Fiji, for example, the Indians preferred the term 'girmit' to represent themselves in the coolie trade. They adopted this term as a distortion of the word 'agreement' or agreement/contract, in order to reject the alienating aspect of the contract, so the word 'girmit' really signalled something bordering on despair or cognitive dissonance.

What the Indian wanted was an agreement, but the contract was not designed to respect his presence or value in the socio-economic world into which he was inserted.

In relation to the former slave, the Indian worker represented himself as a hired contract worker who was protected, as a citizen, as much subject to the law as his recruiter, even if this was based on an underlying deception. And it was this self-representation as subject to habeas corpus that would give rise to petitions and disputations on the part of the coolie, because he made use of royal commissions and local tribunals, etc. to challenge abuses. He would resist not so much by escaping, as the slave had done, but by practicing vagrancy or committing suicide out of melancholy or frustration.

The contract was basically an act of deception …

If we fail to grasp the way in which the contract fixed a specific form of representation, then we fail to grasp the meaning of coolitude which revolves around a twin centre, the contract and the ocean. If there is no contract, in theory, you cannot embark …

In this process of circulation, I would also point out the administrative arsenal that was used to manage the Indian worker, thus producing a truncated representation of the worker in a colonial context, and also revealing an unprecedented level of organization in the transfer of populations to meet international labour needs.

One initial point must be noted here: with slavery, there was not, strictly speaking, any list of names nor any biometric or anthropomorphic data as during the circulation of coolies.

The contract required an established identity, legal and administrative oversight, and bookkeeping onboard the ship because the coolie trade needed the accounting indicators of a commercial trade with profits and losses. For this reason, it was necessary to ensure that the maximum number of people who had embarked could also disembark in the host country, with a name, a contract, and subsequently, a pass.

The circulation of the coolie fitted into an administrative, legal and policing framework.

Before the emergence of indentured labour, the slaves received a name on arrival. What the coolies received was the adulteration of their surname.

This inclusion of the coolie in the symbolic order and colonial law also responded to the abolitionist leagues' demands that forced or semi-forced labour be humanized. They wanted to represent 'free men' working after the abolition of slavery.

But was this the reality of the coolie trade?

The biometric coolie and his circulation

As we were saying, first of all there would be documents drawn up at the point of departure, contracts to be signed and departure lists, which were kept by the police or other authorities. Then, onboard the coolie ships there were logbooks or shipping lists.

Goolam Vahed, the South African historian, details this complex process that was established for the trade in coolies: 'Each volume is composed of tables, with entries inscribed by hand on horizontal lines stretching consecutively across both open pages from left to right. Under column headings – including an assigned registration number, father's name, age, sex, village of origin, caste and identifying "body marks" – are the corresponding details of each named individual. In every colony that received indentured workers from India, officials recorded personal information for identification and to track their employment.' This gave rise to a mass of administrative red-tape, the 'paper regime' of indentured labour which succeeded the Code Noir, the sole legal document that governed the circulation of subjected bodies in the colonies.

Names and personal details connected to indenture formed the basis for all future identity systems, following the biometric standards that still exist today and making us all, effectively, heir to the representation and circulation of indentured labour, through our passports and identity cards.

Deployed for very particular reasons, these markers also established the specific political status of the empire's second-class citizens, representing a means of social control in empires that were driven by the insatiable need for cheap, easily controlled labour.

It should be noted that it was under the terms of Act 15, drawn up in 1842, that the British government set up an Emigration Commission in order to institutionalize the transport of Indians to Mauritius, the Caribbean, America, Fiji and East Africa. There were 450,000 coolies in Mauritius and 416,000 in the West Indies. Globally, almost 1.2 million men, women and children emigrated from India under contracts of indenture, comprising a vast diasporic movement that transformed the cultural and demographic composition of the receiving societies.

In 1859, South Africa, inspired by the legal provisions of Santa Lucia, which imported coolies in 1858, established the conditions of circulation for coolies under Laws 13, 14 and 15. The planters defined the terms of indenture, the length of contract, the minimum wages, transport and health conditions to be respected by the employer and employee.

The immigration officer was required to register the name of every coolie who entered the colonies using public money. Every coolie was listed by number, starting from 1, followed by age, sex, place of origin and name of the boat that transported them. Not surprisingly, a whole written system of administrative bookkeeping began to emerge, designed to exercise control over the circulation of coolies, before, during and after arrival.

In addition to this, the migrant's sponsor or patron was required to keep a separate register and 'issue a note of registration in the name of each contract worker, including the coolie's name, number, father's and mother's names; sex; age; stature; marks; caste; country and village of origin; name of ship and captain who had brought them; whether they were accompanied by children and, if they were, their name, sex and age as well; length of contract; and any other observations. Once this was issued, it would indicate that its holder was a fully 'processed' migrant, whose health and general well-being after the journey had been confirmed on inspection and whose availability for work had been successfully assigned to an employer' (idem).

This made it possible to control the migrants, and the sponsor's office would continue to follow the coolies once they had gone ashore, recording births, deaths, wages that they had either received or sent abroad, illnesses, changes of employer, deaths and any injuries observed. This information also brought abuses to light.

Administrative and political accounting such as this, which created a group of second-class citizens, marked out by the adulteration of their original family names, has left Mauritius with the largest number of indenture records. At the same time, it should be noted that this graphic representation of coolie trade circulation also offered the possibility of a new life.

Some coolies used flaws in the circulation system (e.g. changing surnames) to create a new identity and a new self-representation. The transoceanic voyage also allowed some Indians to escape hardship in India and a past they wanted to forget and make a new life. In this respect, the rather 'approximate' nature of official registration allowed for new identities. On the subject of these records of movement, Uma Dhupelia-Meshtrie has noted that names in registers marked the passage to invented, imagined lives, that had been created from scratch. The inventive nature of coolitude is at work here.

Thus the coolie was not just on the receiving end of geographic displacement, he also used it for personal ends to create new identities and self-representations.

It should be noted that the most exhaustive document of coolie immigration remained the passenger list, a record of indentured labour that was truly oceanic. Various different indicators converged here: the contract documentation issued by the police and that of the colonial administration, which differentiated itself from the system of slavery, as the English conducted what they called 'the Great Experiment', displacing entire populations of their colonies so that they could cultivate the products that were in demand across the empire. They generated a flood of written documentation that laid the foundations for contract labour, producing a large body of colonial record-keeping which coolitude appropriated and deflected to form an imaginary space that was open to the world.

From this biometric coolie, then, as the privileged vector of colonial biometrics, what ultimately emerged was a double writing of coolitude.

If we define biometrics as 'a writing of the body', that is, an administrative transcription of physical and behavioural characteristics, such as fingerprints, photographs and scars, etc., designed to construct a person's identity, coolitude reframes these writings, with a view to creating a 'manifesto' of coolie liberation.

Coolitude writes itself from the very body of these administrative representations of the coolie, in order to give coolitude a voice and an imagination of its own, a form of writing that is anchored in a poetics, a reconstructed memory, a liberated History and a literature, so that the coolie is no longer written by the other, in the manner of an indentured worker whose speech and movements are controlled.

On the anthropometric coolie

We must not forget that all this labour of control that both preceded and accompanied writing, constituted, as we have already noted, a Great Experiment, a process of administrative record-keeping, the recognition of the migrant as a distinctive figure, who moved around the world. We are all heirs to this graphic representation of the body which preceded the circulation of the coolie, and to border police techniques too.

You will certainly have seen these photos of coolies on their passes and entry permits, at the point when they enter the rich iconography of the coolie trade. There are plenty of them in Mauritius.

I would like to focus specifically, for a moment, on the use of photos which took hold in the first three decades of the nineteenth century.

With the coolie trade, we have the first transnational use of photography in the service of controlled migration, in a way that would subsequently become widespread. In this respect, I repeat, we are all, men and women, heirs to this graphic representation of the coolie, for the purposes of circulation. We are heirs to the employment contract too. Both of these were first tried on the coolies, and this occurred at an international level. Their identity photos (the forerunners of our modern passports) were used in order to recognize them, but also to control them.

For example, when a coolie wanted to venture outside his camp, he was asked to produce a photo permit, which he had to renew. The picture cost £1 at the time, while his monthly salary was around ten shillings. So the coolie would elect to stay in the camp, free but under surveillance as if he were on probation.

Anthropometry techniques were developed by Alphonse Bertillon in the 1880s. Bertillonnage is the basis for computerized biometric systems, including facial recognition and hand geometry.

However, alongside the photo, which remains emblematic of these anthropometric and biometric documents, the thing that created a direct link between a physical body and an imprint on paper, was the fingerprint.

It was Francis Galton who formalized these imprints between 1889 and 1901. He was assisted in this development by Azizul Haque and Chandra Bose. This became a widespread practice principally among the coolie populations of empire and other target groups, often fraudsters and criminals. It would be fair to say that fingerprints were the specific ink mark of the coolie, binding him to a coercive system of labour, just as the chains signalled the status of the slave. And we should add here, that the digital chains which bind the citizens of today ever more tightly also derive from the coolie trade.

Biometric rule was created in India and brought to South Africa in 1900 by Indian Colonial Service bureaucrats. Its identification techniques were used in the Transvaal against Indians, including coolies. Gandhi led international protests against fingerprint registration and this issue became a landmark moment in his opposition to colonial policy. These protests put an end to wide-scale identification operations of this kind in India itself. South Africa, however, went on to become the empire's laboratory for technologies of racial segregation. Hannah Arendt referred to these forms of identification as bureaucracy wedded to despotism. It was very much along these lines of thought, that the South African Galton transferred anthropometric data to the field of eugenics, declaring that blacks were created to be slaves. Let us note at this point, that this anthropometric technique, especially as used against the coolie, forged Gandhi's concept of satyagraha and drove him to write against modernism in Hind Swaraj (Indian Home Rule). Despite Gandhi's protests, however, Indians were subjected to discriminatory, racial control up until the 1960s.

Between trace and writing

We should note at this point, that the fingerprint is not a piece of writing as such, but a trace. From the perspective of coolitude, it plays a part in reducing the Indian to the coolie, the coolie to a pair of arms and these arms to a mark left by the body. Surrounding this imprint were anthropometric components which indicated, first and foremost, the subjugation of a body that was recognized as a subject at the point when the contract was 'signed', but which also, paradoxically, linked this body to a system of physical labour which weakened any claim to be a subject of the empire in one's own right. This fingerprint, and indeed the entire apparatus of surveillance, were designed to move the body from one country to another, labelling it in terms of the statistical, political and economic needs of the coolie trade, in order to promote the circulation of the body, but also to reduce its social and geographical mobility. This demonstrates the imperial desire to use biometrics in a way that would ultimately produce an enslaved, subordinate subject, on the margins of empire. The paper empire of the coolie trade was coercive, not liberatory. It controlled, registered and restricted circulation.

So let us note here, that this imposition of graphic representation on the coolie was invented with a view to organizing the migratory flow of coolies in the Indian Ocean and elsewhere, not for the purposes of free circulation. And we must remember that all these statistical data relating to the body are tainted by Galton's world of eugenics (1850s), which extended and developed into our current systems of circulation control. In effect, they joined forces with 'Bertillonage' (1890) which is the global system of criminal identification. It was Bertillon, a keen statistician working in Paris, who managed to link the human body to written records. He applied body statistics (anthropometry) to the process of identifying incorrigible 'vagrants'. His system created the means of tracking a criminal over time. It is his statistics which provide the practical basis for biometrics, allowing eleven parts of the body to be 'measured'. This system drew heavily on phrenology and craniometry, which were influenced by racist beliefs. The aim was to assess personality and intellect, by measuring the size and shape of the skull. These ideas remain embedded in modern bureaucracy. Galton's fingerprinting corrects the statistical weaknesses to be found in Bertillon's anthropometry. The coolie is a direct victim of the collusion between these two researchers. The aim was to corral the 'free' employee into a zone where his citizenship and freedom could be restricted.

Galton published Hereditary Genius in 1869. For him, the black man was inferior and must be corrected with a 'righteous whip'. He was stupid and deviant. Observing the development of English cities, which he considered spaces of decline and decay, Galton warned the British against the movement of 'inferior' populations, basing his argument on the fate of the Athenians. To police all of this, he developed the concept of the fingerprint which had first been posited by the colonial government in India and would subsequently be brought to bear in relation to the circulation of coolies.

One thing must be emphasized here: with the help of a global network of courts and prisons for the enforcement of indenture, private debt, land revenue and long-term criminal sentences, the Indian colonial state created a global infrastructure for coercive identity registration. However, there is a paradox here too: the fingerprint was invented to remedy the absence of writing. Galton, who coined the term eugenics to refer to improvement of the race, directed his fingerprinting method at colonized peoples, from his earliest writings onwards. He claimed illiterate 'blacks' and 'yellow people' could not write or sign their own name. And he further claimed that Europeans could not tell them apart physically. And as they were certified fraudsters, they must therefore be 'recognized' by other physical means.

Historically, fingerprinting was developed in Bengal where conflicts arose after the 1857 Rebellion, between landlords, tenants and the colonial state. The Magistrate of Hooghly district, Sir William Herschel, began logging the handwritten signature of a local contractor in an attempt to curb attempts at fraud. So it was, notably, Herschel who imposed these iniquitous indigo contracts between landlords, peasants and manufacturers, and this was in a context where written documents and receipts were worth nothing. This work around contracts foreshadowed what would later be applied to coolies. Herschel developed a methodology for fingerprinting. He checked administrative records between 1892 and 1895 and concluded that fingerprints did not change with time. His observations were of interest to the colonial authorities and the Chief Constable of Bengal, Edward Henry. The latter understood how useful this could be for reinforcing colonial bureaucracy.

How did Galton develop a system that replaced other methods then in use, including Bertillon's anthropometric photography? He set about eliminating all ambiguity from fingerprints and finding a way to preserve them for rapid consultation. Focusing on the grooves that cut across the intricate tracery of the imprint, he derived a single method of interpretation. And this was the key to the extraordinary power of the Galton system, the determining factor that he managed to isolate. He established a taxonomy of the fingerprint that rested on 8 different components.

From that point on, his method of fingerprint classification provided a simple mechanism for converting the previously unclassifiable qualities of the body into a textual object, subject to the normal indexing procedures that were widely used by document-based bureaucracy. Galton had ultimately found a means, a unique mathematical identifier, a unique marker of the body, for each individual, a biometric marker. And in so doing, he put paid to another source of ambiguity. A surname designated a person's identity but it could also refer to someone else. It could be a homonym. But the fingerprint always refers back to the person it belongs to. There is a one in 64 billion chance that two fingerprints would be the same; this was Galton's contribution to science, established through studying a sample of 8,000 fingerprints. A note of caution is in order here though as there were flaws in the Galton system. In the 1890s, Chief Constable Henry had Azizul Haque and Chandra Bose work on a fingerprint classification system that simplified Galton's. Their findings were published in 1900 and form the system that is still used today. Galton was an ardent advocate of this system, as it allowed the empire to exercise greater power over circulation.

Irregular identities

The fingerprint could be described as a kind of surname or family name that had been impressed upon the thumb, the grooved, serrated and cut-out visual image of a name that could not be forged or written down wrong, playing its part in reducing the Indian, or the poor-man, to representing not just a pair of arms, but a fingertip in the statistics of imperial bureaucracy. It was the metonymy that reduced a pair of arms to a thumb in a legal system that prevented the coolie from circulating while at the same time giving him hope that he was protected by his contract under the imperial system of control. In Cale d'étoiles-coolitude, I highlighted the jagged 'serrated' nature of the surname, this injury to the surname, which unites the tracery of the fingerprint with the pages of the ship's logbook, held together by a serrated metal clip. When you remove the clips, the metal teeth leave a groove mark, metaphorizing damaged identities. These grooves also recall the ancestor of the coolie contract, the indentured or indented contract, the name of which invokes the 'tooth mark' or the groove, in those documents which connected indentured servants to their 'owners' in the Americas and the Caribbean of the seventeenth century. The contracts were torn in half to prove their authenticity: 'Residents of New England in 1650 would have been very familiar with the idea of indenture. When contracts of service were drawn up, the legal document was ripped in two, each part keeping their half as proof – the "indents" of the torn edge would only fit with the other half. This gave rise to the word "indenture"' <https://www.futurelearn.com/cursos/batalha-de-dunbar-1650/0/passos/66260>.

In both cases, before the process of circulation began, the body was marked, cut, injured, reduced to being merely an instrument, a tool, with a signature, an identifier that would remain recognizable from the beginning to the end of the circulation process. A particular representation of the coolie was also being profiled here. He was 'screened' by a system designed to flush out and track criminals and thieves.

Before 1907, the fingerprint system was used in South Africa primarily to control coolies, particularly the Chinese in the Witwatersrand gold mines. They were 'difficult' to identify, speaking no language that was understood in the region. They all had 'identical' names. The authorities were unable to tell them apart. So they used Bertillon's 'talking' portraits, which combined a photo, an anthropometric description and brief notes on a single card, all included in a statistical data system. In seven years, there was not a single misidentification of the 70,000 coolies in the region.

They wanted to do the same with coolies in India, but the Indians resisted because fingerprints were associated with a system that was degrading. Criminals were identified by fingerprints in India.

Gandhi led the insurrection. After six years of fighting against this measure, Gandhi wrote, as I said, Hind Swaraj, rejecting the technologies that created racism and discrimination.

We should point out here that the Mahatma wrote this work while he was on the boat, himself experiencing the circulatory passage of a subject controlled by the empire that was opposing his desire for free circulation.

It is interesting to note that even free Indians in South Africa, businessmen, had to carry a 'pass' with a photo and biometric information, and not just the coolies, indicating that it was not the word 'coolie' that oppressed the Indians, but an entire racial and hierarchical system that made any Indian a second-class citizen, whose mobility was controlled from start to finish.

This new fingerprint identification system was, as Gandhi wrote to The Times, 'a pass and identification system applicable only to criminals'.

Gandhi protested against a colonial representation that made all Indians either a Sammy, a pejorative term for South African Indians, or a coolie.

'The criminalization of Transvaal Indians was an affront not only to '300 million Indians' but also to 'the entire body of Indian officials to which I and most of the members in this delegation belong, and who are insulted along with the natives of India', declared Gandhi.

A coolie was asked to show a 'pass' as was any Indian, purely on the evidence of his difference. Let us emphasize once again here that it was not the term coolie itself that drove this degradation, but the ongoing racial and political hierarchy of South Africa and, to a lesser extent, the empire. The 'other' was not represented as civilized, but as a savage or a native. Gandhian protests were also fuelled by the national humiliation felt at the need for 'passes' and 'ten print registration'. Gandhi was extremely clear on this point, declaring that it was not only forgers of identity who would fall victim to these discriminatory measures, but even Indians in good standing. Gandhi protested against this law of enslavement targeting the whole 'Indian race', stating that governments of the day were, in effect, demonstrating that they wanted to control entire populations and to do so more efficiently.

These anthropometric and biometric elements are absolutely essential to understanding the key issues of circulation and representation in the coolie trade. It is no exaggeration to say that it was because of the coolies of South Africa, the way they were represented and the restrictions of movement imposed on them by a racist state, that Gandhi developed his techniques of non-violence. He was a lawyer and knew how to read, interpret and challenge texts and control mechanisms. However, he wrote: 'The real change was in understanding the nature and purpose of the state.' Prior to 1908, he had seen the state as an instrument of harmony, shaped by science and law, and he had understood his own practices as an extension of that power. Subsequently, he viewed the 'administrative machine', with its technological modes of operation and its telos, as an instrument of destruction. This newfound awareness was not unrelated to the fact that he had worked out exactly how the representation and circulation of the coolie operated under empire.

Conclusion: A systemic view of circulation and representation in indenture

Let us quickly return, in conclusion, to the organization of circulation in the coolie trade.

Methods of registration were obviously invented and put into widespread use, such as Bertillon's 'talking portrait' system, and the anthropomorphic and biometric documents of Galton and Henry, indexed on contracts, 'passes' and travel permits.

Nothing short of a revolution in the surveillance and monitoring of migratory flows took place during the coolie trade, involving collaboration between local and colonial authorities, agents, recruiters, the juridical and legal arsenal and transnational migration police. The latter, along with port authorities and local administrators, listed and filed ships' logbooks and any other documents that would allow them to track the movement and allocation of contract workers.

This system readily explains how labour circulation came to be organized, from a logistical, legal and administrative point of view, with the advent of biometric and anthropomorphic documentation, which was closely intertwined with the way in which coolies were represented.

The coolie, as we have seen, derives from the figuration of the 'other', which emerged from the orientalist perspective of the European empire, where the colonized was represented as a native or an inferior being. Furthermore, in this process of circulation, coolitude clearly indicates that the world of the plantation operated according to a logic of concentration, with a particularity all of its own. The coolie had to be confined to his workplace, while being given the impression that he could circulate freely, thus distinguishing the coolie trade from slavery because ultimately, the servant could return home at the end of the indentured contract. We know that this illusion rapidly fell apart, however, when the coolie realized that he was often in debt after five years, owing to a whole covert system of 're-indenture' resulting from the cumulative effect of fines that were imposed by the 'double cut' (coupure double), alcoholism and the plantation shop. The shop would sell food to the coolies in such a way that they were often deliberately forced into debt so that they could not leave the plantation debt-free. Debt, in this phase where the labour practices of empire acquired international specificity, was a tool designed to impede the coolie's social mobility and physical circulation, effectively binding him to the plantation or the railroad.

This 'free yet constrained' workplace, which emerged after the world of the slave plantation, produced a very particular type of society, where settlers, farmers, former slaves and indentured workers coexisted. It was the matrix of coolitude. Here, the indentured labourer joined the margins of a hybrid society, an internally conflicted society built on creolization. This system of control, which extended from the warehouse to the boat, to the plantation, relied on an almost fail-safe machinery of spatial monitoring which was connected to the police as well as the colonial and imperial authorities. One particular component was emblematic of this apparatus for controlling circulation, and this was Jeremy and Samuel Bentham's panopticon. In the eighteenth century, the two brothers had developed a system which allowed one single person to control an entire prison system from a central vantage point, with circular cells built around a single watchtower. Incidentally, the Benthams were very interested in the penal system – how to reform prisoners by locking them up – and in work. They were inspired by factory planning, and the efficiency and coordination of workers. So what we are seeing here is a space where representation and circulation come under surveillance, affording coolitude across the globe, modes of circulation and representation that are regulated by a transnational system under empire.

And within this scheme, the coolie was represented in very precise ways. He was literally depicted as a beast of burden, pulling a cart, carrying luggage on a dock, or working as a serf in the fields, the mines or the railways.

Ultimately, he was a typical representation of the submissive Asian, a common figure in the colonial period, where the need for cheap manual labour was insatiable. The indentured labourer was a subaltern characterized by contract work, a lowly 'foot soldier' whose services could be bought for a pittance.

By extension, the coolie was the bearer of empire, the casual labourer par excellence. The empire, on the other hand, presented itself as the bearer of light and 'universal' progress. The empire was doing him a great service (this was the dominant discourse of colonialism's 'good conscience') in assigning him to 'liberating' work overseas that would 'free him' from his cripplingly backward traditions. The notions of drudgery and exploitation were entirely absent from this idealized imperial construction of the coolie; ultimately, we might say that this image of the resilient coolie corresponded to a very specific inhabitant of the margins: the person created by nature to bear the white man's 'burden', in the literal sense and figuratively too …

By 'freeing' him from his natural habitat, to put him into circulation in the empire, the master made the coolie into a native who had signed up to a system of labour and could be shaped and influenced at will in the international 'taylorization' phase of the colonial era. The coolie was not a slave in chains, he was a 'free' being, able to move across borders and seas. They have created the documentation for him to do that.

Indeed, as we have seen, this circulation was regulated and subject to levels of border policing and administrative surveillance that were unprecedented in the empire. It went far beyond the apparatus of surveillance that was used to manage slavery, where written documents did not even bother to note the surnames of the enslaved. Rather they simply recorded the individual 'units' that were being loaded and unloaded, and later sold in the slave markets or transferred to masters who bought them. It was pretty much a transfer of goods, a process of pure accountancy. With the coolie trade, the written documentation was 'mercantile' in nature, involving a contract, a legal document, listing names and other details, increasing the power of the immigration police, and internationalizing anthropometric and biometric systems. It marked the beginning of the 'paper empire'.

The invention of the contract and its accompanying documentation allowed a transoceanic transport business to emerge and established a legal framework. The contract was a bit like the code noir of indentured labour. It provided the contract worker with the legal framework of habeas corpus, although this was often violated in practice. In effect, the passes and permits needed to travel outside his own work area made the coolie a marginal figure, a pariah, a second-class citizen, subject to constant surveillance, regarded as an indentured wage labourer whose freedom was thoroughly overseen.

So the coolie was compliant with a colonial mode of representation which ranked men hierarchically by their skin colour and indigenous status. Even if he is considered in his own context, he remains a prisoner of those representations of slavery that preceded the coolie trade. These representations had a direct impact on the coolie's geographical circulation and social mobility in colonial societies.

In short, as I see it, the coolie was a slightly improved version of the slave, in so far he had a contract and a poverty wage.

As a being of lesser importance who was not the master's equal, he could not, for example, ever travel first class (Gandhi was excluded even when he bought a first-class train ticket).

The empire placed the white man at the centre of its hierarchy. It was he who reigned over the panopticon of the colonial plantation: his house occupied the throne at the centre of the surveillance system. It sat in an elevated position, allowing the coolies from the huts below to take in this fact, both mentally and physically.

The coolie trade was contrasted with the barbarism of slavery that preceded indentured labour, and it was supposed to embody the values of human progress (there was a contract after all). But the coolie trade was always, in fact, the dupe in this farcical staging of imperial power.

The indentured labourer could not live where he wanted, he could not travel as he wished, he could not choose who he worked for. If he wanted to report abuses, he would write to the authorities. As I explored thirty years ago in my work on 'Coolitude', he was a petitioner or a pamphleteer, remaining caught in the terms of contestation set by a globalized legal system, even if he ran away (like the marooned slave) or committed suicide in despair.

Ultimately, the coolie was simply the reduction of a human being, a metonym, a pair of arms that was 'free' to work for the specific needs of an empire promising him the world, but offering nothing but a field, a mine or a railway to toil over …

This metonymic representation best sums up the indentured labourer: 'a pair of arms', the manual labour of the colonies. And where the circulation of this manual labour was subject to an arsenal of controls, the pair of arms became a set of fingerprints, and the labourers' rights of citizenship were effectively truncated at the heart of empire. The metonym was then further fragmented and broken down in this policing of movement across the empire.

This was the 'passport' that allowed him to migrate from the periphery of his home country to a periphery somewhere else in the world, condemning him to survive on the margins of colonial societies that were eager to consume the products of manual labour.

Even if he had a first-class ticket, like Gandhi who became close to the coolies, his movements were still delimited.

In the end, being seen as a coolie or a sammy reinserted him in the value system of a racial hierarchy. He may have seen himself as coming from an ancient civilization (India or China) in the first class of History, but the empire inevitably confined him to the third class of the train. He travelled in a class that was just above the livestock and the tools. This was how the disciplinary management of the indentured subaltern or coolie worked. He would not be able to speak out or engage in any literary, artistic or human act of writing until he had first taken his own coolitude back into his full possession …

JOÃO PAULO BORGES COELHO

14 Narrating the Indian Ocean from Mozambique

The writing of a story always starts in the real world. And this real world includes the physical space we call geography. Even when things seem as far removed from the real world as they possibly could be, such as Raymond Roussel's wild *Impressions d'Afrique,* the fantastical island of Casares's *Invención de Morel,* or the inside of Xavier de Maistre's famous bedroom, it is always the real world that sets in motion the journey into fiction.

The course this then follows is not always predictable and it can take a variety of forms. I see it as a kind of prior claiming of spaces – not always straightforward, not always consciously done – so that the speaking voice can find a place to implant itself and tell the story from there. Rather as with architecture, there has to be some pre-existing idea of the space that the project is going to occupy.

I call these spaces 'Narratable Spaces', heterotopic[1] places which are constructed *a priori* (pace the apparent paradox here) providing the main fictional theme – the story or the plot or whatever we want to call it – with its basic structural coordinates and a large part of its material. This narratable space is the structural element that affords the link with reality in the creative act that we call 'writing fiction'. It is not so much a plot or a storyline but rather, as I see it, a *place* where plots and storylines can be constructed.

Narratable spaces are not just settings. Settings are bound up with the events that happen within them, so they expire when the narrative comes

1 Heterotopia is a concept which Michel Foucault derived from Geography and also much used in architecture, which combines space, knowledge and power. It refers to alternative spaces which retain some kind of link with the real (such as the reflection of an image in a mirror, for example).

to an end. Narratable spaces, in contrast, nourish the act of narration, but they also remain intact when the narrative has ended. They stand ready to host and nourish new acts of narration. They are like the 'houses' of fiction; they are houses built to last, more so than the stories they accommodate. They ultimately change too (for nothing is immune to time) but they change more slowly.

Narratable spaces are very diverse (we may find them, for example, in the cities of a particular territory, or on different scales, like a geographical map, or through the prism of childhood or in our dreams, etc.). The Indian Ocean constitutes a narratable space, a place, a house of fiction. It may variously manifest itself as a single entity, as the Indian Ocean of Mozambique, as the bay I can see from my beach, the sea of shipwrecks and cyclones, the sea with offshore gas prospects or the sea known to fishermen or even, as we will see later, the sea of poets.

Not only are narratable spaces very diverse in themselves, but they also accommodate a range of different forms and shapes, which exist in tension with each other, frequently tending towards conflict and rupture. But at the same time, they also favour relations of continuity, sufficient to establish the cohesion which moulds narratable space into a singular, specific entity. The mortar that binds all these dimensions together is provided, I imagine, by *genius loci*, the spirit of place.

The Indian Ocean in all its diversity, with all its different dimensions and internal tensions, from Burma and Iran to Yemen, Australia and the Antarctic, including the east coast of Africa, is clearly being treated, in this book, as a unique and singular entity.

This unique and singular entity is, in fact, the product of rich, multifaceted forms of contact between the different ocean shores. Indeed, Neville Chittick considered the Indian Ocean to have been the largest cultural continuum in the world, between the first millennium and the middle of the current era. As a result, historians note that the extremely multi-ethnic nature of Muslim maritime crews was evident as far back as the thirteenth century.

By the same token, the Indian Ocean, with all its complexities and tensions, is very much its own entity, a narratable space, a *cosa mentale*, to use Leonardo da Vinci's famous phrase. My intention here is to give a brief overview of how I personally perceive some of the many shapes and forms this space takes, and to identify some of the ways, among many, in which they connect with literary fiction.

The ancient Indian Ocean

Africa's relationship with the sea has not historically been a peaceful one. In the encounter between colonizer and colonized, the sea was always, in the eyes of the colonizer, an obstacle to be overcome, to get to wherever their curiosity was leading them; for the people on land in this encounter, the sea represented a kind of non-being, a void, where alien beings would simply materialize on the beach, like Argonauts coming ashore from strange-looking ships which had come from the moon, carried by winds and tides.

There is no doubt that the Indian Ocean, like many seas and oceans, had its own, ancient and very particular modes of navigation. The Pacific was renowned for its light-framed ships and navigators who covered enormous distances with razor-sharp accuracy guided by the stars; the Mediterranean style of seamanship was always alert to winds and tides (thence the advent of terms such as Mistral, Tramontana and Sirocco); the Indian Ocean of Africa had a sophisticated Arab-Swahili culture of navigation driven by the monsoons, a weather system of regular winds which produced their own particular currents and tides (blowing from the Arabian Peninsula to the African coast of Mogadishu or Mombasa from November onwards and in the opposite direction between April and September).

But the great navigational arts of the Indian Ocean always belonged to a culture that had itself come from overseas, like the triangular or lateen sail shared with Mediterranean seafaring, which would fill with wind and drive *dhows* southwards along the coast. It was a culture which influenced only a narrow strip of land, albeit intensely so. Indeed, research tells us that prior to the nineteenth century, no mosques were built more than a mile inland from the east coast of Africa.

In contrast to the New World, where for a large part of the population the sea is a horizon of hope in the quest for ancestral origins, the sea symbolizes loss for Africa. It is the void which engulfed successive generations of slaves in chains.

Bagamoyo, on the coast of Tanzania opposite Zanzibar, is probably a good example. It was once an entrepôt in which huge caravans of slaves converged from the interior of the continent, coming from Cazembe and other places. Legend has it that these slaves were made to walk blindfolded in circles around the Customs House yard, so that they would lose all notion of where they had come from, and then they would be loaded onboard the ship. Only then, when they were far from land were the blindfolds removed so that they were suddenly confronted with a vast expanse of nothingness, the open sea, and were stricken with terror.

This is perhaps one reason, symbolically at least, for the absence of a maritime tradition in Sub-Saharan Africa. There were no great maritime voyages. Fishing dates back to ancient times, but it is limited and it uses small canoes. Even today, (no doubt also for colonial and postcolonial reasons) national navies and native fishing fleets are small and insignificant (shipbuilding in Africa makes up less than 1 per cent of global tonnage and it is estimated that in the last five decades, the continent of Africa has lost more the 100 billion dollars to poaching). The sea is treated as a place of danger and a source of storms. This abstract notion recently became a concrete reality in the form of two huge cyclones, Idai and Kenneth, which trailed death and devastation in their wake.

Indian Ocean: Histories and identities

The importance of the Indian Ocean in the configuration of Mozambique is undeniable, nonetheless. Mozambique consists of highland plateaux in the interior and one huge coastal plane. The coast which is more than 2,500 km long is a major defining element in so far as it opened the country's doors to the outside world and to the East, bringing a degree of cosmopolitanism to the interior, which distinguished it from its neighbours.

Where the upland plateaux retain echoes, stones and myth complexes from the great kingdoms and empires (the Undi, the Mwenemotapa, Gaza), the sands of the beaches attest to life in smaller units, multiple languages, cultural mixing and acculturation – with fragments of Chinese pottery, the ruins of old Arab mosques, Portuguese fortresses, roof tiles from Marseilles and Murano glass beads brought from Trieste. Indian influence still stretches along that coast and up the Zambezi River deep into the interior. One of the greatest contributions that Indian Ocean traffic made to Mozambican identity was, as we know, the connection with Goa, which left its mark on religion, administration, fishing and spice gastronomy, among other things. All of this, which goes far beyond geography, means that Mozambique is profoundly and intrinsically a product of the Indian Ocean, as Alexandre Lobato, has amply demonstrated. And yet there are many different facets to the Indian Ocean. Since it also constitutes the southern limit of Arab influence, Mozambique is the only country which can genuinely claim to belong to two separate regions of Africa – East Africa and Southern Africa – two regions which differ radically, from a cultural, geopolitical and even geoclimatic point of view. However, the Indian Ocean is not just an exceptional case with a high level of cosmopolitanism. It also accounts for the considerable diversity which is clearly manifest in the interior of the country, reflecting the distance that separates the coasts of Quissanga and Ponta do Ouro. As Edward Alpers observes, the coast of Mozambique possesses coral reefs, beaches, islands, mangroves, estuaries and deltas.

The Indian Ocean of the colonial era

The Indian Ocean was profoundly important for the history of colonial Mozambique. It had a decisive, structural influence on one of the three main pillars of its economy: the railway-port system, which designed the colony to act as a gateway to the landlocked countries of the interior and therefore built its main cities on the coast. The ports and railways reinforced the existing directionality of the countries' rivers, making Mozambique a country where the lines of communication run horizontally. Essential in this respect are the axes that link LM-South Africa, Beira-Rhodesia and Nacala-Malawi. By the same token, the country's vertical lines of communication remain underdeveloped (national highway no. 1, which still runs somewhat intermittently, only very recently joined south to north when a bridge was built over the Zambezi, in Caia).

The importance of this role as a maritime gateway to the interior extended beyond the economy to influence colonial society too, creating a culture of beach tourism for the white populations of Rhodesia and South Africa, iconically symbolized by sun, beer, prawns, inter-ethnic contact (frequently sexual) and the LM-B radio station broadcasting in English.

The geo-politics and economics of the Indian Ocean

With the advent of independence, Mozambique inherited a significant part of this structural dependency from the past. In the twenty years that followed, a profound crisis in relations with the countries of the interior, namely Rhodesia, South Africa and even Malawi, led to drastic reductions in railway-port operations, which were already weakened by the Portuguese exodus, and which the ensuing Civil War almost brought to a complete standstill. If we add to all this, a political culture of the new regime that was forged in a rural, interior context, and the natural priority given to reinforcing vertical cohesion, which was essential to consolidating a sense of nationhood, it becomes easy to understand how the sea became even more marginal, in spite of the country's extensive coastline and the potential importance of a maritime economy. This marginalization also characterized the whole southern region, devastated as it was by savage conflicts which emanated from the interior, rather than the coast, and concentrated energy and attention there.

On the wider scale, the end of the Cold War substantially altered the balance of power around the Indian Ocean, specifically the Soviet versus North American bipolarization which had prevailed up to that point. For a short while, the USA reigned supreme, but their strategic interests were progressively forced to confront increasing multipolarity, in the form of institutions such as IORA (the Indian Ocean Rim Association) which was established to develop economic, commercial, investment and research cooperation between the countries along the coast. This multipolarity has now evolved a stage further into a new bipolarity protagonized by India and China. Examples of this latest competition would be IONS (the Indian Ocean Naval Symposium, an Indian-inspired organization which incorporates almost forty countries) and the Chinese strategy known as Belt and Road.

Although we are formally members of these bodies, we do not enjoy full participation in this arena either. Over the last twenty-five years, southern Africa as whole has entered this new environment under the leadership of South Africa, the only country which actually has a maritime vision and the capacity to pursue it. The RSA is the only country in the region which has any real aspirations to a leading international role in the Indian Ocean. It is the only country that has a blue economic programme. The RSA also determines the maritime vision and security of SADC, the Southern African Development Community. It is the main, indeed almost exclusive, guarantor of security there (in 2010, e.g. when the coasts of Mozambique seemed to be under threat, South Africa placed permanent sea and air resources in Cabo Delgado, in an attempt to stop Somali piracy spreading southwards)

For longstanding reasons, as well as more recent ones, Mozambique did not exploit the opportunities offered by local and regional regeneration in the early 1990s, to reinvent its relationship with the sea. Society was not mobilized, no maritime thinking emerged on a national scale, and no consistent maritime policies were developed. The old colonial capacity for cabotage, which was already weak, practically disappeared. Fishing, both industrial and semi-industrial, is operated predominantly by foreign firms, and the tourist industry, associated mainly with the beach areas, is also relatively small scale compared to what it could be.

We do not yet have the communal 'beach for all' as defined by Alain Corbin. With the exception of a few luxury hideaways, which are marketed mainly to foreigners and minimally developed by national operators, the beach is still mainly a place of work where canoes dock to deposit their catches of *marora, magumba*, and African sardines (*hilsa kelee*), the main source of protein for East Africa. Or it provides the principal setting for Zionist religious ceremonies which are ritually connected to the sea.

In short, there is almost no interest in the sea, even though the Civil War years increased the already well-established trend of a rising coastal population. Of the 130 boats fishing tuna off our shores in 2013, only one was Mozambican. And if there has recently been an apparent mega-investment in fishing and maritime security, it has now become apparent that this was merely a cover concealing a corrupt mafia-style operation at the highest level of the state-political apparatus, and causing a national trauma that has only deepened our sense of alienation from the sea. This is a highly regrettable situation at a time when the political and economic hopes of the nation are turned towards the sea, with the possible exploitation of large offshore gas reserves.

Tracking the Indian Ocean through literature

I have already mentioned the Indian Ocean of ancient times, which was mysterious and in many ways hostile. And I have said something about the Indian Ocean of today which remains distant from us. But there remains much to be said about the Indian Ocean in literature, which has, in certain respects, and somewhat against the grain, redeemed Mozambique's relationship with the sea, especially in the work of poets such as Knopfli and Sant'Anna, Lacerda and Lemos, Mendes, Craveirinha, Patraquim, White, Carrilho, Okapi and others.

They achieved what they did in times of colonial repression, by questioning the false harmony of luso-tropicalism and invoking a cosmopolitan history centred largely on Mozambique Island. It was not always a peaceful history nor a happy one. It was a tale of crossings, voyages and sailings, and also of love, raising the island above its peripheral status to something more than just a stopping point.

The poets I mention above continued in this vein after independence, creating spaces and ideas which questioned the simplification of history as well as the stereotypes associated with homogenous, cohesive and originary identity.

This entire process, summed up in Virgílio de Lemos's phrase 'making sea' (as noted by Jessica Falconi, for example), comes close perhaps to what I have defined as a 'narratable space', that is established so that a story can be created out of it.

If we read it metaphorically, the sea is perhaps the ultimate narratable space – an open space, a blank page with no marks on its surface, since all journeys are erased the moment they are completed. It is a page on which everything constantly starts anew. Memories of the voyage are left in this wide open space, always ready to receive new voyages. In this sense, it resembles eternity (and anyone who has visited Mozambique Island will certainly have experienced that sense of eternity which defines narratable spaces). I believe this is the feeling that Jorge Luis Borges is trying to express at the start of his famous poem about the sea:

Before the dream (or the terror) wove
Mythologies and cosmogonies
Before time was minted into days,
The sea, the eternal sea, already existed and was there

And yet, this simulation of eternity, as we know all too well, does not make narratable space into a universal construct that is neutral, eternal or immune to time. On the contrary, since it represents or interprets the world, a narratable space cannot be other than subject to those traces which the passage of time leaves upon it and like the world itself, this is a lesson we are now learning all too painfully. It is subject to profound transformations, some of them largely irreparable.

Things we once considered immutable certainties – houses, ancient trees, mountains and the sea itself – are becoming vulnerable and provisional, as we look on in fear.

The signs are so obvious that we cannot just keep rationalizing and denying them, as the dark forces that live within and among us do. The changes are cutting ever deeper, a clear sign that nature itself is hurtling to its doom. It is as if nature itself had been bled dry and was showing us that it can no longer match the dizzying pace of what we call growth and progress. Elias Canetti used to say that if one thing sums us up, it is the *drive to mastery*. We do more and more, we do it faster and faster. 3G, 4G, 5G piled onto an exhausted world already in its death throes. We are like a deadly cloud of locusts sweeping over a field of corn.

And inevitably in the future that will follow, much of the debate will focus on how we respond to these signs of gathering doom. There will be increasingly drastic responses that are anxious and hasty, clashing with each other in a cacophony of rules and regulations, accusations and shouting, sanctions, changed behaviours and frustrated dogma.

How do we get beyond this confrontation between man and nature that is so central to Western culture and has now become global? This is a good moment to paraphrase the poet Manuel António Pina:

> *It is neither yet the end nor the beginning of the world, fear not, it is merely a little late.*

Aldo Leopold (*A Sand County Almanac*) back in 1949 referred to the need for ecological awareness based not just on the imposition of new laws but on the assumption of new values, 'on altering out intellectual priorities, our loyalties, our tastes and our convictions'. This also means, to come extent, setting new priorities in art and literature too.

Obviously, this does not have to imply the crude instrumentalism of using literature, which is sovereign in its own operations, as a means of ecological consciousness-raising; what it means is deploying literature to unearth fresh, creative ways of relating to nature and its narratable spaces, in this current context where the signs of impending doom grow ever more frequent.

Pushing speed to the limits, pursuing novelty for novelty's sake and cultivating the outlandish will not necessarily capture the spirit of the avant garde, even in art. Dissolution, purely for its own sake, may only, in the end, be yet another symptom of confusion and anguish.

Claude Monet built a water garden in Giverny so that he could paint the waters at different times of the day; so that he could paint light, essentially. There is more of truth in Claude Monet's water garden than there is in any of the monstrously overblown temples of modern architecture scattered across the world today, showcasing an art that has more to do with making war, than deciphering its mysteries.

Returning to the Indian Ocean of Mozambique as a narratable space, contrary to what the millionaire tourist getaways of palm trees and golden sands would have us believe, it is not timeless or eternal either. It too is susceptible to this twilight age, with the extinction of ancient forms of canoe fishing, precarious, informal cabotage and shipwrecks, various forms of trade and piracy, over-fishing, cyclones and ultimately terrorism. And it is also an Indian Ocean where maritime exploitation of fossil fuels is beginning, on a large scale and somewhat belatedly, promising progress and well-being for all, but also threatening false hope, pollution, disorder and catastrophe.

And yet, the Indian Ocean, as we land-dwellers understand it, still offers the possibility of escape from a world in decline. It still represents the lure of the quest, a promise of new lands beyond the horizon, a guarantee that the imagination will survive and a sense of utopia, ultimately, the journey into fiction. It is more than just a continuation of land onboard the ship. The sea is a place where old earthbound principles, such as frontiers, propriety and order are called into question, paving the way for new grammar and forms of syntax.

Literature can help us renew our age-old contract with the sea and its cosmopolitan spirit of openness. This principle is all the more pertinent the more the sea, in its fluidity, becomes illusory and hard to capture, with its multiple moods, tonalities and momentary flashes of light, where the dilution of frontiers forms a vivid contrast with the rigidly defined identities of the present day.

We need to see the Indian Ocean, not as a huge, inhospitable void, but as a central core of meaning and potential, connecting all its different shores and making them vibrant.

A place of freedom, rebellion and risk, a place to try out new beginnings. The last frontier before the horizon, the last untameable category.

In spite of the extinction of fish and coral which this century will surely see, in spite of the huge tides of plastic and the oil platforms which are springing up like filthy metal islands, in spite of the aircraft carriers and the sinister lone radioactive submarines lurking in its submerged landscapes, it is an Indian Ocean which still re-appears, every morning, made entirely new.

And by way of conclusion, the words of Borges serve me once again.

Who looks upon it, sees it always for the first time.
With that wonderment in the wake of things perfect,
Elemental, the evening's
Beauty, the moon, the bonfire's flame.

LUÍS CARLOS PATRAQUIM

15 Poems

Omuhípiti, The Sibilant

Over this night, let the shadows sink down
The whisper of the wind smoothing the skin of the palm trees
After the street

Over this night   eternal,
Let her lie in the arms of the light

Perfume Visible
Quivering
Naked

For João Paulo Borges Coelho

The Indian Ocean is an inner sea
A babel of rafts
The gargoyle that spits out in fury
The shipwrecks of light.

And we
Still fearing the song in the flesh of the maize
Of the islands the liquid part becoming
houses whipped by the wind

Drunken birds is what we are
Where the moon
Weeps in her womb
Mother enfolding the fear

Chiveve, Manga, Munhava, Maquinino,
The Búzi river rising unbidden
The names all
Her children
Begging the waters of the Beginning
Shining like the face of morning
After the end

'Bar Tropical'

to Diniz, Camilo, Funcho, Licínio, Paulo and 'Che'

evening finds us at the table of the bar
the light with memory in its crystal glass
and the profiled foam

sharp suburban stakes
upended in us no falling tear
scalds the news that we spread
before the stars in the city

Desolate temple without heroes
who from a dark oracle flow
there we drink

when the blood is anchored
and oceans of india grow wild
we soar to an uncertain Olympus
of secret names that circle us

Muhípiti

For you, with the island; to Rui Knopfli

It is where I lay down all the weapons. A palm tree
Soothing our sleep. The shadow.
Where I am. Slow and naked. Over
the eternal waves. Where I have never been and the angels
play at boats with books as hands.
Where we eat the acidic last bud
of useless rhetoric. This is where we have no use.
Pure natural objects. A palm tree
Of beads with the sun. Singing.
Where in the night the Island gathers every isthmus
and the voices whisper. The statuary in the groin.
Gushing forth. Maconde woman not turned to stone.
It is where I am in this poem and and I have never been.
Your name which I shout as I laugh at the name.
By my name cancelled. The voices that announce you.
And I lose myself. And I am naked. Slowly. Within the body.
A palm tree opening up to the silence.
It is where I know the jaw that bleeds. Where the leopards
Drown. Time. The cicada rattling
in the lungs. The earth soaked. Gushing forth. Scarlet.
It is where I becomes confused with you. A boy bowed
under the weight of being a man. A palm tree in blue
forehead damp. The memory of the infinite
The repose that doubts itself.
Listen.
The watch patrol and no plane has taken off. It is where we are.
Where the birds are birds and you sleep.
Through sobs of syllables I wander. Where
I flee from this poem. A palm tree of fire.
On the Island. Burning our names.
(*Vinte e tal novas formulações e uma elegia carnívora*, 1991)

KHAL TORABULLY

16 Poems from the book *Voices at the Aapravasi Ghat-Indentured Imaginairies*

Book I – *The rape of the flock* (fragments)

1.
I am woman deposed
in a depot of dhobis
in a ghat of coolies
the plantation was lost cause
Old Caribbean poise and prose
told me of Mauritius Fiji jahajis
Kala give me pani kala
from Calcutta
before baitka and madrassa
Make my dhal massala in Guadeloupe Trinidad Guyana
Yes jahaji Mauritian mix with chilli
pappad and dhall puri
see, I want more mango chutney
then jilébi and red rose alouda
No dark waters to drink
but clear waters to think
Yes jahaji Trinidian flavour with more chilli
Blissful curry from Reunion with vanilla
Some roti for you Ghataji
-Who wants to blink
for a woman at the depot ?
O cook the curry and the dhal pita mangela

2.
The enquirer inquired
from the Queen's own office :
Britannia rules the waves
and i'm only poor human foam
resisting dispersion and plankton.
Not yet a diaspora but only indentured
Tooth against booth
Nail against hell and hair upon tooth
India reversed, revulsed in my vulva.
Arkati and mestri said them voices :
So status of female coolies is far from female slaves ?
How can an object of Desire be a subject of Empire?
How Sahib ?
How can I be freer than a woman ?
I mean a coolie and colonized woman from a paria subcontinent ?
Twice subjected in this dominion.

Book II – *Coral imaginaries: The fakir's aesthetics* (fragments)

1.
Then from an army of defeated sepoys Came a fakir as lean as the moon is high.
He rose his voice to carve a sigh
And told all pariahs of the sea their plight. "Whatever you think is your right,
You must defend. Your star must remain bright. But allow me to give you a piece of advice,
You sailing in a world of lies and vice.
I am the man of imbalance and plasticity. In me the gestures of the ocean are free.
I may stand on my elbow or on my knee. Whatever is in the world must flow easily.
I am the fakir of Bangalore, I have connections galore. My network is beyond boundaries and imaginaries.
I compute I dispute and I refute
Words from distant cultures and of great repute.
I have a binary net in my mosaic head.
I am lean, don't think I am transparent.
I am the opacity of people who are conversant With the idioms of coral and currents".

2.

I have known the amber of Muscat And enjoyed the curry of Surat. My poems are delicate
As the flimsy Rubbaiyat.
I have kept the beauty of the rose
And have perfumed words with a cause. Life on the kala pani
Can also be the voyage to ultimate reality.
Keep crossing jahaji
Keep steering dear coolie
Life is a ceaseless journey
From your broken dream to a new day.

JESSICA FALEIRO

17 Merman

In the beginning, there was the sea.
It was between us,
then, around us,
roaring in our veins.
In our brains.

The first time you said my name aloud,
waves crashed through me.
The first time you whispered it softly in my ear,
I felt them still.

Your fingertip drawn across my skin.

You thought the world had forgotten you, when we met, and you preferred it that way. I wonder what you would think now if you knew that I could still feel discombobulated by the memory of your index finger stroking the inside of my elbow. Twenty years have passed in a moment, and I've thought about you everyday. Did you return to the sea that you came from, drawing away on the same current that brought you to my shoreline?

I've graduated, moved away, gotten married and have grown children. I buried my father yesterday.

But still,
whenever I smell salt in the air,
the inside of my elbow tingles.

I lift my eyes to the horizon,
like I did all those years ago,
and
my heart
is full.

Notes on contributors

MARTA BANASIAK is a postdoctoral researcher at the Department of Literary Theory of the University of Campinas (Brazil) and holds a research grant from the São Paulo Research Foundation (FAPESP) – research grant no. 2020/03902-9. She has a degree in Czech philology (2009) and a master's degree in Portuguese philology (2013) from the Charles University in Prague (Czech Republic) and PhD in Romanesque Studies with a specialty in African Studies (2019) from the University of Lisbon (Portugal). She has researched and published in the areas of Portuguese-speaking African literature, particularly Mozambican, cinema from Portuguese-speaking countries and comparative African literature. In addition to the areas mentioned, her recent research interests focus on theories of world(-)literature, (semi-) peripheral literatures and Indian Ocean Studies. Her publications include 'On allegories and interventions in the (semi)periphery – alternatives to historical writing in Campo de trânsito by João Paulo Borges Coelho and Life & Times of Michael K by J. M. Coetzee' (*Via Atlântica*, 2020), 'On the transition times – ethical writing in Crónica da Rua 513.2 by João Paulo Borges Coelho and in Disgrace by J. M. Coetzee' (Abriu, 2021), 'Combining the Uneven: Literatures of the Lusophone Indian Ocean in the Context of World-Literature – Proposal for a Theoretical Approach Applied to Mozambican Literature' (*Portuguese Studies*, 2021).

ELENA BRUGIONI is Associate Professor in Comparative Literature (African literatures and Postcolonial theory) at the Department of Literary Theory of the University of Campinas UNICAMP, and Lecturer at the Graduate Program in Literary Theory and History of Unicamp. She graduated in Modern Literatures at the University of Bologna (MA, 2004) and has a PhD in African Literatures (2009) from the University of Minho. She is co-coordinator of the KALIBAN – research group in postcolonial studies and world literature (CNPq) at Unicamp and Associate Researcher of CESA – Centre for African and Development Studies of ISEG, University of Lisbon. She publishes and researches in the fields of comparative literatures, African literatures, Indian Ocean studies, postcolonial theory and world literature.

JOÃO PAULO BORGES COELHO is a Mozambican writer and historian. He studied History at the University Eduardo Mondlane in Maputo and got his PhD in Economic and Social History at the University of Bradford (UK). He has been Professor of Mozambican and African History at Eduardo Mondlane University for over thirty years, and he is now retired. To date, his literary works comprise nine novels, two novellas and two volumes of short stories.

JESSICA FALCONI is an assistant researcher at the Centre for African and Development Studies of the University of Lisbon funded by the Portuguese Foundation for Science and Technology of Portugal. She is Assistant Visiting Lecturer at the Faculty of Arts of the same university. She holds a PhD in Iberian Studies from the University of Naples 'L' Orientale'. In 2018 she was Visiting Lecturer at Universitat Autónoma de Barcelona and coordinator of the Portuguese Language Center Camões. In 2017 she concluded a postdoctoral fellowship funded by the FCT. She co-edited eight books and published various articles and book chapters within the field of Lusophone African Literatures and Cultures.

JESSICA FALEIRO has an MA in Creative Writing with commendation from Kingston University, UK. Her work has been published in various anthologies, journals and magazines, including *Asia Literary Review, Forbes, Indian Quarterly, Himal Southasian, Bengaluru Review, Coldnoon, Mascara Literary Review* and the *Times of India*. Her debut novel *Afterlife: Ghost Stories from Goa* (2012) about a Goan family and their 'ghostly' encounters was followed by *The Delicate Balance of Little Lives* (2018), a collection of interlinked stories about five women coping with loss. Jessica was the commissioning editor for the *Joao*

Roque Literary Journal for three years and won the Joao Roque Literary Award 'Best I Fiction 2017' for her short story 'Unmatched'. She won a place on the Helter Skelter writing residency in Goa, working with author Janice Pariat and also on the University of East Anglia-Ashoka University Creative Writing Workshop, working with author Amit Chaudhuri. She lives in Goa, India, where she also teaches creative writing.

UTE FENDLER is Chair of 'Romance Literary and Comparative Studies', University of Bayreuth (Germany) and Deputy Spokesperson of EXC Africa Multiple (since 2019). Her research interests include literatures and film cultures of the Caribbean, West Africa, the Indian Ocean and South America, comparative romance literary and film studies, iconographies, transmediality, popular culture and memory. Her current research project includes 'Multiple Artworks – Multiple Indian Ocean' (Ute Fendler/Rémi Tchokothe/Clarissa Vierke). Some of her recent publications are Ute Fendler/A.-M. Kohl/C. Odhiambo/G. Shang Ndi/C. Vierke: *Of Worlds and Artworks. For a Relational Perspective on African and Afro-Diasporic Artworks*. Leiden: Brill (in print, 2022); 'Hacia una tercera-estética – Ecos polirrítmicos Sur-Sur en novelas antillanas e africanas.' In: Ineke Phaf-Rheinberger et al. (eds.): *Más allá de la Línea – Convergencias, enfrentamientos y cruces en las culturas ibéricas*. 2022, 22pp.; with Elena Brugioni: 'Islands, Theory and the Postcolonial Environment: Reading the Work of Khal Torabully'. In: *Portuguese Studies* Vol. 37, N°2, 2021, 165–177.

PAMILA GUPTA is Research Professor at the University of the Free State in Bloemfontein, South Africa, affiliated with the Centre for Gender and Africa Studies (CGAS). She was formerly Full Professor based at WiSER at the University of the Witwatersrand (2008–2022). She holds a PhD in Anthropology from Columbia University. Her research and writing interests include Portuguese colonial and Jesuit missionary history in India; diasporas, islands, tourism, heritage and design in the Indian Ocean; photography, tailoring and visual cultures in East Africa; and architecture, infrastructure and affect in South Africa. She is the co-editor of Eyes across the Water: Navigating the Indian Ocean (with Isabel Hofmeyr and Michael Pearson, UNISA 2010) and author of three monographs: *The Relic State: St. Francis Xavier and the Politics of Ritual in Portuguese India* (Manchester UP, 2014) and *Portuguese Decolonization in the Indian Ocean World: History and Ethnography* (Bloomsbury Academic, 2019); *Heritage and Design: Ten Portraits from Goa* (India) (Cambridge UP, 2022).

ANA MAFALDA LEITE is Associate Professor of African Literature at the University of Lisbon and an associate researcher at the Centre for African and Development Studies at the same university. Her areas of research include Indian Ocean studies, African cultures and literatures, oral literatures and postcolonial studies. Her recent publications include *Narrating the Postcolonial Nation: Mapping Angola and Mozambique* (ed. 2014), *Speaking the Postcolonial Nation: Interviews with Writers from Angola and Mozambique* (ed. 2014) and *Cenografias Pós-coloniais. Estudos sobre Literatura Moçambicana* (2018).

JOANA PEREIRA LEITE is an economist and historian. She is an retired assistant professor of Lisbon School of Economics and Management (ISEG), University of Lisbon (1990–2019) ISEG | Lisbon School of Economics & Management. She is an integrated researcher for CESA Centre for African and Development Studies (since 1990), and her research areas include economic and social history of East Africa and Mozambique, XIX–XX centuries; Portuguese colonial history, XIX–XX centuries; and history of the Indian diasporas in the West Indian Ocean. She hold a PhD from Ecole des Hautes Etudes en Sciences Sociales (1990, EHESS, Paris).

MANUEL LOBATO is an integrated researcher at the History Centre of the Faculty of Letters of the University of Lisbon. He has published multiple book chapters and articles in specialized journals. He has edited/co-edited books and participated in numerous events abroad and in Portugal. He works within the area of Humanities with an emphasis on History and Social Science, especially related to South East Asia, Estado da India, Malacca, Mozambique, Moluccas, Islam, Christianity, Timor and Philippines.

JOANA PASSOS is a senior researcher at CEHUM – The Centre for Humanistic Studies of the University of Minho, and she is a lecturer at the PhD Program on Compared Modernities. She was a team member of the following projects: Womanart (University of Minho), Nilus (CesA, University of Lisbon), Pensando Goa (University of S. Paulo, Brazil) and Prémios Literários (University of Minho). She published a book on Goan literature and co-edited several others. She has published over fifty articles in English and Portuguese, and she has translated several academic texts by authors such as Rosi Braidotti, Edward Said and Amitav Ghosh. Her research interests combine gender and postcolonial studies, comparative literature and the visual arts.

LUÍS CARLOS PATRAQUIM (1953) is a Mozambican poet, playwright and journalist. Refugee in Sweden in 1973, he returned to Mozambique in 1975, where he went to work for the newspaper *A Tribuna*. After the newspaper was closed, he joined the founding group of the Mozambique Information Agency (AIM) under the direction of Mia Couto. From 1977 to 1986, he worked at the National Institute of Cinema of Mozambique (INC) as the author of scripts and screenplays and as editor of the cinematographic newspaper *Kuxa Kanema*. Together with Calane da Silva and Gulamo Khan, he coordinated, between 1984 and 1986, the *Gazeta de Artes e Letras of Tempo* magazine. He has resided in Portugal since 1986. He collaborates in the Mozambican and Portuguese press, on film scripts and writes for theatre.

KAMILA KRAKOWSKA RODRIGUES is Assistant Professor at Leiden University Centre for the Arts in Society (Netherlands). Her research addresses identity negotiations in arts and cultural practices in the globalized Portuguese-speaking world, with a special focus on African and Afrodescendant communities. Her PhD thesis (2014, Coimbra University, Portugal) on the motif of travel in Brazilian and Mozambican literature was awarded with the international Fernão Mendes Pinto prize and published by the Institute Camões (2019). She participated in several FCT-funded collaborative research projects at the University of Lisbon which resulted, among other outputs, in two co-edited volumes of interviews with writers and filmmakers from the Portuguese-speaking Africa and its diasporas: *Speaking the Postcolonial Nation* (2014) by Leite, AM, S. Khan, J. Falconi and K. Krakowska and *Voices, Languages, Discourses. Interpreting the Present and the Memory of Nation in Cape Verde, Guinea-Bissau and São Tomé and Príncipe* (2019) by Leite, AM, J. Falconi, K. Krakowska, C. Tindó Secco and S. Khan. She was a co-investigator in the HERA-funded grant 'Night Spaces: Migration, Culture, and Integration in Europe' (2019–2022) and is currently leading the ERC project 'City Tales: An Art-Based Participatory Framework for Studying Migration-Related Diversity' (2024–2028).

MEG SAMUELSON is Associate Professor in the Department of English, Creative Writing, and Film at the University of Adelaide, and Associate Professor extraordinary at Stellenbosch University. She works in the oceanic humanities and with literatures of the South and is particularly interested in relating Indian Ocean, African and other southern situations to planetary thought. She has published widely in these fields, including recent essays on the 'African Anthropocene and Mia Couto's poetics of the planet', 'amphibious form', 'blue focalisation and thinking from the Cape' in K Sello Duiker's writing, 'the blue Anthropocene and the oceanic South', 'coastal form', 'coastal thought', JM Coetzee's 'literatures of the south', 'littoral world-making' in Abdulrazak Gurnah's fiction and the oceans in world literature. Other recent publications include the book *Claiming the City in South African Literature* (Routledge, 2021). Meg also co-edits the Palgrave Macmillan series 'Maritime Literature and Culture' and, with Mandy Treagus, is CI on the Australian Research Council project 'Between Indian and Pacific Oceans: Reframing Australian Literatures'. Her co-authored essay in this volume is drawn from an ongoing collaboration with Pamila Gupta that engages with photographic culture and images of Zanzibar.

CARMEN LUCIA TINDÓ SECCO is Professor Emeritus of African Literatures in the Portuguese Language at the Federal University of Rio de Janeiro-UFRJ. She holds a PhD in Letters from the Federal University of Rio de Janeiro (1992). She completed a Postdoctoral Internship at the Fluminense Federal University – UFF and at the Polytechnic University of Mozambique (2009–2010). In 1993, it set up the Sector of African Literatures in the Portuguese Language at UFRJ. He is a member of the Jorge de Sena Chair for Luso-Afro-Brazilian Literary Studies. She is a consultant for FAPERJ and CAPES, B1 Researcher for CNPq and State Scientist for FAPERJ. She is the founding editor of the *Revista Mulemba da UFRJ.* She is a corresponding member of the Angolan Academy of Letters in Brazil and a member of the Honor Committee of the Fernando Leite Couto Foundation in Mozambique. Her publications include *Antologias do mar na poesia africana* (Rio de Janeiro: F. Letras/UFRJ, 1996, 1997, 1999. 3 v.). *A Magia das letras africanas: ensaios escolhidos sobre as literaturas de Angola e Moçambique.* São Paulo: Editora Kapulana, 2021 and *Como se o mar fosse mentira* (with Rita Chaves and Tânia Macedo). Luanda: Chá de Caxinde, 2006. In 2013, together with Maria Geralda de Miranda, she organized the book Paulina Chiziane: *Vozes e rostos femininos de Moçambique* (2013). She also published *Pensando o cinema moçambicano* (2018); *Cinegrafias moçambicanas: memórias e reflexões* (2022), with Ana Paula Tavares, Ana Mafalda Leite and José Octávio Serra Van-Dúnem.

GIULIA SPINUZZA, member of CESA, Centro de Estudos sobre África e Desenvolvimento (CSG, ISEG, University of Lisbon), was a research fellow in the same institution under the Project *NILUS* Narrative of the Indian Ocean in the Lusophone Space funded by FCT, Fundação para a Ciência e a Tecnologia (PTDC/CPC-ELT/4868/2014). She received her PhD in Literary and Cultural Studies, specialization in African Studies in 2017 from Faculdade de Letras, University of Lisbon.

KHAL TORABULLY is a critically acclaimed poet, essayist, film director and semiologist. He is the holder of a PhD in the Semiology of Poetics at the University of Lyon II. He is the author of more than twenty-five books. Torabully has a significant part in UNESCO'S International Indentured Labour Route Project. He pioneered indentured studies.

CLARISSA VIERKE is Professor of Literatures in African Languages at the University of Bayreuth. She is an expert in Swahili poetry and manuscript cultures, and has been working on travelling texts in East Africa – both in Anglophone Kenya and Tanzania, and Lusophone Mozambique. Together with colleagues working on Francophone and Lusophone literature, she is currently running a research project on literary entanglements in the Indian Ocean, in which she herself focuses on popular Swahili poetry. She is the principal investigator of the Cluster of Excellence 'Africa Multiple. Reconfiguring African Studies' and the spokesperson of the research section Arts and Aesthetics. Together with colleagues from Leipzig and Cologne, she set up the project 'Recalibrating Afrikanistik' funded by the Volkswagen foundation with the aim to critically consider the future of the study of African languages and literatures.

GABRIELA BEDUSCHI ZANFELICE is a PhD student at the Department of Romance Languages and Literatures at the University of Chicago and holds a master's degree from the Department of Literary Theory of the University of Campinas (DTL/UNICAMP) and a bachelor's degree in Languages and Literatures from the same university. She co-edited the book *A obra literária de João Paulo Borges Coelho: Panorama crítico* (Editora da Unicamp, 2022) with Elena Brugioni and Fernanda Gallo. She translated the work *Combined and Uneven Development: Towards a New Theory of World-Literature* (WREC, 2015) to Portuguese, published by Editora da Unicamp as *Desenvolvimento combinado e desigual: Por uma nova teoria da literatura-mundial* (WREC, 2020). Her research focus is within the fields of African literatures and comparative literature, with emphasis on Mozambican literature, Indian Ocean studies, world literature and postcolonial theory. She is a member of Kaliban – The World Literature and Postcolonial Studies Research Centre (CNPq).

Index

RECONFIGURING IDENTITIES IN THE PORTUGUESE-SPEAKING WORLD

Edited by

Paulo de Medeiros and Cláudia Pazos-Alonso

The series publishes studies across the entire spectrum of Lusophone literature, culture and intellectual history, from the Middle Ages to the present day, with particular emphasis on figurations and reconfigurations of identity, broadly understood. It is especially interested in work which interrogates national identity and cultural memory, or which offers fresh insights into Portuguese-speaking cultural and literary traditions, in diverse historical contexts and geographical locations. It is open to a wide variety of approaches and methodologies as well as to interdisciplinary fields: from literary criticism and comparative literature to cultural and gender studies, to film and media studies. It also seeks to encourage critical dialogue among scholarship originating from different continents.

Proposals are welcome for either single-author monographs or edited collections (in English and/or Portuguese). Those interested in contributing to the series should send a detailed project outline to oxford@peterlang.com.

Vol. 1 Ana Margarida Martins: Magic Stones and Flying Snakes: Gender and the 'Postcolonial Exotic' in the Work of Paulina Chiziane and Lídia Jorge.
ISBN 978-3-0343-0828-1. 2012

Vol. 2 Ana Mafalda Leite, Hilary Owen, Rita Chaves, Livia Apa (eds): Narrating the Postcolonial Nation: Mapping Angola and Mozambique.
ISBN 978-3-0343-0891-5. 2014

Vol. 3 Ana Mafalda Leite, Sheila Khan, Jessica Falconi, Kamila Krakowska (eds): Speaking the Postcolonial Nation: Interviews with Writers from Angola and Mozambique.
ISBN 978-3-0343-0890-8. 2014

Vol. 4 Francisco Bethencourt (ed.): Utopia in Portugal, Brazil and Lusophone African Countries.
ISBN 978-3-0343-1871-6. 2015

Vol. 5 Ana Luísa Amaral, Ana Paula Ferreira, Marinela Freitas (eds): New Portuguese Letters to the World: International Reception.
ISBN 978-3-0343-1893-8. 2015

Vol. 6 Fernando Beleza and Simon Park (eds): Mário de Sá-Carneiro, A Cosmopolitan Modernist.
ISBN 978-3-0343-1885-3. 2017

Vol. 7 Eleanor K. Jones: Battleground Bodies: Gender and Sexuality in Mozambican Literature.
ISBN 978-1-78707-317-3. 2017

Vol. 8 Maria do Carmo Piçarra and Teresa Castro (eds): (Re)imagining African Independence: Film, Visual Arts and the Fall of the Portuguese Empire.
ISBN 978-1-78707-318-0. 2017

Vol. 9 Maria Manuel Lisboa: A Heaven of Their Own: Heresy and Heterodoxy in Portuguese Literature from the Eighteenth Century to the Present.
ISBN 978-3-0343-1962-1. 2018

Vol. 10 Doris Wieser and Ana Filipa Prata (eds): Cities of the Lusophone World: Literature, Culture and Urban Transformations.
ISBN 978-1-78874-251-1. 2018

Vol. 11 Paulo Pepe and Ana Raquel Fernandes (eds): Beyond Binaries, Sex, Sexualities, and Gender in the Lusophone World.
ISBN 978-1-78707-615-0. 2019

Vol. 12 Ana Mafalda Leite, Jessica Falconi, Kamila Krakowska, Sheila Khan, and Carmen Tindó Secco (eds): Voices, Languages, Discourses: Interpreting the Present and the Memory of Nation in Cape Verde, Guinea-Bissau and São Tomé and Príncipe.
ISBN 978-1-78707-585-6. 2020

Vol. 13 Ana Mafalda Leite, Hilary Owen, Ellen Sapega and Carmen Tindó Secco (eds): Postcolonial Nation and Narrative III: Literature & Cinema. Cape Verde, Guinea-Bissau and São Tomé e Príncipe.
ISBN 978-1-78707-581-8. 2019

Vol. 14 Elena Brugioni, Orlando Grossegesse, and Paulo de Medeiros (eds): A Companion to João Paulo Borges Coelho.
ISBN 978-1-78707-986-1. 2020

Vol. 15 Kristian Van Haesendonck (ed.): The Worlds of Mia Couto.
ISBN 978-1-78874-594-9. 2020

Vol. 16 Peter J. Maurits: The Mozambican Modern Ghost Story (1866–2006): The Genealogy of a Genre.
ISBN 978-1-78997-541-3. 2022

Vol. 17 Paulo de Medeiros and José N. Ornelas (eds): Saramago After the Nobel: Contemporary Readings of José Saramago's Late Works.
ISBN 978-1-78707-894-9. 2022

Vol. 18 Nazaré Torrão and José Cândido de Oliveira Martins (eds): The Power of the Image in the Work of Lídia Jorge.
ISBN 978-1-80079-911-0. 2023

Vol. 19 Emanuelle Rodrigues dos Santos: The Late Postcolonial Condition: Twenty-First-Century Reconfigurations in the Literatures of Portuguese-Speaking Africa.
ISBN 978-1-78707-450-7. 2025

Vol. 20 Ana Mafalda Leite, Elena Brugioni, Jessica Falconi and Marta Banasiak (eds): The Indian Ocean and the Portuguese-Speaking World: Literary and Cultural Intersections.
ISBN 978-1-80079-096-4. 2025

www.ingramcontent.com/pod-product-compliance
Ingram Content Group UK Ltd.
Pitfield, Milton Keynes, MK11 3LW, UK
UKHW021015120926
8556IPUK00040B/1153

9 781800 790964